Geomedicine

Editor

Jul Låg, Dr. Agric.
Professor
Department of Soil Science
Agricultural University of Norway
Ås, Norway

CRC Press
Boca Raton   Ann Arbor   Boston

Library of Congress Cataloging-in-Publication Data

Geomedicine/editor, Jul Låg.
   p.   cm.
  Includes bibliographical references and index.
  ISBN 0-8493-6755-7
  1. Medical geography.   I. Låg, J. (Jul)
RA792.G48   1990
616.9'8--dc20

90-1945
CIP

Direct all inquiries to CRC Press, Inc., 2000 Corporate Blvd., N.W., Boca Raton, Florida, 33431.

© 1990 by CRC Press, Inc.

International Standard Book Number 0-8493-6755-7

Library of Congress Card Number 90-1945
Printed in the United States

PREFACE

Nearly all over the world a very rapid increase in environmental problems has taken place over the last 20 to 30 years. The literature dealing with these topics has grown enormously.

An important part of the environmental question concerns the effects of factors on the health of man and animals. Initiative was taken in several countries in the early 1970s to organize activities in this scientific field. Some distribution patterns of specific diseases had been known for a long time, but it was now necessary to systematize existing knowledge and to gain new knowledge.

Different names have been used for this group of problems. In Scandinavia we have applied the term geomedicine and have prepared a definition. Some related concepts, e.g., environmental geomedicine and health, only cover the effects of chemical factors and thus are more limited.

Certain aspects of geomedicine have been studied since ancient times. However, it is only recently that the subject has acquired a separate status, and we are evidently only at the very beginning of development in this scientific field.

In this book some geomedical knowledge is presented. We hope that this may be useful in teaching and as a starting point for continued investigation. In addition to focusing on environmental problems, the prevention of disease in man and animals has drawn great attention. Comprehensive geomedical activity as a basis of prophylaxis may be expected in the near future.

Unfortunately, despite a clear agreement, the manuscript on geomedical problems in eastern Europe was not delivered. I wish to thank all the contributors for their help in compiling the manuscript of the book. Further, I am grateful to the publishers for their cooperation. Professor Alina Kabata-Pendias, author of the CRC book, *Trace Elements in Soils and Plants*, has given valuable advice. Among the Norwegian research workers who have helped me, I am especially grateful to Professor Knut Westlund, National Health Screening Service, Oslo.

Agricultural University of Norway
J. Låg

THE EDITOR

Jul Låg, Dr. Agric., Professor emeritus, is chairman of the Geomedical Committee of the Norwegian Academy of Science and Letters, and of the working group, Soils and Geomedicine, of the International Society of Soil Science.

He has a candidate degree (1942) and a doctor degree (1949) from the Agricultural University of Norway. He was professor in the period 1949-1985 and for most of that time he was head of the department of soil science in this institution. He was rector (president) of the University from 1968 to 1971.

Dr. Låg has been a member of the Norwegian Academy of Science and Letters from 1953, and president for 5 years. He is a member of academies in Denmark, Finland, Poland, and Sweden, and has received the Copernicus Medal from the Polish academy.

In Norway Dr. Låg has been elected to a number of positions, e.g., chairman of the Agricultural Research Council of Norway, member of the Royal Norwegian Council of Scientific and Industrial Research, member of the Norwegian Research Council for Science and Humanities, vice-chairman of the Norwegian Council of Parliaments and Scientists, and chairman of the Norwegian Committee of ICSU program SCOPE.

Among international engagements, he has been Chairman of the Scandinavian Committee of Agricultural Research, chairman of the Soil Section of the Scandinavian Agricultural Research Workers Association, member of the Council of the International Society of Soil Science, and member of the editorial board (consulting editor) of *Soil Science* (U.S.), *Agrochimica* (Italy), *Geoderma* (Netherlands), *Ambio* (Sweden), and *Alexandria Science Exchange* (Egypt). He is honorary member of the Soil Science Society of the Soviet Union, and corresponding member of the Forestry Society of Finland and of the Soil Science Society of West Germany.

Dr. Låg has published more than 150 papers in soil science and related subjects. He has written 5 books in Norwegian, and has been editor of 6 books in connection with symposia: *Geomedical Aspects in Present and Future Research* (1980), *Basis of Accounts of Norway's Natural Resources* (1982), *Geomedical Research in Relation to Geochemical Registrations* (1984), *Geomedical Consequences of Chemical Composition of Freshwater* (1987), *Commercial Fertilizers and Geomedical Problems* (1987), and *Health Problems in Connection with Radiation from Radioactive Matter in Fertilizers, Soils and Rocks* (1988). (The distribution office is Norwegian University Press.)

CONTRIBUTORS

Jan Aaseth, M.D., Ph.D.
Department of Occupational Medicine
University Hospital
Tromsø, Norway

U. Aswathanarayana, D.Sc., F.N.A.,
F.A.Sc., F.G.S.
Department of Geology
University of Dar es Salaam
Dar es Salaam, Tanzania

Alf Björklund, Ph.D.
Department of Geology
Åbo Akademi
Åbo, Finland

Brian E. Davies, B.Sc., Ph.D.
Department of Environmental Sciences
University of Bradford, West Yorkshire
England

Eystein Glattre, M.D., Ph.D.
Cancer Registry of Norway
Oslo, Norway

Olav Hilmar Iversen, M.D., Ph.D.
Institute of Pathology
University of Oslo
Oslo, Norway

Jul Låg, Dr. Agric.
Department of Soil Science
Agricultural University
Ås, Norway

Charles Lawrence, Ph.D.
Department of Water Resources
Melbourne, Victoria
Australia

Jetmund Ringstad, M.D., Ph.D.
Department of Occupational Medicine
Tromsø University Hospital
Tromsø, Norway

H. W. Scharpenseel, Ph.D.
Institute of Soil Science
University of Hamburg
Hamburg, West Germany

Eiliv Steinnes, Ph.D.
Department of Chemistry
University of Trondheim, AVH
Dragvoll, Norway

Jan Alexander, M.D., Ph.D.
Department of Environmental Medicine
National Institute of Public Health
Oslo, Norway

Peter Becker-Heidmann, Dr. rer. nat.
Institute of Soil Science
University of Hamburg
Hamburg, W. Germany

Bjørn Bølviken, M.Sc.
Geological Survey of Norway
Trondheim, Norway

Arne Frøslie, Dr. med. vet.
National Veterinary Institute
Oslo, Norway

Francis D. Hole, Ph.D.
Department of Soil Science
University of Wisconsin
Madison, Wisconsin

M. W. Johns, Ph.D.
Richmond, Victoria
Australia

Pertti W. Lahermo, Ph.D.
Department of Geochemistry
Geological Survey of Finland
Espoo, Finland

S. S. Randhawa, M.V.Sc. (Vet. Med.)
College of Veterinary Science
Punjab Agricultural University
Ludhiana, Punjab, India

Kaare Rodahl, M.D., D.Sc.
(Retired Professor)
Oslo, Norway

Per Slagsvold, Dr. Vet. Med.
Professor Emeritus
Bekkestua, Norway

P. N. Takkar, Ph.D. (Soils)
Indian Institute of Soil Science (ICAR)
Madhya Pradesh, Bhopal, India

To the memory of my wife Ingrid Låg (1926-1979) who suffered from multiple sclerosis,

and to her physician, Professor Dr. Sigvald Refsum, Emeritus President of the World Federation of Neurology.

TABLE OF CONTENTS

Chapter 1

GENERAL SURVEY OF GEOMEDICINE

J. Låg

TABLE OF CONTENTS

I. ANCIENT SUGGESTIONS AND MODERN DEFINITION

As long as medical science has existed, there has been knowledge of certain human illnesses related to particular geographical areas. Hippocrates, the founder of the subject, has mentioned such examples.[1] Other Greeks, too, tackled these questions more than 2000 years ago.

Empirically based knowledge on specific animal diseases also originated long ago. However, most of such observations were lost because they were not written down.

Publications on medical history contain a lot of information on valuable knowledge reached by experience (e.g., see Lyons[2]). The old basis for successful use of medical plants is in many ways astonishing.

As the science grew, many of the previously unknown relations of causes began to be understood. To find reasons for the geographical distributions of certain diseases appealed to the investigators.

New types of problems have, however, appeared in connection with the development of the industry. Well-known elements have been applied in new ways, and rare elements have come into use. Products like plastics, special steel alloys, metallic glasses, superconductors, etc. may be mentioned. Mankind is exposed to new products.

Geomedicine is now defined as the science dealing with the influence of ordinary environmental factors on the geographical distribution of health problems in man and animals. Accordingly, this is a complicated subject. Health is a very comprehensive notion (explanations from the World Health Organization of the United Nations). Contributions from essentially different scientific fields are often needed when geomedical problems are to be solved.

According to this definition, medical problems caused by so-called indoor factors are not included in geomedicine. They belong to the subject of occupational medicine. There are many borderline cases which are difficult to classify. Harmful radiation due to radioactive processes under ground has to be examined as a geomedical problem. If the radiation originates from objects in the house, it is reasonable to assume that this is a question for occupational medicine.

Geomedicine can be divided into subgroups, e.g., according to types of environmental factors.[3,4]

Maps are often convenient for geomedical accounts and presentation. We should, however, avoid premature conclusions when maps showing patterns of disease distribution and natural conditions are compared. The maps may be similar without any casual connection.

II. SOME CLASSIC, INTERNATIONALLY WELL-KNOWN EXAMPLES

A. GOITER DUE TO IODINE DEFICIENCY

A hypothesis showing that iodine deficiency results in goiter was presented early in the 19th century. In France exact chemical analyses led to a definitive conclusion and to the recommendation of using extra iodine in the goiter districts. These important scientific results were, however, overlooked for a long time.

Soon after the beginning of this century the prevention of goiter by the addition of iodine was commonly recognized. Mixing of iodine compounds in table salt for purchase is now a recognized practice in many countries.

Continued research has revealed that certain compounds can depress the effects of iodine in food and feed. Further, different types of goiter have been discovered. The biological effect of iodine is connected to the thyroid hormones. Very large intakes of iodine can lead to poisoning.

Even if the relationships are somewhat more complicated than formerly believed, the connection between goiter and iodine deficiency is nevertheless a classic geomedical example. In many developing countries goiter is quite common. More comprehensive use of existing knowledge could be of benefit for mankind.

A general survey of the research on goiter is, e.g., given by Underwood.[5]

B. EXCESS AND DEFICIENCY OF FLUORINE

In the 1920s many research results showing harmful effects of fluorine compounds were published. Earlier, too, some serious poisonings were reported.[6,7]

After eruptions of the volcano Hekla (Iceland) in 1693, 1766, and 1845, detailed descriptions of fluorosis were presented.[6] Acute poisoning was described, and damage to teeth was reported. In a publication on the eruption in 1693, the special expression "ash-teeth" is used for teeth with black marks which are connected to the precipitation of volcanic ash.

Roholm[7] has presented descriptions and results of chemical analyses of bones from sheep being affected by fluorosis. The concentration of fluorine was as high as 0.44 to 2.06%. He carried out experiments which resulted in similar damage in sheep and other animals when they received feed with intermixture of a fluorine compound.

Since World War II, Hekla has had eruptions in 1947, 1970, and 1980, and a number of analyses of fluorine have been performed. It is unequivocally stated that the volcano had delivered huge quantities of fluorine, and farmers have been advised which precautions to take to avoid damage in animal husbandry as far as possible. Immediately after the eruption in 1970, a concentration of 4300 ppm fluorine in the grass was found, and in 1980, 1000 ppm.[8] However, the concentration decreased quite rapidly.

Even some years before World War II, attention was pointed to the preventive effects of fluorine on caries. Early in the 19th century the content of fluorine in teeth and bones was pointed out. A long time elapsed, however, before such knowledge was applied to medical problems.

The use of various types of additives of fluorine to prevent dental caries is now common in many countries. Vehement discussions on the addition of fluorine compounds to drinking water have often taken place. In many parts of the world the incidence of caries has decreased considerably, and the use of fluorine is thought to be of importance.

Damage to man and animals because of too high natural concentration of fluorine in water and plants is also reported from many other countries. The ability to take up fluorine varies between the plant species. For instance the tea plant can be very rich in fluorine. In several places, e.g., in Africa, there is high fluorine concentration in the water and fluorosis in man and animals.

Contamination has often led to poisoning by fluorine compounds. Domestic and wild animals have suffered from such contaminations in the vicinity of aluminum factories. In some cases the raw phosphates for the production of commercial fertilizers have a too-high concentration of fluorine.

C. SELENIUM POISONING

Serious illnesses in animals, reported in the U.S. in the middle of the 19th century, were (early in the 1930s) proven to be the results of selenium toxicity. Since then, selenium poisoning has been found in several other places in the world.

The names "blind staggers" and "alkali disease" were used in the first descriptions of the diseases in America. More detailed classifications are to follow. Anemia and deformation of hooves and other organs normally containing the amino acids with sulphurus (methionine, cysteine, cystine) are common indications of selenium poisoning. Lameness and failure in the respiratory organs may occur in acute attacks.

There seems to be a positive correlation between selenium supply and dental caries in children.

Different plant species take up varying amounts of selenium. Some leguminous plants have a special ability to absorb this element.[9]

Some time ago selenium was suspected to cause cancer in certain situations, but this hypothesis was later abandoned. In fact, an inverse suggestion was that some special cancer types are due to selenium deficiency.

III. LESS KNOWN, OLD EXAMPLES FROM SCANDINAVIAN COUNTRIES

A. PHOSPHOROUS DEFICIENCY AND OSTEOMALACIA

Norwegian farmers have, for a long time, had knowledge of the extraordinarily high frequency of osteomalacia in domestic animals in certain districts. They had also developed a procedure to help combat against the disease: crushed bones were used in the feed of the animals. Some farmers thought that a special pasture plant caused osteomalacia.

A Norwegian official Jens Bjelke (1580-1659), with an interest in botany and knowledge in foreign languages, gave the suspected plant the Latin name *Gramen ossifragum* ("the grass that breaks bones"). The name has also been written *Gramen Norwagicum ossifragum*. In modern botanical nomenclature the species name *ossifragum* has been retained. The famous Swedish botanist von Linnaeus adopted it, and the British scientist William Hudson made a new description of the species which now has the name *Narthecium ossifragum* (L.) Huds.

One hundred years ago the geologist J. H. L. Vogt investigated the bedrock in a region where the deficiency was common. He knew about the farmers' experience of using crushed bones for the animals. When he found very small amounts of the mineral apatite in the rocks, he drew the logical and correct conclusion that deficiency of phosphorus was the reason for osteomalacia.[10]

Another Norwegian geologist, Esmark,[11] had previously pointed out that the vegetation was extra sparse over the bedrock which Vogt later found very poor in apatite.

Once the reason was stated, it became, in principle, easy to prevent the damage. Phosphorus fertilizer is a simple and effective remedy.

B. THE SHEEP DISEASE "ALVELD"

As mentioned above the reason for osteomalacia was not the plant *Narthecium ossifragum*, but the deficiency of phosphorus. On the other hand it seems that a curious sheep disease "alveld" can be traced back to this plant species.

The name alveld ("elf's fire") is a Norwegian dialect word. Another dialect term used in Norway is "hovudsott" ("sickness of the head"). Typical symptoms include a rash which resembles burns, particularly round the ears, which causes them eventually to hang down. A large number of lambs die of alveld every year in Norway.

This disease has been known in Norway for a long time, and the name indicates that superstition may be connected to the explanation of the cause.

Now alveld is explained as a photosensitive disease.[12,13] Stabursvik[13] has found, by analyzing *Narthecium ossifragum*, matter which gives alveld-like symptoms in lambs when the compound is mixed in the feed.

There are also other plant species which may cause similar diseases in animals. Species of the genus *Hypericum* are suspected to have such qualities in Scandinavia.

C. SCURVY

For a long period of time scurvy (in Latin *scorbutus*) was a dreaded nutrient disease, and was mentioned in Scandinavian literature as early as the 10th century.[14-15] The expression is said to be derived from the Norwegian name for sour milk (skyr). The term seems to be connected to the suspicion that the use of such old milk, instead of fresh milk, caused the illness.

Remedies against scurvy were tried long before knowledge about vitamin C existed. It is easy to understand that questions about scurvy have been particularly recurrent in the polar and subpolar regions, with the short summer, and in connection with navigation.

In a description of Norway, printed in 1632, Peder Claussøn Friis explains that far to the north of the country, people are generally healthy except they suffer from scurvy.[16] He tells that the plant species now named *Angelica archangelica*, *Rubus chamaemorus*, and *Cochlearia officinalis* are used as remedies against scurvy. At some places along the coast large quantities of *Cochlearia officinalis* are found, and people harvest the plant material in summer for use during the winter time. (The English name for the last mentioned plant is scurvy-grass.)

IV. SOME CONCEPTS WITH RELATION TO GEOMEDICINE

Cartographical presentations of the distribution of different human diseases have been in use for a long time. Expressions such as medical geography and geographical medicine have been employed. The term geomedicine seems to have been introduced by Zeiss,[17] in an effort to expand the field and place more stress on causality (cf. difference between the Greek words graphos and logos).

Zeiss used geomedicine and geographical medicine as synonymous concepts, and said that this is a branch of medicine where geographical and cartographical methods are used to present medical research results. He further maintained that medical geography belongs to the subject geography. It seems that Zeiss looked at the relation to the words geopolicy and geojurisprudence when he introduced the term geomedicine.

Rimpau[18] supported Zeiss in his introduction of the new concept, but he said it was not convenient to continue to stick to the old terms geographical medicine and medical geography. He divided geomedicine into two main groups, geostatic and geodynamic, and put geographical medicine as a part of the first one.

Zeiss had also pointed out that geomedicine is a comprehensive subject. In his publication in 1931 he said that medical doctors, veterinarians, and botanists ought to collaborate with geographers, meteorologists, soil scientists, entomologists, and geologists. He may have been of the opinion that veterinary medicine and plant pathology should be included in geomedicine. However, he continued to use examples from human medicine. An exact definition of geomedicine has not been found. It seems that Zeiss has been greatly influenced by his teacher, Ernst Rodenwaldt. Jusatz,[19] too, gave a lot of attention to the work of Rodenwaldt.

The expression "geomedicine" was not widely used in the first decades. In 1972 it became known that geomedical research activities had been carried out in a center of the Heidelberg Scientific Academy for a period of 20 years.[20] As mentioned, Zeiss[17] suggested collaboration with other scientists in order to solve geomedical problems. But the continued research work in Germany seems to have been concentrated directly on human medical questions.

Since 1931, there has been a considerable growth of knowledge in human medicine, veterinary medicine, and environmental factors. To combine results from investigations in human and veterinary medicine has often proved to be advantageous. Experiments with animals are comparatively easy to carry out. Originally, medical doctors and veterinarians concentrated their work on curing the illnesses. Since then, prophylaxis has been given more attention; for instance, nutritional questions have been brought into focus. Both pathological and nutritional problems are regarded under the health concept.

In defining the subject of geomedicine, new types of knowledge were considered (see Section I). Satisfactory explanations and accurate definition will hopefully benefit the advance of this science.

It has been suggested that plant pathology be included in geomedicine. The expression geophytomedicine was once mentioned.[21] However, plant pathology is a separate unit, and it does not seem convenient to expand the concept of geomedicine so far. On the other hand, close relationships between plant pathology and human and veterinary medicine problems do occur. If deficiency or surplus of an element results in plant diseases, this situation may lead to difficulties when the plant material is used as food or feed. The problems may be transmitted higher up in the nutritional chains.

The word epidemiology in human and veterinary medicine was originally used in connection with infectious diseases. Illnesses with physiological background were included later. In environmental epidemiology many geomedical problems are discussed.[22]

The expression "environmental geochemistry and health" has a somewhat more limited meaning than geomedicine. Early in the 1970s organized activities were built up in the U.S., Canada, and Great Britain, having geochemistry as a main basis.[23] In the Soviet Union the geochemical subject was for a long time in a strong position, and relationships to health problems have been dealt with in a fruitful way (see, e.g., References 24, 25).

In Scandinavian countries we have preferred to use a broad perspective on such problems. As in the Soviet Union, geochemical research has been regarded as relatively important in Norway (cf. Reference 26). Using the expression "geomedicine" we have, for instance, included physical factors together with chemical ones. When the committee of the international program "Man and the Biosphere" started work in 1971, the question of research in geomedicine was taken up in Norway. We had some time earlier discovered a serious case of natural lead poisoning of the soil,[27] and important influences of the chemical climate. Investigations of forest humus, sampled by cooperation between the Agricultural University of Norway, the National Forest Survey, and the Geological Survey of Norway, had shown relationships between the chemical composition of the precipitation and the soil, relationships supposed to have biological consequences.[28,29]

Ecotoxicology, a relatively new concept, deals with some of the geomedical problems. The name is made up of the first part of the word ecology with the addition of toxicology which is the science of toxic matter and toxic effects. Ecotoxicology may be said to be the science of poisoning effects in ecological connections, or harmful effects on ecosystems by chemical matter. A more comprehensive definition is introduced by SCOPE (Scientific Committee on Problems of the Environment): "Ecotoxicology is concerned with the toxic effects of chemical and physical agents on living organisms, especially on populations and communities within defined ecosystems; it includes the transfer pathways of those agents and their interactions with the environment.[30]

Other concepts, with some relationships to geomedicine, are environmental toxicology and environmental chemistry. The activities in these fields have expanded rapidly due to the increase of pollution problems. A great number of new books and many new journals dealing with such questions have appeared during the last decades.

V. PROGRESS IN ANALYTICAL CHEMISTRY AND PHYSIOLOGY

For a few decades many challenging geomedical problems had to be given up because methods of chemical analyses were insufficient. The questions were often tied up with matter which occurred in very small concentrations.

Since World War II we have had a nearly revolutionary development in analytical techniques. By means of atomic absorption spectrometry, X-ray fluorescence analysis, polarometry, ion chromatography, ICP technique, neutron activation analysis, PIXE technique, etc. even very small concentrations can be determined accurately.

A comparison of the figures published by Clarke in 1924,[31] and more recent geochemical literature such as Wedepohl,[32] can demonstrate the progress in analytical techniques.

Important physiological processes have gradually been elucidated. Complicated enzyme and hormone reactions have been explained. On the basis of new knowledge in physiology, the use of refined analytical methods can be used more purposefully, and they then serve as valuable means of producing extensive information. Cooperation between persons working in physiology and analytical chemistry has led to very good results.

The comprehensive textbooks and handbooks in nutrition and in human and veterinary medicine contain plenty of facts for further geomedical studies. For examples of books dealing with interesting chemical compounds see References 31 to 35.

VI. GENERAL GEOCHEMICAL DISTRIBUTION PATTERNS

Geochemistry is the science of the chemical composition and actual or possible chemical changes of the earth. The term seems to have been introduced in 1838.

The geochemical literature is now very comprehensive. At the beginning most of the publications had a descriptive character. Later this changed, and, e.g., regularities in the distribution of the different elements began to be studied intensively. The Norwegian scientist V. M. Goldschmidt (1888-1947) made a pioneer contribution in this field. A number of natural laws regarding the composition of minerals and rocks were discovered. The size and charge of the ions were found to be very important.

The processes in the soil are more complicated than in the bedrock, and because of this they are not so well known. Circulation of matter between soils, plants, animals, and man, with feedback of waste products to the soils is very important (Figure 1). In modern societies pollution has to be taken into account in this connection.

The properties of the soil are determined by the following five natural factors or groups of factors: climate, living organisms, mineral material, topography, and time of soil formation. Among some of them there exist interrelationships (see Figure 2) and, in addition, there are influences by man. It is not possible to elucidate exactly all these factors and to describe in detail all the processes which are going on.

For geochemical survey literature, refer to chapters written by other authors in this book.

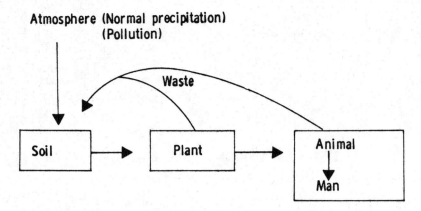

FIGURE 1.   Main circulation processes in nature.

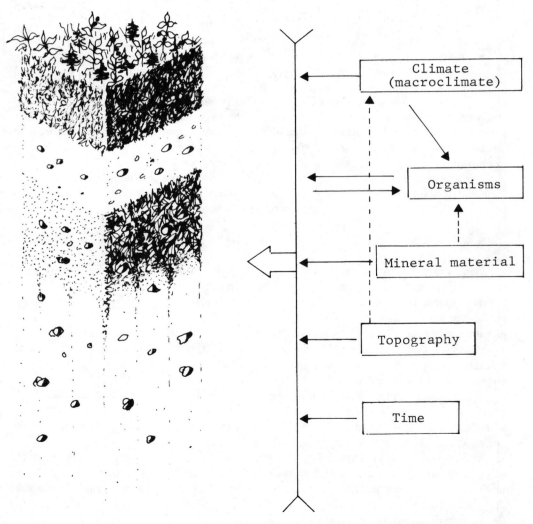

FIGURE 2.   Schematic presentation showing the influence of soil-forming factors. (Redrawn from Låg, J., Reference 36.)

VII. THE CHEMICAL CLIMATE

A. INFLUENCE OF CLIMATE ON SOIL FORMATION

Development of modern soil science started when the relationship between climatic conditions and geographical distribution of great soil groups was discovered a little more than a hundred years ago. The Russian V. V. Dokuchaev was the first to clearly recognize this regularity.

The climate has direct influence on soil formation, e.g., through transport of matter by water in the soil profile. The living organisms are also dependent on climatic factors. In this way the climate has both a direct and indirect influence. Figure 2 shows, in a schematic way, relations between soil and soil-forming factors.

Temperature, precipitation, and wind are very important for the development of the soil, but chemical variations in the climate occur, too. We may talk about both physical and chemical climatology. When the analytical methods were improved sufficiently, such climatic variations were registered. Later there were also possibilities of tracing the chemical differences from precipitation to the soils and further to the plants.

B. ANALYSES OF PRECIPITATION
1. Determination of Nitrogen

In the 19th century, scientists were intensely concerned with the plant's supply of the nutrient nitrogen. The famous German agricultural chemist Justus von Liebig[37] maintained, however, that the content of nitrogen in the precipitation should be sufficient for the vegetation, but he had calculated with too high concentrations. His patented artificial fertilizer did not contain nitrogen, and when it was used it fell short of expectations.

At Rothamsted experimental station, England, nitrogen analyses of precipitation and field experiments were carried out. There, it was proved that agriculture plants, except the leguminous species, had to be fertilized with nitrogen. Later, determination of nitrogen in precipitation was conducted in many places around the world. Before the industrial fixation of the free nitrogen was a reality, some scientists thought that scarce resources of this nutrient would limit the world production of food.

In addition to the traditional questions on plant nutrition, the problems of pollution by nitrogen compounds came into focus after World War II. An investigation begun in Sweden, and greatly expanded under the geophysical years 1957/58, brought a lot of new information. The analytical figures proved that concentration of nitrogen compounds in precipitation increased in the direction of towns and industrial communities.

In connection with the threat of dying forests, especially in some areas in Mid-Europe and North America, there is a greater interest in nitrogen analyses of precipitation besides determination of sulfur dioxide and ozone. Some forest damage has been supposed to be connected with excessive nitrogen concentration.

2. pH of Precipitation

Distilled water in equilibrium with a normal atmosphere containing 0.03% CO_2, has a pH about 5.6. The soil is, we understand, regularly influenced by precipitation with some acidity.

Increasing acidity has, however, caused intense discussion since the late 1960s. By using huge quantities of fossil fuel in industrialized societies, large quantities of acid matter are released into the atmosphere. Many types of damage have been found. Monuments and buildings have been affected, and influence on biological systems has been discussed. A decrease of the forest production was postulated at an early stage. Fish death in many watercourses has been proven.

The additional acidity of rain and snow is caused by burning coal and oil, thus generating sulfuric acid and to some degree nitric acid.

TABLE 1
Analyses of Precipitation from Norwegian Meteorological Stations

Station	Year	Precip. (mm)	Yearly amount (mg/m²)								pH (avg.)
			S	Cl	NO₃-N	NH₃-N	Na	K	Mg	Ca	
Ås	1955-62	719	616	697	146	156	488	137	94	536	5.3
Vågåmo	1955-62	292	294	135	30	46	134	108	50	406	6.2
Lista	1955-62	1025	1871	25,742	345	276	14,831	839	1734	1381	4.9
Ytterøy	1957-62	640	393	3,246	57	75	1,820	283	273	561	5.9
Tana	1958-62	336	409	1,613	31	69	966	138	132	413	6.0
Gjermundnes	1957-62	991	501	4,570	53	106	2,785	185	364	584	5.9
Stend	1957-62	1116	907	4,365	148	209	2,491	245	336	643	5.4
Fortun	1957-62	622	366	404	51	98	253	123	60	528	6.0
Fanaråken	1957-62	616	320	424	55	86	363	120	53	264	5.8
Trysil	1957-62	673	465	187	88	107	126	81	42	394	5.7
Kise	1957-62	543	417	141	92	126	104	75	45	491	5.7
Dalen, Telemark	1957-62	768	504	340	112	90	240	165	66	759	5.9

From Låg, J., *Acta Agric. Scand.*, 18, 148, 1968. With permission.

Comprehensive investigations on the harmful effects of acid precipitation have been carried out, especially in Europe and North America. After protracted negotiations, many states have agreed to reduce the release to the atmosphere of sulfuric compounds by 30% before the year 1993.

The fish death is an example of a special veterinary medical problem caused by acid precipitation. Possibilities of several human medical difficulties have been mentioned. For example, the increased acidity can result in a higher concentration of soluble aluminum and heavy metals, which may cause health problems.

3. Other Compounds in Precipitation

As minute analyses of rain and snow continued, it was clear that the soil received considerable amounts of several elements, and that important geographical variations occurred.

In Norway, large climatic differences in comparatively small areas can be clearly demonstrated. Often the precipitation comes with air currents from the west, giving high rainfall and snowfall in the western areas and low rainfall and snowfall to the east of the mountains. The chemical composition reflects to some degree the distance from the ocean[29] (Table 1 and Figure 3). The station Vågåmo is situated in a so-called rain-shadow of the mountains.

Increasing interest in pollution problems has resulted in analyses of organic micropollutants, and some of these compounds found in the precipitation may be carcinogens.

In the near future we can expect rapidly growing activity in the investigation of the chemical composition of the precipitation.

4. Movement of Matter in the Atmosphere Independent of Precipitation

Huge amounts of matter are transported in the atmosphere as gas and dust. Such processes play an important part in geological circulation.

Transportation of iodine as gas in the atmosphere has been discussed for a long time. Goldschmidt[26] refers to older literature, and he explains his opinion that air transport must be of importance for the iodine supply, especially for distribution in the soil. He further points out that in Quaternary glaciated areas the iodine content of the soil is low because of the short period of time for the supply.

Other volatile matter can behave somewhat similar to iodine. In this connection we ought to remember that both elements and compounds should be taken into account.

During the study of acidification through pollution of the atmosphere, registration of

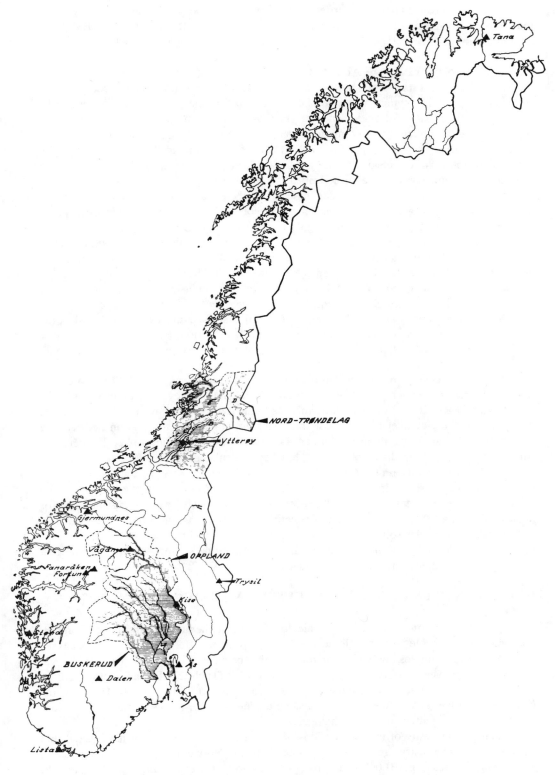

FIGURE 3.   Sampling stations and areas investigated. (From Låg, J., *Acta Agric. Scand.*, 18, 148, 1968. With permission.)

so-called dry depositions has been carried out. In particular, deposition of sulfuric compounds independent of rain and snow has been noticed. Values for sulfur as high as 30% of the total amounts have been reported.[38]

C. DIFFERENT TYPES OF AIR POLLUTION

Natural disasters may for short periods lead to abnormal atmospheric composition. Volcanic eruption is a typical example. The release of fluorine is mentioned in Section II. Hydrogen-sulfide, sulfur dioxide, and carbon dioxide are common gases under volcanic activities, and also many other compounds can occur in gases from volcanoes.

Some historical reports, e.g., from the eruption of Vesuvius in the year 79 A.D. give drastic descriptions of effects of poisoning. Many volcanic eruptions have, however, resulted in serious poisoning without being described.

Increasing content of carbon dioxide in the atmosphere may result in higher temperatures due to the so-called ''greenhouse effect'', with important biological consequences. Fires in forests and grass steppes, started after lightning, can give strong air pollution locally.

However, the most comprehensive pollution of the air is done by mankind, and is primarily connected to industrializing and urbanization of the society. In some towns the air is periodically so polluted that it represents direct health dangers. For example, serious examples are described from London and Tokyo. In some industrial areas the atmosphere is permanently impure and gives chronic health problems.

Most air pollutants sooner or later reach the earth's surface and will be included in the soil. In this way noxious matter can enter the biological circulation. Even if the concentration in the atmosphere is low, continuous addition to the soil may give accumulations with harmful effects.

Many attempts have been made to quantify the different types of pollution. To obtain reliable figures is, of course, difficult. However, a valuable survey was recently published by Nriagu and Pacyna.[39] They conclude that human activities now have a major impact on the global and regional cycles of most trace elements, and that there is an accelerating accumulation of toxic metals in the human food chain. Supply to atmosphere, water, and soils, respectively, from different sources, has been calculated.

The duration of different types of pollution varies greatly. Most of the pollution substances stay only a short time in the atmosphere. In the water a lot of the matter will be absorbed by the sediments. In the soil, however, harmful effects of pollutants may last for hundreds of years. With the exception of negatively charged ions, most of the heavy metals will be absorbed in the soil and to some degree go into the circulation processes: soil — plant — animal(man) — soil. In some areas the soil may be so polluted after a time that growing of food plants ought to be dissuaded.

D. INFLUENCE OF CLIMATE-CHEMISTRY ON CHEMICAL COMPOSITION OF SOILS

When figures showing large geographical variations in chemical composition of precipitation appeared (Table 1), it was logical to look for possible effects on the soils. Analyses of forest humus samples showed clear relationships between the distance from the ocean and the ratio between different exchangeable cations[28,29] (Table 2). The amount of sodium and magnesium in relation to calcium decreased from ocean to inland areas. The elements chlorine, bromine, and iodine showed the same tendency. Later it was found that selenium had a distribution pattern like these halogenes.[40-42]

When we are aware of the long periods of time the climatic influences have lasted, the effects on soil chemistry are easy to understand. In the Norwegian forests the period has been, roughly speaking, 10,000 years, that is to say since the end of the last glaciation. In other parts of the world, outside the areas of Quaternary glaciation, the processes of soil formation can have run for a much longer time.

TABLE 2
Analyses of Humus Samples, Podzol

District	Number of samples	Loss on ignition	pH	Degree of base saturation	Ca (%)	Mg (%)	K (%)	Na (%)
Nord-Trøndelag	198	74.2	4.1	25.3	52.8 ± 0.94	31.8 ± 0.70	12.0 ± 0.30	3.4 ± 0.24
(F) Ytre Nord-Trøndelag	22	79.7	4.1	23.2	43.6 ± 1.88	41.0 ± 1.43	10.4 ± 0.52	5.0 ± 0.43
(E) Midtre Namdalsbygdene	46	77.6	4.1	24.4	49.4 ± 1.75	34.6 ± 1.18	11.8 ± 0.65	4.2 ± 0.49
(D) Sørli-Nordli-Røyrvik	33	67.9	4.3	28.8	59.7 ± 1.86	26.2 ± 1.24	12.4 ± 0.73	1.7 ± 0.15
Oppland	487	60.2	4.2	23.2	72.9 ± 2.33	15.3 ± 0.42	10.4 ± 0.24	1.4 ± 0.12
Buskerud	296	67.2	4.1	20.6	74.2 ± 2.94	14.9 ± 0.37	9.5 ± 0.24	1.4 ± 0.06

From Låg, J., *Acta Agric. Scand.*, 18, 148, 1968. With permission.

From the figures in Table 1 we can derive, e.g., that the supply of sodium at the Lista station has been as high as approximately 150 kg per m² in the course of a period of 10,000 years. Corresponding calculations can be made for other elements and other stations.

A comparison of the amount received by precipitation with the total content of the element in the soil may be of interest. A normal soil profile in mineral material often may have a content of sodium of 20 to 40 kg per m². During postglacial time the precipitation has brought an amount of about five times as much as that. The quantities of magnesium and calcium in rain and snow are lower at the Lista station, and may be approximately half of the total content of these elements in the soil. If the soil profile is shallow or rich in humus, the relative part derived from the precipitation will of course be larger.

The availability for the plants of most elements in the mineral matter is very low compared with those in the precipitation water.

The alkali and alkaline earth ions are more or less absorbed in the soils, especially those with two charges. Ions with electronegative charges such as chlorine and sulfate easily follow the percolating water, and are brought into the water-courses.

The total concentration of chlorine and sulfur in the soils is low. The concentration is for the most common rocks 0.05%. In normal soils we may often find a content of the order of magnitude of less than 1 kg per m². During a period of 10,000 years the soil surface at Lista has received about 250 kg chlorine and 20 kg sulfur per m². Even at Vågåmo the supply is approximately 1 and 3 kg, respectively.

The weight of the two elements sodium and chlorine together, brought by precipitation to Lista during postglacial times, accounts for more than the weight of a normal plough layer. The stations Stend, Gjermundnes, and Ytterøy, situated in fjord regions, have received approximately 20 to 30% of the bulk of a 20 cm deep cultivated layer over a period of 10,000 years.

As a geological period the time since the Quaternary deglaciation is short. In many parts of the world the chemical climate has influenced soil formation much longer, and thus the soil properties are correspondingly stronger.

The large amount of matter brought down by rain and snow, and the pronounced geographical differences will, of course, have an influence on soil chemistry as well as on the ecosystems.

Analyses of barley and wheat showed a larger content of iodine and bromine in samples from coastal areas than from central parts of Norway.[3,43] Chlorine did not behave in the same way, and the reason seems to be that large quantities of this element are added to the soils by fertilizers.

The supply by precipitation of the nutrient magnesium is, as Table 1 shows, much larger in coastal than inland areas. In an experimental farm in the island Smøla, deficiency of most nutrients except magnesium has been stated, and the explanation is the high concentration in the precipitation water.

E. EXAMPLES OF GEOMEDICAL CONSEQUENCES OF CLIMATE-CHEMISTRY

The relationship between chemical climatology, iodine content in soils and plants, and distribution of goiter are easy to understand. In Norway goiter was common in the eastern central part of the country up to the first part of this century. Rare use of saltwater fish was for a long time supposed to be the reason. Now we know that also the plant food and feed produced in this district are poor in iodine.

More recently, the element selenium has received a lot of attention in the medical field. As mentioned in Section VIII, the content of selenium in forest humus decreases with increasing distance from the ocean, and with decreasing precipitation. In districts which have been shown to have low selenium content in the soil, veterinarians have previously used selenium preparations to prevent muscular degeneration in domestic animals on an empirical basis.

The ratio between calcium and magnesium is interesting in connection with some medical questions. Great variations in the amounts supplied from the atmosphere occur, and more comprehensive studies may give valuable results, e.g., concerning cardiovascular diseases.

Differences in precipitation acidity can affect the concentration of, e.g., soluble aluminum and manganese in soils and water, thus having geomedical consequences.

Because it is only a short time since chemical climatology was brought up in serious discussion, we may expect that valuable discoveries will appear in the near future.

VIII. DIFFERENT REQUIREMENTS OF PLANTS AND ANIMALS

Animals and man need even more elements than the plants. The nutrients are commonly classified, in accordance with the quantities needed, in macro- and micronutrients. The expression "trace elements" is used for those which occur in minute concentrations, both nutrients and elements not needed by the organisms.

We have the following macronutrients for the plants: C, O, H, N, P, S, K, Ca, and Mg. As micronutrients are regarded: Fe, Cl, Mn, B, Cu, Zn, and Mo. Some scientists would also include Si and V in the last group.

The animals, including man, need the same elements except most possibly B. In addition Na, I, Co, Se, Cr, and F are nutrients. Further, in some textbooks, Si, V, Ni, and Sn are also included.

Perhaps, in the future, still more elements will prove to be necessary for plants and animals.

Many elements are toxic, even in small concentrations. Up to now it has been common to speak about lead, mercury, cadmium, and arsenic as the toxic ones. However, many other elements can give serious poisoning effects. In very small concentrations poisonous matter such as cadmium and arsenic may have a positive influence. Some nutrients can give poisoning in small amounts. Boron in plants and selenium in animals may be mentioned as examples.

For the nutrition of man and animals it is beneficial if the plants are sensitive toward matter being dangerous higher up in the nutrition chain. If the plants are killed by lower concentrations than what is harmful for man and animals, the last group can escape the threat of poisoning. The situation is in this way favorable concerning, e.g., chromium and cobalt, whereas cadmium, mercury, and thallium can more easily pass the plants and then harm animals and man.[44]

IX. INFLUENCE OF PHYSICAL FACTORS

Climatic conditions like temperature, precipitation, wind, and air humidity can have both a direct and indirect influence on the health situation. Radiation of different kinds is another important group of physical factors.

It is easy to understand the effects of temperature after drastic reactions like sun-stroke and frostbite. On other occasions the explanations are more difficult to find. As an example may be mentioned the geographical distribution of the disease multiple sclerosis. We don't know the reason why this illness does not occur in warm climatic regions.

The temperature is of course one of the decisive factors for the distribution of the so-called tropical diseases. Such infectious illnesses are serious plagues to man and animals in the tropics and subtropics. High temperature is necessary for the development of the infectious organisms or the intermediate hosts they need.

The humidity of soil and air is of importance for the distribution of many diseases. For instance, many intermediate hosts both in the tropics and in cooler regions are dependent on open water or wet soils. Some illnesses in the respiratory organs are related to the degree of humidity in the air.

Ultraviolet radiation may cause cancer of the skin, but it is a beneficial factor for calcium metabolism. Questions on changing radiation through the atmosphere have been discussed vigorously in connection with possible disturbance in the layer of ozone.

Specific problems are connected to radioactive radiation. After the discovery of the occurrence of such radiation in 1896, the medical effects have been studied thoroughly. Incidences of diseases after the explosion of the atom bombs in Japan in 1945, and special disasters like the one in Chernobyl in the Soviet Union in 1986 have been starting points for comprehensive investigations.

X. INCREASING KNOWLEDGE OF THE EFFECTS OF DIFFERENT TRACE ELEMENTS

Important functions for many of the macronutrients were comparatively easy to discover. The situation was different for many trace elements. Modern research methods have, however, given very interesting results.

Many micronutrients are central ions in enzymes. Cobalt is included in vitamin B 12, and iodine in the hormone thyroxine. Metabolic processes, of primary importance to life, are in these ways dependent on such necessary trace elements.

It has gradually become clear how many minor nutrients act in different enzyme systems. Some elements, especially iron and zinc, take part in many different reactions, while others are more specific. Intercellular processes are to a great extent dependent on these elements.

Attempts have been made to calculate the exact amounts of essential elements in the human body. A person weighing 70 kg has approximately 20 mg selenium, 5 mg chromium, and 5 mg molybdenum. Even with these minute amounts each cell has hundreds of thousands of atoms if the matter is evenly distributed.[45] For these three elements the figures should be 1.5×10^6, 6×10^5, and 3.2×10^5, respectively.

XI. DEFICIENCY, POISONING, AND BALANCE SITUATIONS BETWEEN DIFFERENT ELEMENTS

A. INTRODUCTION

Nutrition problems have a central position in the subject of geomedicine. Hunger is always a serious question. Hundreds of millions of people in the world are suffering. But the general hunger problem will not be examined in this connection.

Acute cases of deficiency and poisoning are anxiously observed, because they are serious and, as a rule, easy to recognize. However, chronic instances are of high importance because they are quite common.

Antagonism and synergism between different substances should be taken into account. It is often impossible to explain causal connections without taking note of these relations.

There are special difficulties in the general evalution of these problems, caused by the varying quality of information on health situations in different parts of the world. Even in countries having well-developed statistics, the registration of cases of deficiency and poisoning in man and animals is usually rather incomplete. In this connection, we have to remember that we have derived clear diagnoses for some of these important illnesses only a short while ago.

Large parts of the world have only inadequately developed health statistics. Deficiency and poisoning are likely to be quite frequent in the developing countries.

When we try to estimate future development, we should be aware that scientists have studied, for example, cadmium and selenium problems for only a comparatively short period, and we can expect that important new discoveries will be made in rapid succession. We may now be on the verge of a valuable development.

B. DEFICIENCY OF MACRO- AND MICRONUTRIENTS

An inadequate supply of calcium and phosphorus will cause a faulty skeleton. The amounts of available calcium and magnesium are thought to be of importance for cardiovascular diseases. A very large number of people in the world suffer from a lack of protein. Insufficient intake of nitrogen compounds is a reason for many nutritional problems.

A large number of other health problems connected to deficiencies of macronutrients have, of course, been under investigation. However, research activities have concentrated on micronutrients. Knowledge of the functions of trace elements in nutrition has been growing rapidly. Thus, a basis is formed for solving serious nutritional questions. Exact statistics both of health problems and related environmental factors are prerequisites for reaching beneficial results.

C. SCARCE SUPPLY OF VITAMINS AND OTHER ESSENTIAL ORGANIC COMPOUNDS

The geographical distribution of diseases which are caused by deficiency of vitamins (avitaminosis), shows instructive geomedical examples. Often the natural factors have been decisive, but on some occasions man's pretreatment of food and feed has had unfortunate results.

Populations living in intimate contact with nature have often by experience gained impressive knowledge of complicated nutritional problems. However, when they are influenced by people from industrialized nations, they often meet difficulties.

In developed societies, changes in living habit and food treatment have introduced special problems. The history of vitamin research presents a number of such examples.

Deficiency of essential amino acids can have very serious consequences. In developing countries a large number of inhabitants suffer from this nutritional condition. Other organic compounds, e.g., some fatty acids, are also found to be essential.

D. POISONING BY INORGANIC AND ORGANIC MATTER

The toxic effects of arsenic compounds and mercury have been known since ancient times. Lead poisoning has also been known for a long time. Exact knowledge on cadmium as a toxic element emerged for the first time late in the 1950s.

The questions on poisoning are of course tied up to the concentration and amount of the substance concerned. If the amount is large, most of the elements can cause toxication.

Classic examples of explaining serious poisoning concern fluorine and selenium (see Section II, B and C). A large number of other elements have on special occasions caused poisoning.

Nitrate and nitrite poisoning in man and animals is often a result of excessive use of commercial fertilizers.

Toxic effects of plant material have been known for a very long time. The Greek philosopher Socrates was, according to tradition, executed by poison from the plant *Conium maculatum*. Comprehensive surveys on dangerous vascular plants have been prepared (see, e.g., Reference 46). Nevertheless, every year tragic disasters happen because children have eaten poisonous plant matter.

Among the fungi, a large number of species are toxic. Some bluegreen algae are dangerous producers of poison.

Some useful organic compounds can also be poisonous when consumed in too-large quantities. A number of hyper-vitaminosis cases have been reported.[34]

E. MUTUAL EFFECTS BETWEEN DIFFERENT ELEMENTS

The biological effect of one specific element often depends on other elements in the environment. The concept of antagonism and synergism has had a central position in physiology for a long time. In trying to explain many of the complicated processes in the soil, mutual reactions between different elements have to be regarded. As an instructive example, the relation between potassium and magnesium in soil and in animal nutrition may be mentioned. When large quantities of potassium fertilizers are used, the plants take up less magnesium, even if the magnesium content of the soil is not changed. Feed with low concentrations of magnesium can result in a deficiency of this element in the blood, followed by an attack of tetanus. Other elements, too, affect the magnesium concentration of the blood.

Among the trace elements a number of nutritional interactions are known. Common examples are those concerning cadmium. Antagonism between cadmium and several other elements has been demonstrated.

Between copper and molybdenum important interactions occur. Molybdenum deficiency can be a contributing factor in copper poisoning in animals, and in the opposite way copper deficiency can promote molybdenum poisoning.

Selenium interferes in the metabolism of several elements. Relationships between selenium and vitamin E (tocopherol) are shown. The content of selenium, and also of cobalt, seems to decrease in the vegetation in connection with use of phosphorus fertilizer.[47]

XII. REGIONAL INVESTIGATIONS

Knowledge on the regional distribution of health problems and of important environmental factors can give a valuable starting point for geomedical investigations.

In many countries, fairly detailed surveys on geographical distributions of important groups of diseases have been carried out. The results of the registrations have partly been presented by maps. In particular, cancer and cardiovascular diseases have been mapped. Special records have been compiled for many other illnesses. The quality of such presentations varies. In many developing countries this type of material is very scarce.

We hope that exact registrations of important diseases will increase rapidly. Comprehensive investigations are especially valuable in many developing countries. Further comparison of old and new diagnostic criteria will be of interest in countries which have had statistics for longer periods.

Compilation of data concerning environmental factors of importance in geomedicine can be valuable as basic material. Questions covering nutritional needs will have a central position in this connection. Many other problems, e.g., radiation, should also be taken into account.

Some examples of regional investigations are reported in this book by other authors.

XIII. PROPHYLAXIS

To cure diseases has been a primary duty of physicians and veterinarians. For a long time it was nearly the only task for many of them, but the scope of work gradually expanded. Now it is just as important to prevent diseases as it is to care for the sick. The limited activity in prophylaxis can have had several reasons. Treatment of diseases has been regarded as more urgent than prevention. Knowledge in broader scientific fields will, as a rule, be more necessary in prophylaxis than in curing.

A rapid growth will most possibly take place in the prevention of diseases in the near future, both in developing and industrialized countries. An important basis for successful activities will often be well-founded knowledge of environmental factors.

In many developing regions, natural conditions which lead to erroneous nutrition are frequent. Dangerous poisonings may occur. Many different parasites are troublesome.

In developed countries, pollution is a threat in many places. The medical consequences of such situations have been realized through time.

The acute cases of deficiency and poisoning have been concerns for a long time. More comprehensive investigations will probably show that more stress should be laid on the chronic illnesses. In preventive medical activities, both drastic cases and less-pronounced long-lasting effects should be taken into account.

A number of diseases are not regarded as so serious because they normally do not end with death, for example, rheumatism and asthma. When preventive medicine is more developed, such groups of illnesses will receive more attention.

Industrialization has caused great changes in geomedical problems. Meeting some nutritional requirements has become relatively easy; on the other hand, pollution has created new serious difficulties.

The future food supply of mankind has been a basis for anxious discussions. The FAO (Food and Agriculture Organization of the United Nations) has provided a lot of statistics in this connection. Modern technology has made it possible to increase food production quite rapidly, but many of the reclamation enterprises have had important biological consequences. Every year large areas, especially of forest and grassland, are cultivated for agriculture (see, e.g., Reference 48). At the same time, earlier cultivated land is abandoned due to soil erosion and spreading of towns. Intensive cultivation by irrigation, heavy fertilization, and pesticides can pollute the surroundings. A considerable reduction in the rapid growth of world population is needed if the food supply is not to turn into a catastrophic situation within some decades.

Developed countries have continued to load the environment with pollutants. A division between contamination and pollution has been tried,[49] and the last-mentioned concept should in this connection only be used when a harmful biological effect is a reality. Generally, however, the two words have been regarded as synonymous.

A special manner of evaluation compares the amount of matter which yearly circulates in nature with the amount brought out by mining. An element is regarded as a potential pollutant if mining brings at least ten times as much as the natural circulation.[49] Based on such calculations Bowen specifies the following pollutants: Ag, Au, C, Cd, Cr, Cu, Hg, Mn, N, Pb, Sb, W, and Zn.

The degree of toxicity varies quite a lot. Some elements can be pollutants even if they are not present in large quantities compared with the regular circulation, for example, arsenic, beryllium, and nickel.

The burning of huge amounts of coal and oil has increased the content of carbon dioxide and released a range of pollutants into the atmosphere. Anxious discussions have dealt with the harmful effects of acid precipitation. Fish death is a drastic example. However, such pollution may also have long-lasting effects which are more difficult to interpret. The precipitation annually supplies the soil with unwanted matter. The amounts are, as a rule, very small, but after a long period they may have negative biological effects. The ratio between available nutrients of plants and animals can be changed by pollution.

Polluted precipitation usually contains nitrogen compounds. The vegetation will often react positively to this addition of matter. However, partial addition of nutrients, as a rule, leads to other necessary elements in the soil being exhausted. Many experiments have shown interesting mutual effects between various nutrients. A large part of the nitrogen compounds in pollution gases are formed during the burning processes.

Local enterprises have in many places loaded the environment heavily. Many types of metallurgic industry have poisoned their surroundings with heavy metals. Arsenic and fluorine are pollutants from special factories. Some industries spread poisonous organic compounds. Municipal combustion plants have on some occasions caused difficulties. Mining has been an important source of toxic elements; on some occasions serious poisonings have taken place. The disaster in Bhopal, India, in 1984 may be mentioned as a recent example.

Means of transit, especially cars, pollute a lot. Some investigations on the dust in large towns have revealed high figures of dangerous elements like cadmium and lead. For example, in Tokyo an annual supply of 109 mg lead, 98 mg manganese, and 10 mg cadmium per m^2 was recorded.[50] During a period of 25 years, a common plough-layer has increased its cadmium concentration by about 1 ppm. The content of lead increases more than 10 times as rapidly.

Consideration of the size of the harvest has generally been a dominating concern in agriculture. Alteration of cultivation techniques, however, can change both quantity and chemical composition of the crops. Demands for good quality of agricultural products are now more pronounced than before. High concentrations of toxic metals, or low content of nutrients, are serious defects. In prophylactic activity in the future, these questions will be studied intensively.

Changed processes in food industry can result in changed quality. An example is the occurrence of beriberi disease after the introduction of a new rice preparation technique late in the 19th century. With the new method valuable matter was removed, which later proved to be rich in vitamin B1 (thiamine).

Knowledge of change of common environmental factors is of importance when prophylactic activities are planned. In this preventive work the causes of health troubles must be considered, i.e., physical, chemical, nutritional, infectious, and psychological factors. Resistance to factors which have a negative influence on health must be studied. Improving the environment is an important task in prophylaxis.

XIV. FINAL REMARKS

As explained above, a number of isolated discoveries were made a long time ago. Many problems were solved by using existing knowledge both on diseases and environment. It is easy to understand that Hippocrates and other ancient Greeks studied environmental factors in order to explain the causes of deficiencies. For a long time medical doctors worked closely with the natural sciences.

Many physicians also had comprehensive knowledge in sociological questions; such information was often valuable when epidemic problems were to be solved. A connection between the water supply and the cholera epidemic in London in 1854 was determined on this basis. The water pump in Broad Street became world-famous because John Snow was able to trace the epidemic back to this place.

Fundamental microbiological discoveries in the last half of the nineteenth century introduced a new epoch in the fight against many illnesses. It was said that diseases could begin to be studied directly, without looking at the social situations. This way of working entailed, however, a narrowing of the field of medical activity. Recently, many universities and colleges have also given relatively less emphasis on nutrition in the education of medical doctors.

We are much better equipped now to find solutions to geomedical questions. The total sum of knowledge is much greater, and possibilities of carrying out experiments better. On the other hand, mastering all the pertinent information is more difficult for a single person.

The desire to strengthen the prophylactic activities concerning man and animals increases the need for taking account of environmental factors. Thorough systematization and explanation of the factors of greatest geomedical importance are monumental. Highly improved analytical techniques are very valuable in this connection.

Specialization is necessary in all sciences, but this can result in isolation between different scientific fields. In order to be able to solve many of the complicated questions it is necessary

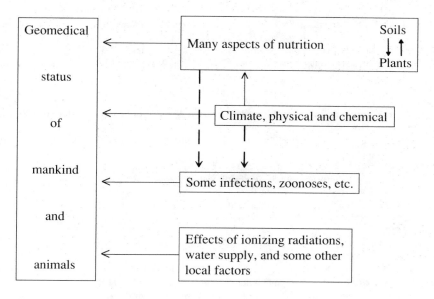

FIGURE 4. Schematic presentation showing influence of some important factors on geomedical conditions.

that specialists from essentially different subjects take part. Geomedicine is a typical example of a field where such collaboration is needed. (See Figure 4.)

Increasing geomedical research is very important. Particular expectations are connected to those investigations which form a better basis for prophylactic activities.

Some organizational efforts have resulted in cooperation on geomedicine (see Section IV.) One example is the Working Group "Soils and Geomedicine" of the International Society of Soil Science. It may be beneficial to set up a broadly based organization to promote geomedical activities.

New challenges are expected; e.g., biotechnology, now in an early stage of development, will most probably create geomedical problems previously unknown.

It is possible that we are now at the beginning of a fruitful development in geomedical research. By purposeful work we should be able to progress and render valuable service to mankind.

REFERENCES

1. **Jacobi, G.,** *Das goldene Buch des Hippocrates. Eine medizinische Geographie aus dem Altertum.* Schriftreihe des Natura, Vol. 5, Stuttgart, 1930, 75.
2. **Lyons, A. S. and Tetrucelli, R. J.,** *Medicine. An Illustrated History,* Abrams, New York, 1978, 616.
3. **Låg, J.,** Soil science and geomedicine, *Acta Agric. Scand.,* 22, 150, 1972.
4. **Låg, J., Ed.,** *Geomedical Aspects in Present and Future Research,* Universitetsforlaget, Oslo, 1980, 226.
5. **Underwood, E. J.,** *Trace Elements in Human and Animal Nutrition,* Academic Press, New York, 1977, 545.
6. **Palsson, P. A., Gregorsson, G., and Petursson, G.,** Fluorosis in farm animals in Iceland related to vulcanic eruptions, in Låg, J., Ed., *Geomedical Aspects in Present and Future Research,* 1980, 123.
7. **Roholm, K.,** *Fluorine Intoxication,* Nyt Nordisk Forlag, H. K. Lewis & Co., London, 1937, 364.
8. **Fridriksson, S.,** Fluoride problems following volcanic eruptions, Proc. Intern. Fluoride Symp., Logan, Utah, May 1982, 339, 1982.
9. **Rosenfeld, I. and Beath, O. A.,** *Selenium, Geobotany, Biochemistry, Toxicity, and Nutrition,* Academic Press, New York, 1964, 411.

10. **Vogt, J. H. L.,** Norske ertsforekomster. V. Titanjernforekomsterne i noritfeltet ved Ekersund-Soggendal, *Archiv for Mathematik og Naturvidenskab.,* 12, 1, 1888.
11. **Esmark, J.,** Om Norit-Formation, *Magazin for Naturvidenskaberne,* 1, 205, 1823.
12. **Ender, F.,** Undersøkelser over alveldsykens etiologi (English summary) *Nordisk Veterinaermedicin,* 7, 329, 1955.
13. **Stabursvik, A.,** *A Phytochemical Study of Narthecium ossifragum (L.) Huds.,* Norges Tekn. Vitenskapsakademi, 1959, 2.
14. **Reichborn-Kjennerud, I., Grøn, F., and Kobro, I.,** *Medisinens Historie i Norge,* Grøndahl & Søns Forlag, Oslo, 1936, 328.
15. **Nicolaysen, R.,** *Arktisk ernaering.* Fridtjof Nansens Minneforelesninger. VII. Det Norske Videnskaps-Akademi, Oslo, 1970, 26.
16. **Storm, G.,** *Samlede Skrifter af Peder Claussøn Friis,* Kristiania. 1881, 493.
17. **Zeiss, H.,** Geomedizin (geographische Medizin) oder medizinische Geographie? *Münchener med. Wochenschr. Nr.,* 5, 98, 1931.
18. **Rimpau, W.,** Geomedizin als Wissenschaft, *Münchener med. Wochenschr. Nr.,* 25, 940, 1934.
19. **Jusatz, H. J., Ed.,** *Fortschritte der geomedizinischen Forschung,* Geographische Zeitschrift, Beiheft, 1974, 164.
20. **Doerr, W.,** Grussworte . . . am 9. Oktober 1972, in Jusatz, H. J., Ed., *Fortschritte der geomedizinischen Forschung,* Geographische Zeitschrift, Beiheft, 1974, 1.
21. **Weltzien, H. C.,** Geophytomedizin. Aufgaben und Möglichkeiten, in Jusatz, H. J., Ed. *Fortschritte der geomedizinischen Forschung,* Geographische Zeitschrift, Beiheft, 1974, 110.
22. **WHO,** *Environmental Health Criteria 27, Guidelines on studies in environmental epidemiology.* World Health Organization, Geneva, 1983, 351.
23. **Thornton, I., Ed.,** Proceedings of the 1st International Symposium on Geochemistry and Health, Oxford, 1987, 264.
24. **Vinogradov, A. P.,** *The Geochemistry of Rare and Dispersed Chemical Elements in Soils,* 2nd ed., Chapman and Hall, London, 1959, 209.
25. **Kovalskij, V. V. M.,** *Geochemische ökologie. Biochemie.* Berlin, 1977, 353.
26. **Goldschmidt, V. M.,** *Geochemistry,* Claridon Press, Oxford, 1954, 608.
27. **Låg, J., Hvatum, O. Ø., and Bølviken, B.,** An occurrence of naturally lead-poisoned soil at Kastad near Gjøvik, Norway, *Norges Geol. Unders,* 266, 141, 1970.
28. **Låg, J.,** Undersøkelse av skogjorda i Nord-Trøndelag ved Landsskogtakseringens markarbeid sommeren 1960. (English summary.) *Medd. fra Det norske Skogforsøksvesen,* 18, 107, 1962.
29. **Låg, J.,** Relationships between the chemical composition of the precipitation and the contents of exchangeable ions in the humus layer of natural soils, *Acta Agric. Scand.,* 18, 148, 1968.
30. **Butler, G. C., Ed.,** Principles of Ecotoxicology, SCOPE 12, John Wiley & Sons, New York, 1978, 350.
31. **Clarke, F. W.,** *The Data of Geochemistry,* U.S. Geological Survey, Bull. 770, 841, 1924.
32. **Wedepohl, K. H., Ed.,** *Handbook of Geochemistry,* Vol. 1, Springer-Verlag, Berlin, 1969, 442.
33. **Schormüller, J., Ed.,** *Handbuch der Lebensmittelchemie,* 9 vols., Springer-Verlag, Berlin, 1965-1970.
34. **Rechcigl, M., Ed.,** *Nutritional Disorders,* Vol. 1, Effect of Nutrient Excesses and Toxicities in Animals and Man. Vol. 2, Effect of Nutrient deficiencies in animals, CRC Press, Boca Raton, FL, 1978.
35. **Friberg, L., Nordberg, G. F., and Vouk, V. B.,** *Handbook on the Toxicology of Metals,* 2 vol., Elsevier, Amsterdam, 1986.
36. **Låg, J.,** Klimaets humiditet og jordsmonnutviklingen. Skogen og klimaet. Bergens Museum, Bergen, 1948, 87.
37. **von Liebig, J.,** Die organische Chemie in ihrer Anwendung auf Agricultur und Physiologie, Vieweg, Braunschweig, 1840, 353.
38. **Overrein, L., Seip, H. M., and Tollan, A.,** Acid precipitation — effects of forest and fish. Final report. Oslo-Ås., 175, 1980.
39. **Nriagu, J. O. and Pacyna, J. M.,** Quantitative assessment of worldwide contamination of air, water, and soils by trace metals, *Nature,* 333, 134, 1988.
40. **Låg, J. and Steinnes, E.,** Soil selenium in relation to precipitation, *Ambio,* 3, 237, 1974.
41. **Låg, J. and Steinnes, E.,** Regional distribution of halogens in Norwegian forest soils, *Geoderma,* 16, 317, 1976.
42. **Låg, J. and Steinnes, E.,** Regional distribution of selenium and arsenic in humus layers of Norwegian forest soils, *Geoderma,* 20, 3, 1978.
43. **Låg, J. and Steinnes, E.,** Halogens in barley and wheat grown at different locations in Norway, *Acta Agric. Scand.,* 27, 265, 1977.
44. **Sauerbeck, D.,** Welche Schwermetallgehalte in Pflanzen dürfen nicht überschritten werden, um Wachstumsbeeinträchtigungen zu vermeiden? Kongressband 1982, 108. Vorträge. 94 VDLUFA-Kongress, Münster, 1983.

45. **Kieffer, F.,** Metalle als lebensnotvendige Spurenelemente für pflanzen, Tiere und Menschen, in Merian, E., Ed., *Metalle in der Umwelt,* Verlag Chemie, Weinheim, 1984, 118.
46. **Frohne, D. and Pfänder, H. J.,** *Giftpflanzen,* 3rd ed., Wissenschaftliche Verlag, Stuttgart, 1987, 344.
47. **Halpin, C., Caple, I., Schroder, P., and McKenzie, R.,** Intensive grazing practices and selenium and vitamin B_{12} nutrition of sheep, in Gauthorne, J. M., Mowell, J. M., and White, C. L., Eds., *Trace Element Metabolism in Man and Animals,* Springer-Verlag, Berlin, 1982, 222.
48. **Wolman, M. G. and Fournier, F. G. A., Eds.,** *Land Transformation in Agriculture,* SCOPE 32. J. Wiley & Sons, New York, 1987, 531.
49. **Bowen, H. J. M.,** *Environmental Chemistry of the Elements,* Academic Press, London, 1979, 333.
50. **Kitagishi, K. and Yamane, J., Eds.,** *Heavy Metal Pollution in Soils of Japan,* Japan Scientific Society Press, Tokyo, 1981, 302.

Chapter 2

DEFICIENCY OF MINERAL NUTRIENTS FOR MANKIND

Jetmund Ringstad, Jan Aaseth, and Jan Alexander

TABLE OF CONTENTS

I. INTRODUCTION

Living organisms consist mainly of hydrogen, carbon, nitrogen, and oxygen. The concentration of these elements in biological matter can be expressed in grams per kilograms (g/kg), and in humans they are required in grams per day in the forms of fat, sugars, proteins, water, vitamins, and oxygen (O_2).

Another important group of elements found in living matter are the macrominerals calcium, phosphorous, sulfur, potassium, sodium, chlorine, and magnesium. They serve as necessary components of tissues and body fluids and are essential for the function of all cells. Their concentration in living organisms is somewhat lower than that of the main elements, but can still be expressed as grams per kilogram, and the requirement as grams or fractions of a gram per day.

Over 99.9% of all animal matter is made up from the abovementioned 11 fundamental elements and macrominerals, mainly as water, proteins, fat, and carbohydrates. The remaining 0.1% of elements found in living matter consist of a heterogenous group of minerals called trace elements. They occur as milligrams (mg) or micrograms per kilogram (μg/kg) of tissue (Table 1). These elements could not easily be quantified by earlier analytical techniques, hence, the name "trace".[1]

A. ESSENTIALITY

Trace elements may still be divided into subgroups: those which are nonessential and those which are essential to all cells. The nonessential elements may occur as environmental hazards or contaminants (lead, mercury, cadmium, aluminum, etc.) or even as drugs (gold, platinum); further reference will be made to them in Chapter 4.

The essential trace elements, however, are of importance in several biological processes. Though they are required in minute concentrations, inadequate intakes cannot be compensated for and will consistently result in impairment of a function from optimal to suboptimal. Supplementation of physiological levels of the missing element will prevent or cure this impairment. Essentiality is generally acknowledged when it has been demonstrated by more than one independent investigator and in more than one animal species.

The number of essential elements still seems to be increasing. There are two main reasons for this. First, comprehension of the biochemical functions, pathogenic defects, and pathobiochemical mechanisms of trace elements has gained in depth. Today, scientists from a large number of disciplines in the medical and natural sciences focus on the essentiality of the different elements. Second, the development of specific, precise, and reliably reproducible analytical techniques has allowed trace elements to be measured in body fluids and tissues with the same precision as the macroelements. The lack of such techniques was an enormous problem only a few decades ago and led to rather confusing results, as illustrated in Figure 1.

The following trace elements are now considered as essential for man and animals: iron, zinc, copper, iodine, chromium, manganese, molybdenum, cobalt, selenium, fluoride, vanadium, nickel, tin, arsenic, (boron), and bromine. The latter six or seven elements are designated "ultratrace" elements[2] because their estimated dietary requirements usually are less than 1 μg/g, and often less than 50 ng/g dry diet. In fact, their essentiality for mankind is not generally accepted, and boron is probably only essential for some plants. Nutritional deficiencies have not yet been described for any of the suggested ultratrace elements. However, this area of nutritional research is in rapid progress, and it is likely that some of these elements will prove to be more important in higher animals than is now generally acknowledged.

B. DOSE AND RESPONSE

Figure 2 shows the dependence of biological function on tissue concentration or intake of a nutrient.[3] Two conclusions can be inferred from this figure: (1) for each element there

TABLE 1
Physiological Functions and Estimated Total Amounts (in Grams) of Some Essential Macro- and Microelements

Element	Amount	Function
Calcium	1000	Enzyme activator, electrolyte
Phosphorus	700	Nucleic acid element
Sulfur	150	Proteins (SH bridges)
Potassium	140	Enzyme activator, electrolyte
Sodium	95	Balance of water and electrolytes
Chloride	80	Balance of water and electrolytes
Magnesium	35	Enzyme activator, electrolyte
Iron	5	Metallo-enzyme, enzyme activator
Zinc	4	Metallo-enzyme, enzyme activator
Copper	0.15	Metallo-enzyme
Iodine	0.03	Thyroid hormone element
Fluoride	?	Enzyme inhibition (glycolysis)
Manganese	0.02	Metallo-enzyme, enzyme activator
Molybdenum	0.02	Metallo-enzyme
Vanadium	0.02	Redox system
Selenium	0.015	Enzyme protein
Nickel	0.01	Metallo-enzyme
Cobalt	0.003	Metallo-enzyme, enzyme activator Vitamin component

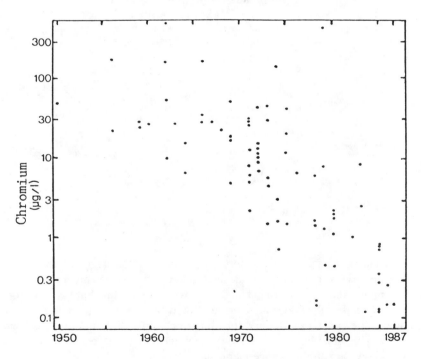

FIGURE 1.   80 ''normal'' values for chromium in human blood, serum or, plasma published between 1950 and 1987. (Note: logarithmic scale)

is a safe and adequate range of exposures and (2) deficient as well as excess intakes may lead to depressed cell function and death. The range of safe and adequate intakes is usually fairly large for most trace elements, compared to other pharmacologically active substances or nutrients. For instance, the intake of zinc may be increased 100-fold before symptoms of toxicity occur. On the other hand, one should beware that nonessential trace elements as

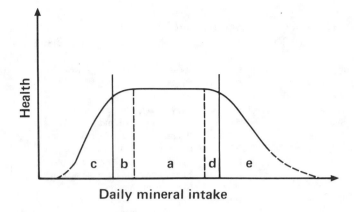

FIGURE 2.   The dependence of health on adequate intakes of trace elements. (a) Safe and adequate intake; (b) marginally deficient intake; (c) deficiency; (d) marginal overexposure, may have pharmacological effects; (e) excess.

TABLE 2
Recommended Safe and Adequate
Dietary Intakes for Adults

Element	Intake (mg/d)
Calcium	800
Phosphorus	800
Magnesium (males)	350
Magnesium (females)	300
Iron (males)	10
Iron (females)	18
Zinc	15
Manganese	2.5—5.0
Fluorine	1.5—4.0
Copper	2.0—3.0
Molybdenum	0.15—0.5
Chromium	0.05—0.2
Selenium	0.05—0.2
Iodine	0.15

well as other nutritional factors may interact with the essential trace elements at a cellular level so that the dose-response and the dose-effect curves may be shifted.

Table 2 shows the recommended daily allowances in adult humans for some trace elements as presented by the Food and Nutrition Board of the National Academy of Sciences.[4] The recommendations are well defined for iron, zinc, and iodine. For fluorine, chromium, manganese, copper, selenium, and molybdenum, the ranges have been estimated, while ranges are unknown for the ultratrace elements.

C. BIOLOGICAL ACTION

Trace elements and macrominerals perform key functions in the metabolism of animals and man, either as co-factors in various enzyme reactions or as necessary components of important structural proteins.[5]

This may be illustrated by the transport and metabolism of oxygen in higher animals. Iron transports oxygen by reversible combination to hemoglobin-Fe^{2+} and myoglobin-Fe^{2+}. Iron also participates in biochemical redox reactions involving Fe^{2+}/Fe^{3+}-interchanges, e.g.,

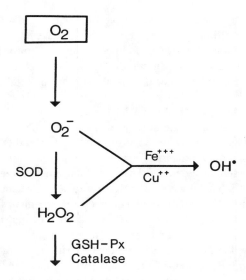

FIGURE 3. The univalent reduction of oxygen is illustrated. Superoxide (O_2^-) may be formed accidentally. It is converted to the less toxic H_2O_2 by cytoplasmic SOD (superoxide dismutase) or by mitochondrial SOD. H_2O_2 is detoxified by GSH-Px (see text), which contains selenium, or by catalase containing Fe^{3+}. Mixtures of H_2O_2 and O^- may give rise to the most toxic radical, OH·. The presence of *ionic* iron or copper catalyzes the formation of the latter radical.

in cytochrom oxidase in the mitochondria. Cytochrom oxidase, which contains both copper and iron, is an electron pair-transferring enzyme system which uses redox reactions to convert oxygen to water. The metabolism of oxygen within the cell is taken care of by enzymes containing copper, zinc, iron, and selenium. The overall *rate* of metabolic activity, and thus the total oxygen consumption, is regulated by the iodine-containing hormones thyroxin and triiodothyronine. Chromium is possibly part of a cofactor (glucose tolerance factor, GTF) that facilitates the insulin-regulated transport of glucose into the cell.[6] When glucose "burns" to CO_2, the escaping electron pairs move along a cascade of mitochondrial enzymes to reach O_2. The cellular production of energy, which is derived by this complex interaction of O_2 with carbohydrates forming H_2O_2 and CO_2, is stored by the use of phosphorus in energy-rich phosphate bindings. The turnover of such energy-accumulating compounds, e.g., ATP (adenosine triphosphate), is regulated by enzymes highly dependent on magnesium. The CO_2 derived from glucose combustion is dissolved and removed from the circulation (and lungs) by the action of a zinc-containing enzyme, carbonic anhydrase. Accidental uptake of a single electron by O_2 will give rise to oxygen radicals (O_2^-) which are toxic to cellular constituents.[7] For instance, an unphysiological dissociation of oxyhemoglobin may give rise to O_2^- and hemoglobin-Fe^{3+} (methemoglobin). Several enzymes containing copper, zinc, selenium, and iron (superoxid dismutase, glutathione peroxidase, and catalase) will protect against the toxicity of unphysiological oxygen species, thereby interrupting the free radical peroxidative chain reactions (Figure 3).

It appears that micro- and macrominerals are of crucial importance to energy and oxygen metabolism in *aerobic* organisms. It is, however, noted from Figure 3 that free ionic (nonenzymatic) iron or copper can give rise to toxic reactions, owing to their ability to catalyze the formation of the extremely toxic radical OH· from H_2O_2 and O_2^-. This demonstrates that essential elements, including oxygen, also possess toxic potentials.

Another example showing how minute amounts of an essential trace element may play a key physiological function, is the vitamin B_{12} complex, which has a chemical structure made up of porphyrines as hemoglobin. Cobalt, however, and not iron, is the important coordinated constituent of this coenzyme. Daily turnover of vitamin B_{12} corresponds to approximately 2 parts per trillion of the adult body weight. Nevertheless, deficiency of this vitamin, with its trace element *cobalt*, results in a serious condition, pernicious anemia and peripheral neuropathy. It is also an intriguing biological fact that this porphyrin-metal complex is indispensable for the synthesis of the visible porphyrin-metal complex, hemoglobin.

D. ABSORPTION, TRANSPORT AND EXCRETION

The preconditions for an adequate supply of trace elements and macrominerals are a sufficient supply of food and an undisturbed absorption. A mixed diet of cereals, meat, fish, milk, and cheese, as commonly consumed in most Western countries, is a good guarantee against deficiencies of essential elements. However, the quantitative amount of a trace or macroelement in the diet is not always an indication of the efficiency of utilization (bioavailability) of the element. Bioavailability depends on the chemical form of the element in the food, the pH of the stomach contents, and the presence of and interactions with other nutrients in the food. Sulfur, cobalt, and chromium are examples of elements that are not bioavailable at all if they don't occur as components of particular organic compounds, viz., sulfur-amino acids, vitamin B_{12}, and GTF, respectively. The intestinal segment of absorption as well as the mechanisms of absorption and mucosal transport differ to a large extent between the minerals. Simple diffusion or anion exchange transport is used by small anions, such as fluoride or selenite, while most cationic elements (e.g., zinc, iron, copper, magnesium, and calcium) use faciliated diffusion or active transport.

Metallo-plasmaproteins (albumin, transferrin, ceruloplasmin, and others) function as specific carriers of the absorbed elements. These carriers are normally not fully saturated and thus present a certain buffering capacity against excesses. Within the cell, iron and zinc are bound to specific storage proteins, ferritin and metallothionein. Copper is also attached to metallothionein. Of the macrominerals, calcium and magnesium are partly dissociated, whereas sodium and potassium, as dissociated ions, represent the osmotic water "binders", extracellularly and intracellularly.

The excretion of trace elements is usually through the intestines and kidneys. Urine accounts for nearly all the excretion of selenium, chromium, fluorides, and iodides, while iron, zinc, and copper are excreted into the intestines. Other excretory routes are losses via skin, sweat,[8] and breath, which can become important in hot climates. In conclusion, the organism has an amazing heterogeneity of mechanisms that regulate trace element and macromineral homeostases and maintain the plateau of optimal function according to the dose-response curve already described.

II. GENERAL ASPECTS OF INORGANIC ELEMENT DEFICIENCIES

As trace and mineral elements are transferred up the soil-plant-animal-human food chain, geographical differences account for large variations in the mineral content of foods. For instance, in The People's Republic of China, daily average intake of selenium ranges from 11 to 4990 μg.[9] Some of the most convincing associations between mineral deficiencies and disease have therefore been registered in the poor part of the world. Here, people depend on trace element concentrations in soil and water and thus are under local environmental influence, much more so than the Western populations with their rich access to varied diets. In the industrialized part of the world, public health measures of increasing mineral intake are feasible. Table salt and bread are fortified with iodine and cheese is commonly fortified with iron. In addition, domestic animals are supplemented with minerals and sales of mineral pills have increased several fold. Therefore, the main nutritional problems in developed countries are the recognition of *marginal deficiencies*, the definition of their consequences, and, if necessary, their prevention. An important question that remains unanswered is whether increasing the intake of certain minerals will decrease the risk of cancer or cardiovascular disease.

Going back to the different elements, deficiencies of iron and iodine are classical, easily recognized clinical entities, well described in standard textbooks of medicine. Iron-deficiency anemia and iodine-deficiency goiter or hypothyroidism thus no longer present particular difficulty in diagnosis. Deficiencies of some other elements (e.g., zinc, magnesium, copper, and selenium) may also give rise to characteristic clinical pictures, which will be described in the following text. As for protein and calorie malnutrition, as well as for iron and iodine deficiency, the reader is referred to standard textbooks of medicine.

A. DEFICIENCY OF ZINC

Clinical and experimental studies suggest that zinc deficiency plays an important role in the pathogenesis of several diseases. The clinical diagnosis of zinc deficiency is, however, difficult because of the complex interactions of pathophysiological factors influencing the serum levels. Zinc is an essential constituent of several vital proteins. More than 70 enzymes have been found to require zinc for their proper functioning.[10] Among biochemical processes requiring zinc-containing enzymes are the synthesis of DNA and RNA, as well as the synthesis of proteins, such as keratin and collagen. The crucial role of zinc in nucleic acid metabolism and protein synthesis may explain the relatively high zinc requirements in man. Disorders of zinc metabolism occur in a broad spectrum of clinical disorders.[11] Intestinal malabsorption as well as nutritional deficiency may lead to hypozincemia. In Iran, Armenia, and Egypt, zinc-deficient patients were reported under the term *"hypogonadal dwarfism"*. A detailed description of this condition, which can be cured by zinc supplementation, is given by Prasad.[12] The dominating traits of this condition were retarded growth, anorexia, and hypogonadism. Patients with *acrodermatitis enterohepatica*, a disease with selective failure in the absorption of zinc, suffer from retarded growth, dermatitis, and alopecia. This condition is well treated with large, oral doses of zinc. Delayed wound healing and low resistance toward infections may accompany zinc deficiency. Impaired sensory functions, including night blindness and loss of the senses of taste and smell, are restituted when zinc therapy is given. Even some cases of infertility in the male seem to be related to deficiency of this trace metal, but further research is necessary.

B. DEFICIENCY OF SELENIUM

The biological role of selenium is considered in the context of the protection offered by this nutrient against oxygen radical-induced damage to cellular structures. Clinical effects of selenium deficiency have been reported from a district in The People's Republic of China where selenium intake is extremely low. Chinese investigators showed that selenium deficiency is one of the principal factors responsible for Keshan disease, a cardiomyopathy mainly affecting children and women of child-bearing age.[13,14] Supplementation with sodium selenite was shown to be effective in reducing the incidence, morbidity, and fatality of the disease. A similar condition was later described in subjects administered total parenteral nutrition without selenium supplementation.[15,16] A large number of selenium-deficient diseases are described in veterinary medicine.[17] Mulberry heart disease seen in pigs, white muscle disease in sheep, and exudative diathesis in poultry brought about great economic loss.

In recent years, the risk of both cardiovascular disease and cancer morbidity and mortality have been observed to be somewhat lower in subjects with a high serum selenium concentration.[18,19] However, whether increasing the intake of selenium will decrease the incidence of these diseases remains unanswered.

C. DEFICIENCY OF COPPER

Copper deficiency might be a problem in malnourished children and in patients on long-term parenteral nutrition. Signs of copper deficiency are anemia, leucopenia, skeletal defects, and demyelination and degeneration of the nervous system.[12] Animals severely deficient in copper are also hypercholesteremic and their hearts and arteries have abnormal connective tissue.[20] Severe copper deficiency in humans is seen in Menkes' kinky hair disease, an X-linked, recessively inherited disease where the absorption of copper is defective.[21] Even if plasma copper can be raised by infusions, these patients usually die before the age of 2 years, probably because of a defective cellular utilization of the element. The biochemical functions of copper are linked to specific cuproenzymes, for example, cytoplasmic superoxide dismutase, cytochrome c oxidase, dopamine-β-monoxygenase, and tyrosinase. Several investigators have pointed out that the average copper intake is marginal or deficient in most of the Western world, compared to the recommended intake of at least 2 mg/d. Based upon results from animal experiments, it has been hypothesized that such marginal intake is of importance in the pathogenesis of ischemic heart disease.[22]

D. DEFICIENCY OF CHROMIUM

In experimental animals, chromium deficiency is characterized by impaired glucose tolerance with fasting hyperglycemia, glycosuria, hypercholesterolemia, corneal opacities, and impaired growth. Strong indices of the essentiality of chromium originate from three studies in man on parenteral nutrition.[23] These patients developed an impairment of glucose tolerance that was reversed by chromium administration. Since a deficiency of chromium may disturb normal cholesterol and sugar metabolism, there may be several risk factors for cardiovascular disease.[24] Although direct relationships have not yet been established, future research might conclude that supplementation of chromium will improve general public health.

E. DEFICIENCY OF MAGNESIUM

The magnesium ion acts as a cofactor and activator of many enzyme systems. A normal magnesium balance is of importance for normal potassium homeostasis. Magnesium is necessary for the function of Na^+/K^+-ATP-ase and thus the transport of K^+ and Na^+ across the cell membrane. Magnesium is linked to energy-requiring processes such as neurochemical transmission, muscle contraction, and protein, fat, and nucleic acid synthesis. Dietary deficiency of magnesium will therefore lead to reduced function within several organ systems.

Especially important is the consequence of magnesium deficiency on intracellular potassium and calcium levels. Deficiency of magnesium will entail elevated levels of intracellular calcium and loss of intracellular potassium, both of which may lead to cardiac arrhythmias, i.e., ventricular extrasystoles, ventricular tachycardias, and even ventricular fibrillation. The increased mortality of ischemic heart disease reported from geographical areas with low levels of magnesium in drinking water is thus of particular interest.[25] Because magnesium is predominantly an intracellular cation, serum magnesium is invalid as an index of magnesium deficiency. This implies that clinical studies on diseases linked to hypomagnesemia must be interpreted with caution.

III. DISCUSSION

Diseases due to iron deficiency, iodine deficiency, and potassium losses as well as protein or calori malnutrition are among the most frequent health problems in the world. Since these illnesses are well described in ordinary medical textbooks, they are not discussed further in the present text. A wide range of other effects due to deficient intake of mineral nutrients have been reviewed above and further described in other overviews.

Of particular geomedical interest are hypotheses that marginal deficiencies of several mineral nutrients might contribute to the development of cardiovascular disease and cancer. In the following discussion, some of these aspects will be referred to, and some comments on analytical problems will be given.

A. CARDIOVASCULAR DISEASES

Life style factors such as low relative intake of polyunsaturated fat (low P/S ratio), low physical activity, and smoking have been shown to promote cardiovascular diseases.[26] Low dietary intake of magnesium and probably copper, chromium, and selenium are also apparently linked to the development of atherosclerosis.[24]

The epidemiologic indications linking magnesium and atherosclerotic disease come from reports of an inverse association between the level of water hardness and cardiovascular mortality.[25] Studies from Finland and other countries have precipitated this hypothesis. However, there are both positive and negative findings with regard to the cross-sectional relationship between magnesium and cardiovascular disease. No difference was found in serum magnesium between people who died of coronary heart disease and the controls in a follow-up study of 10 years.[27] On the other hand, there is no doubt that magnesium deficiency can cause cardiac arrhythmias in patients with acute myocardial infarction and in patients taking diuretics and digitalis. The serum magnesium level may also influence the size of the myocardial infarction.

Severe copper deficiency produces both myocardial and vascular lesions and hypercholesterolemia in experimental animals.[20] Elevated levels of serum cholesterol have also been found in man during copper depletion. This change has been reversed during copper repletion. Data on the association between copper and cardiovascular risk factors in humans are, however, difficult to interpret, as smokers have increased serum copper concentrations.

The data linking chromium to human cardiovascular disease are inconsistent. Results from Finland suggested that coronary heart disease may be associated with a low concentration of chromium in the drinking water,[28] a finding which is not confirmed in other studies. When evaluating the studies on the association between chromium and human disease, one has to remind oneself of the many difficulties in measuring this element.

Serum measurements of magnesium, copper, and chromium are of limited, if any, value in assessing the body status of these three minerals. This may explain why we still lack knowledge regarding the associations between such mineral intake and heart disease.

As has been previously pointed out in this chapter, a dietary deficiency of selenium causes cardiomyopathy in both humans and animals. We have, however, no evidence that selenium deficiency promotes the development of atherosclerotic disease. Nevertheless, an increased risk of stroke and coronary death have been observed in some prospective follow-up studies.[18,29] The reason for this may be thrombotic mechanisms as selenium seems to play a part in platelet aggregability. Selenium may also lead to reduced prostacyclin synthesis, reduced protection against cardiotoxic heavy metals, an impaired active calcium transport, and a reduction in peroxidation capacity, all mechanisms which may be of importance in the etiology of cardiovascular disease.[30]

B. CANCER

Cancer is the second most frequent cause of death in the Western Hemisphere. In the U.S., this disease is responsible for over 20% of the annual deaths. Today, cigarette smoking and heavy alcohol drinking are among the leading known causes of avoidable cancers. Nutritional factors also play a significant role in the process of carcinogenesis. Estimates of the percent of cancers for which dietary factors are responsible range from 30 to 60%.[31] Modification of dietary habits, as well as increased intake of specific mineral nutrients, have therefore been postulated to modify the process of carcinogenesis.

Selenium is even *identified* as a potential modifier of the process of carcinogenesis. Evidence for a role of selenium in cancer is based upon *in vitro,* animal, ecological (correlational), case control, and prospective studies. Selenium decreases the mutagenic activity of various known carcinogens,[32] and it alters the degradation of certain carcinogens to favor less toxic metabolites.[33] Animal studies show that under certain conditions relatively high dietary levels of selenium can have a protective action against a variety of chemically induced, spontaneous, and transplantable tumors in rats and mice.[34] Geographic areas with low selenium levels in the soil or in pooled serum samples generally have higher cancer rates than areas with higher selenium levels.[35] Patients with cancer usually have depressed serum selenium levels,[36] probably due to insufficient dietary intakes. More interesting is the inverse association with total cancer risk observed in cancer patients with low prediagnostic serum selenium concentrations.[37,38] These prospective (or nested case control) studies, adjusted for age, sex, and cigarette smoking, have shown a particularly strong protecting effect of selenium against gastrointestinal, lung, and hematological malignancies in males. The proposed cancer protective *mechanisms* of selenium could be alleviation of carcinogen-induced oxidative damage, alterations in carcinogen metabolism, and selective toxicity to dividing cells.[39]

In general, one has also to remember that mineral nutrients are essential for normal proliferation of tissues. It is therefore most likely that mineral elements are involved in the metabolism and proliferation of cancer cells, too. For instance, zinc and magnesium have crucial functions in cellular growth. Theoretically, *depletion* of these elements might retard the growth of an already *existing* tumor, while an excess might stimulate tumor growth. In fact, this has been shown for zinc,[40] while the role of magnesium in the pathogenesis of neoplasia is not very clear at present. However, this mechanism of growth promotion should be differentiated from true carcinogenecity. It would appear to be paradoxical if an accepted essential nutrient were unveiled as a carcinogen. Nevertheless, an increasing literature is now pointing on the ability of (hexavalent) chromium to induce bronchiogenic cancers.[41] Even iron is suspected to be a lung cancer promoter when it is inhaled as dust.[42]

C. ANALYTICAL ASPECTS

The mineral content of whole blood, serum or plasma, and urine are the most commonly used measurements for trace element status assessments. They are easy to sample, contamination can be controlled, and analyses can be made on stored samples. However, the choice of substrate and the interpretation of data depend on thorough knowledge of the metabolism of the element under investigation. Recent food intake, diurnal variations, influences of hormonal status, chemical transformations, and equilibrium with available body stores are thus factors to be considered.

Another approximation to measuring (or calculating) mineral nutrient status can be made by comparing the actual dietary intake of minerals with the recommended dietary allowances. However, the quantitative amounts and the bioavailability (i.e., efficiency of utilization) of minerals in the diet may vary with both geographical and ecological conditions.

Biochemical functional measurements (e.g., determinations of enzyme activities) may be used when diagnosing deficiency states. One must, however, remember that enzyme activity tends to plateau and that data on enzyme activity is difficult to compare between laboratories.

Apparently, assessments of the mineral status of humans are still not an easy task in clinical laboratories, neither for the "ultratrace" minerals nor for the common elements like magnesium, copper, zinc, and chromium.

IV. CONCLUDING REMARKS

When working with geomedical problems, it is mandatory to have good estimates of dietary intakes and the body status of the different elements. This makes demands on the analytical measurements, as the specificity, accuracy, and precision of the analyses are the foundation for trace element status assessment.

When correlations between disease and local geochemistry are found, they must be interpreted with caution, as knowledge about the metabolism and bioavailability of the elements also is claimed. Nevertheless, geomedical studies can be indicative of a causal relationship, as is illustrated by the Chinese studies on selenium and Keshan disease. Hypotheses derived from observed correlations, however, have to be confirmed by using other approaches and investigations.

The classical examples of geomedical diseases are iodine-deficiency goiter, hypogonadal dwarfism due to zinc deficiency, and anemia in iron deficient regions. In the modern Western world, the constant refinement of diet may lead to suboptimal intake of mineral nutrients. It remains to be demonstrated if such food processing, involving mineral depletion in some cases, will have an impact on the important disease groups discussed above, cardiovascular disease and cancer.

REFERENCES

1. **Underwood, E. J.,** Trace elements in human and animal nutrition, 4th ed., Academic Press, New York, 1977.
2. **Nielsen, F. H.,** Nutritional significance of the ultra-trace elements, *Nutr. Rev.,* 46, 337, 1988.
3. **Mertz, W.,** The essential trace elements, *Science,* 213, 1332, 1981.
4. National Research Council, Recommended Dietary Allowances, 9th ed., National Academy of Science, Washington, D.C., 1980.
5. **Kruse-Jarres, J. D.,** Clinical indications for trace element analyses, *J. Trace Element Electrolytes Health Dis.,* 1, 5, 1987.
6. **Anderson, R. A., Polansky, M. M., Bryden, N. A., Roginski, E. E., Mertz, W., and Glinsmann, W.,** Chromium supplementation of human subjects: effects on glucose, insulin and lipid parameters, *Metabolism,* 32, 894, 1983.
7. **Halliwell, B. and Gutteridge, J. M. C.,** Oxygen toxicity, oxygen radicals, transition metals and disease, *Biochem. J.,* 219, 1, 1984.
8. **Jacob, R. A., Sanstead, H. H., Munoz, J. M., Klevay, L. M., and Milne, D. B.,** Whole body surface loss of trace metals in normal males, *Am. J. Clin. Nutr.,* 34, 1379, 1981.
9. **Yang, G., Wang, S., Zhou, R., and Sun, S.,** Endemic selenium intoxication of humans in China, *Am. J. Clin. Nutr.,* 37, 872, 1983.
10. **Burch, R. E. and Sullivan, J. J., Eds.,** *The Medical Clinics of North America,* Vol. 60, *Symposium of Trace Elements,* W. B. Saunders, Philadelphia, 1976.
11. **Prasad, A. S.,** Clinical manifestations of zinc deficiency, *Ann. Rev. Nutr.,* 5, 341, 1985.
12. **Prasad, A. S., Ed.,** *Trace Elements in Human Health and Disease,* Vol. 1, *Zinc and Copper,* Academic Press, New York, 1976.
13. Keshan Disease Research Group of the Chinese Academy of Medical Sciences, Beijing, Observations on effect of sodium selenite in prevention of Keshan disease, *Chin. Med. J.,* 92, 471, 1979.
14. Keshan Disease Research Group of the Chinese Academy of Medical Sciences, Beijing, Epidemiologic studies on the etiologic relationship of selenium and Keshan disease, *Chin. Med. J.,* 92, 477, 1979.
15. **Johnson, R. A., Baker, S. S., and Fallon, J. T.,** An occidental case of cardiomyopathy and selenium deficiency, *N. Engl. J. Med.,* 304, 1210, 1981.
16. **Fleming, C. R., Lie, J. T., McCall, J. T., O'Brien, J. F., Baillie, E. E., and Thistle, J. L.,** Selenium deficiency and fatal cardiomyopathy in a patient on home parenteral nutrition, *Gastroenterology,* 83, 689, 1982.
17. **Shamberger, R. J.,** Selenium diseases in animals, in *Biochemistry of Selenium,* Plenum Press, New York, 1983, 31.
18. **Salonen, J. T., Alfthan, G., Huttunen, J. K., Pikkarainen, J., and Puska, P.,** Association between cardiovascular death and myocardial infarction and serum selenium in a matched-pair longitudinal study, *Lancet,* 2, 175, 1982.

19. **Willett, W. C., Morris, J. S., Pressel, S., et al.,** Prediagnostic serum selenium and risk of cancer, *Lancet,* 2, 130, 1983.

20. **Klevay, L. M.,** Hypercholesterolemia in rats produced by an increase of the ratio of zinc to copper ingested, *Am. J. Clin. Nutr.,* 26, 1060, 1973.

21. **Menkes, J. H., Alter, M., Steigleder, G. K., Weakly, D. R., and Sung, J. H.,** A sex-linked recessive disorder with retardation of growth, peculiar hair, and focal cerebral and cerebellar degeneration, *Pediatrics,* 29, 764, 1962.

22. **Klevay, L. M.,** The ratio of zinc to copper of diets in the United States, *Nutr. Rep. Int.,* 11, 237, 1975.

23. **Freund, J., Atamian, S., and Fischer, J. E.,** Chromium deficiency during total parenteral nutrition, *JAMA,* 241, 496, 1979.

24. **Virtamo, J. and Huttunen, J. K.,** Minerals, trace elements and cardiovascular disease, *Ann. Clin. Res.,* 20, 102, 1988.

25. **Karppanen, H., Pennanen, R., and Passinen, L.,** Minerals, coronary heart disease and sudden coronary death, *Adv. Cardiol.,* 25, 9, 1978.

26. **Stamler, J.,** Research related to risk factors, *Circulation,* 60, 1575, 1979.

27. **Niskanen, J., Marniemi, J., Piironen, O., et al.,** Trace element levels in serum and urine of subjects who died of coronary heart disease, *Acta Pharmacol. Toxicol.,* 59, (Suppl.), 340, 1986.

28. **Punsar, S. and Karvonen, M. J.,** Drinking water quality and sudden death: observations from west and east Finland, *Cardiology,* 64, 24, 1979.

29. **Virtamo, J., Valkeila, E., Alfthan, G., Punsar, S., Huttunen, J. K., and Karvonen, M. J.,** Serum selenium and the risk of coronary heart disease and stroke, *Am. J. Epidemiol.,* 122, 276, 1985.

30. **Salonen, J. T.,** Selenium in ischaemic heart disease, *Int. J. Epidemiol.,* 16, 323, 1987.

31. National Academy of Sciences, Committee on Diet, Nutrition, and Cancer, *Diet, Nutrition and Cancer,* National Academy Press, Washington, D.C., 1982.

32. **Jacobs, M. M. and Griffin, A. C.,** Effects of selenium on chemical carcinogenesis. Comparative effects of antioxidants, *Biol. Trace Element Res.,* 1, 1, 1979.

33. **Marshall, M. V., Arnott, M. S., Jacobs, M. M., and Griffin, A. C.,** Selenium effects on the carcinogenecity and metabolism of 2-acetylaminofluorene, *Cancer Lett.,* 7, 331, 1979.

34. **Ip, C.,** Selenium inhibition of chemical carcinogenesis, *Fed. Proc.,* 44, 2573, 1985.

35. **Schrauzer, G. N., White, D. A., and Schneider, C. J.,** Cancer mortality correlation studies. III. Statistical associations with dietary selenium intakes, *Bioinorg. Chem.,* 7, 23, 1977.

36. **Shamberger, R. J., Rukovena, E., Longfield, A. K., Tytko, S. A., Deodhar, S., and Willis, C. E.,** Antioxidants and cancer. I. Selenium in the blood of normals and cancer patients, *J. Nat. Cancer Inst.,* 50, 863, 1973.

37. **Salonen, J. T., Alfthan, G., Huttunen, J. K., and Puska, P.,** Association between serum selenium and the risk of cancer, *Am. J. Epidemiol.,* 120, 342, 1984.

38. **Kok, F. J., De Bruijn, A. M., Hofman, A., Vermeeren, R., and Valkenburg, H. A.,** Is serum selenium a risk factor for cancer in men only?, *Am. J. Epidemiol.,* 125, 12, 1987.

39. **Vernie, L. N.,** Selenium in carcinogenesis. *Biochim. Biophys. Acta,* 738, 203, 1984.

40. **Mills, B. J., Broghamer, W. L., Higgins, P. J., and Lindeman, R. D.,** Inhibition of tumor growth by zinc depletion of rats, *J. Nutr.,* 14, 746, 1984.

41. **Langård, S. and Norseth, T.,** Chromium. I. Friberg, L., Nordberg, G. F., and Vouk, V. B., Eds., *Handbook of the Toxicology of Metals,* Elsevier, Amsterdam, 1985, 185.

42. **Elinder, C. G.,** Iron II. Friberg, L., Nordberg, G. F., and Vouk, V. B., Eds., *Handbook of the Toxicology of Metals,* Elsevier, Amsterdam, 1986, 289.

Chapter 3

PROBLEMS ON DEFICIENCY AND EXCESS OF MINERALS IN ANIMAL NUTRITION

Arne Frøslie

TABLE OF CONTENTS

I. GENERAL ASPECTS

A. INTRODUCTION

Like many diseases in man, several disease conditions in animals show a definite geographical distribution. This may be due to geographical variation in climatic conditions, soil type, and vegetation or to regional differences in husbandry practices, variation in resistance against diseases between breeds, etc. Diseases due to such factors could all be included in the broad definition of geomedicine. However, in veterinary medicine, the term geomedicine is often limited to disorders which are related to geochemistry or soil chemistry. Most relevant in this connection are the trace elements (microminerals), in particular the essential ones. Deficiencies of such elements are of main concern in a global context.

As regards animals, the disorders usually appear as true deficiency conditions. Animals are in a situation where they have little choice, they have to make do with the feed they are offered or which they can come across when grazing. Local deficiencies in nutrients are therefore likely to be reflected in their nutritional status, unless they are offered supplementary feed. This contrasts with the situation for people in most regions of the world, where foodstuffs are traded and transferred from place to place, even from continent to continent. Any deficiencies or imbalances in trace elements will tend to be evened out, the likelihood of specific deficiency conditions arising thus being reduced. Such considerations may partly explain why deficiencies of trace elements give rise to more pronounced problems in animals than in humans. In humans, with few exceptions, it has only been possible to demonstrate ''associations'' between the occurrence of certain diseases and the concentration of certain trace elements in the soil.

The animals most exposed to trace element deficiencies are the domestic animals, and in particular the ruminants. Pigs, poultry, and certain other animal species are usually fed on a varied ration, often comprising feed ingredients which are traded internationally. This in itself reduces the risk of trace element insufficiencies, though feed mixtures are nevertheless usually largely based on locally or regionally produced feedstuffs. Deficiency conditions may therefore still appear unless appropriate feed supplements are offered. The provision of such supplementary feeds is, in fact, the normal practice in countries with an advanced livestock industry.

Wild animals are far less subject to deficiency diseases than domestic livestock kept under intensive conditions. Any deficiency disorder appearing in wildlife might be considered one of the many phenomena or factors which contribute to the maintenance of a balanced ecological system. Wild animals are able to graze selectively, eating a little here and a little there of different leaves, herbs, and grasses, and because the ability of different plant species to accumulate trace elements varies, deficiency conditions will thus be avoided.[1] Moreover, wild forage plants are usually poorer in protein and energy than cultivated plants, which means that wild animals have to eat relatively more than their domestic counterparts. Both these factors will contribute to a greater likelihood of trace element requirements being met in the wild than in conditions of intensive production with reliance on monoculture pastures.

B. EFFECTS OF MODERN FARMING

The modern farmer has improved his cultivation techniques, increased the use of fertilizers, and improved his plant material, all with the aim of increasing crop yields. This may well influence the trace element status of feed crops, especially those elements which are either of no significance for plant growth or which are required at lower levels in plants than in animals. Little interest has been shown among crop farmers in supplying plants with such elements. Selenium is a typical example in this connection. However, copper, cobalt, selenium, and magnesium are now among those elements which are recommended for top-dressing of pastures to control deficiency disorders in livestock.[2] The continual increase in the levels of productivity, growth rate, and feed conversion ratio may also well increase the needs for certain trace elements.

C. LOSSES CAUSED BY TRACE ELEMENT DEFICIENCIES

As mentioned, the greatest problems in livestock farming are created by deficiencies of trace elements rather than excesses. On a world-wide basis, large economic losses result from such deficiencies, though it is impossible to give even an approximate estimate of the actual sums involved.[3] This is due not only to lack of statistical data, but also because many deficiency disorders only cause unthriftiness and reduced growth, the loss to the farmer being indirect. Such deficiency conditions, associated with unsatisfactory production and obscure, ill-defined symptoms, may well be confused with other chronic diseases, such as moderate parasitic infestations. Many deficiency diseases are also difficult to diagnose, even when of clinical significance.[4]

Deficiency conditions in ruminants would make livestock farming completely impossible in certain districts were they not corrected. In Scotland alone, measures to prevent and treat copper deficiency cost GBP Pounds (£) 0.6 mill. (approx. 1 mill. U.S. dollars) annually.[5] According to Lewis and Anderson,[4] calculations indicate that, in spite of prophylactic efforts, there is a possible production loss of nearly GBP Pounds 5 mill. (approx. 8.5 mill. U.S. dollars) annually due to copper deficiency alone in Britain. As for the selenium deficiency diseases, the same authors suggest that the diseases diagnosed represent only the tip of the real selenium deficiency iceberg.

D. HISTORICAL DEVELOPMENT OF TRACE ELEMENT RESEARCH

The history of research on the significance of trace elements for animal health goes back a long way, though most progress has been made during the last 50 years or so. In this connection, reference is made to the reviews of Lee[6] and Underwood.[7] Some of the important milestones will be mentioned, as well as some less well-known, older reports from Norway.

The technique of dose-response was the most common method previously employed to reveal deficiency diseases. Such techniques ("diagnostic treatment") are still useful.[8] Subsequently, the method of feeding a controlled diet took over in importance, a technique by which the feed was manipulated so that just *one* or only a few nutrients were lacking. In this way, it became possible to experimentally induce specific deficiency diseases.

However, important as these developments were, it has been the progress made in the field of analytical chemistry which has really increased our knowledge on the role of trace elements for animal health. The most important steps in this progress have been taken during

TABLE 1
Approximate Dietary Requirements and Maximum Tolerable Levels of Some Important Trace Elements in Animals' Feed[2,9-11]

	Requirements (mg/kg dry matter)	Tolerance Levels (mg/kg dry matter)			
		Cattle	Sheep	Pigs	Poultry
Cobalt	0.1[a]	5—10	10	10	10
Copper	4—15[b]	100[b]	25[b]	250	300
Fluorine	—	40—100	60—150	150	200
Iodine	0.1—1.0	50	50	400	300
Iron	30—80	1000	500	3000	1000
Manganese	20—40	1000	1000	400	2000
Molybdenum	0.5[a,c]	6—10[c]	10[c]	20	100
Selenium	0.1—0.3	2	2	2	2
Zinc	30—70[d]	500	300	1000	1000

[a]   Ruminants.
[b]   Dependent on Mo and S supply in ruminants.
[c]   Dependent on Cu supply in ruminants.
[d]   Very dependent on Ca supply in pigs.

the last 50 years. Early analytical methods were not sensitive enough to determine levels of trace elements in plants or animals. Today, sensitive and specific high-capacity apparatus, often permitting multielemental analysis, is commonplace in every laboratory of any size working with trace element analysis. This development, which has taken place during the last 20 years, has revolutionized trace element research and produced an enormous amount of data on trace elements in soil, plants, and animals.

It has thus become very easy to obtain analytical data. However, the interpretation of all the data produced is not always as easy, since there are many factors which can influence the transfer of elements both from soil to plants and from plants to animals. Furthermore, it is still not always possible to define an optimal intake of a particular trace element, let alone determine the optimal relationship *between* various macro- and microelements in the feed. Even optimal levels of trace elements in animal tissues are difficult to define. It may thus pose a problem to differentiate between a marginal insufficiency and a deficiency condition.

Faced with such problems, modern biochemical methods have proved very helpful. As more and more knowledge has been gained on the metabolic processes in which trace elements are involved, it has become possible to determine the body levels of certain specific metabolites or enzymes and thereby obtain a measurable expression of the *physiological* nutritional status. The determination of viramin B_{12} in cobalt deficiency in ruminants is an example, the vitamin becoming deficient with an inadequate supply of cobalt. The measurement of the activity of the selenium-containing enzyme glutathione peroxidase has become a standard method for the diagnosis of selenium deficiency in animals.

E. REQUIREMENTS AND TOLERANCES

One of the difficulties which arise when attempting to interpret the results of trace element analyses is created by the diffuse delimitation and transition between true deficiency with manifest disease, marginal insuffiency, inadequate supply, optimal supply, marginal excess, and absolute overload. The different elements also show great variation in the factor between requirement and tolerance, i.e., the therapeutic index, as well as there being considerable differences between animal species in their tolerance to trace element intake. A good example of this is copper, which has been given as a growth promoter to pigs at levels up to 250 mg/kg feed without significant risk, whereas levels as low as 10 to 15 mg/kg feed can, if the concomitant molybdenum intake is marginal, result in copper poisoning in sheep.

F. INTERACTION

Another problem is caused by interactions between elements during their transfer from soil to plant and from plant to animals. Levels measured in soil are by no means always a reliable expression of the levels of the element biologically available to the animal. Moreover, different methods give different results, and it is therefore difficult to compare results from different countries or laboratories because of the lack of standardization. Furthermore, the physical and chemical properties of the soil, for example, soil pH and drainage, exert a significant influence on the uptake of trace elements by plants.

Many interfering factors also exert their effect during uptake of trace elements from plants to animals, and many pitfalls exist if reliance is placed on individual analyses. A good example in this connection is the relationship between copper levels in plants and animals. In Norway, copper levels above those considered to be safe[2] have hardly ever been found in pasture plants, and yet copper poisoning in sheep is commonly occurring.[12] Thus, faulty conclusions may easily be drawn if only copper levels in the feed are determined. In this case, it is the low molybdenum levels which result in the Cu:Mo ratio being too high, with poisoning as a result.

Interaction between various essential and toxic elements thus plays an essential role in trace element nutrition.[2,13-15] In light of such a complex situation, it is a wonder that imbalances in the supply of minerals to our animals can be avoided at all. However, it has to be noted that animals, and also humans, possess a considerable ability to maintain a homeostatic equilibrium, even under circumstances of marked imbalances of supply. This is true for animals in natural surroundings and for the natural levels of minerals which, after all, fluctuate only to a limited extent in most cases. When elements are removed from their natural context, i.e., their natural mineral sources, however, imbalances may become too great for a natural homeostatic equilibrium to be maintained. This state of affairs may also arise with intensive cultivation of pastures and selective improvement of livestock, the gap between requirements and supply becoming too wide.

G. ESSENTIAL AND TOXIC ELEMENTS

The listing of trace elements varies, depending on the definition of essentiality applied. Underwood[7] considered, in 1977, 26 of 90 natural elements to be essential for animal life. Of these, 11 were macro elements, namely, carbon, hydrogen, oxygen, nitrogen, sulfur, calcium, phosphorus, potassium, sodium, chlorine, and magnesium. Essential trace elements totalled 15, namely, iron, zinc, copper, manganese, nickel, cobalt, molybdenum, selenium, chromium, iodine, fluorine, tin, silicon, vanadium, and arsenic. In addition, boron is essential for plants. Underwood and Mertz,[16] in 1987, added lead and lithium to the list of possible essential trace elements.

There is no disagreement concerning the essentiality of chromium, manganese, iron, copper, cobalt, zinc, selenium, molybdenum, and iodine,[16] though chromium is of minor interest in animal nutrition. Of the trace elements, nine have been associated with clinical problems in livestock farming in the form of *excess*, namely, iron, zinc, copper, manganese, cobalt, molybdenum, selenium, fluorine, and arsenic.[13] In addition, a group of elements exert toxic effects in low doses, namely, lead, cadmium, and mercury.[13] Attention has also recently been focused on aluminium.[17]

H. REVIEW LITERATURE

An enormous amount of literature has accumulated on trace elements and on the problems they create in connection with deficiency diseases and poisonings in animals. It is thus impossible to do proper justice to all this work in a brief review like this. Moreover, the extent of the problems associated with trace elements varies greatly from country to country and between geographical areas, depending on livestock farming practices, climatic conditions, etc. The situation in one country or region will therefore not necessarily correspond to an apparently similar situation in another country. This account will consequently be restricted to general circumstances with illustrative examples. Those seeking a deeper and more thorough treatment of the subject should consult the textbooks and reviews referred to[2,5,18-35] and the specific literature dealing with the individual trace elements.

II. TRACE ELEMENTS OF SPECIAL INTEREST

A. COBALT

Cobalt is an essential trace element, though deficiency conditions occur only in ruminants. The reason for this is that cobalt is a component of vitamin B_{12} (cobalamin), which is synthesized by microorganisms in the rumen. Although all animal species require vitamin B_{12}, the higher animals (and plants) are unable to synthesize the vitamin and depend on a supply of preformed, microbially produced vitamin B_{12} in the diet, rather than cobalt itself. Ruminants must rely on the presence of an adequate amount of cobalt in the rumen for microbial production of vitamin B_{12}, while other herbivores and those mixed feeders that do not consume ruminant flesh receive their vitamin B_{12} through other food chains. Cobalt as such does not seem to be essential to the animal body.[36] Cobalt is not very toxic, and poisoning does not seem to occur except by accident or overdosage in connection with the treatment or prevention of deficiency.[2]

1. Cobalt Deficiency

According to the Scottish agricultural research institutes,[5] the use of the term "pining" or "vinquish" (unthriftiness) to describe a disorder in sheep which was associated with a particular type of soil or special geographic regions in Scotland was first reported in 1807. The first hypotheses that a particular disease ("bush sickness"), which later proved to be cobalt deficiency, could be due to a deficiency factor were put forward in New Zealand around the turn of the present century.[6] The occurrence of a similar disease ("coast disease") had been reported in South Australia in 1883, though it was not until the mid-1930s that Filmer and Underwood, while studying "enzootic marasmus" in Australia, clarified the etiology of the diseases caused by cobalt deficiency (see Lee[6] and Smith[36]).

Cobalt deficiency is widespread in New Zealand, Australia, and Great Britain.[2] The disease has previously been registered in cattle in the coastal areas of Norway, often in conjunction with copper deficiency,[37-39] and is now seen in lambs in both Norway[40] and Sweden.[41] Apart from selenium and copper deficiency, cobalt deficiency is probably the most economically important trace element deficiency disease.

As cobalt is a component of vitamin B_{12}, and is therefore necessary for the synthesis of the substance in the rumen, cobalt deficiency manifests itself as a deficiency of the vitamin. A significant effect of this is that propionic acid metabolism, which is of major importance in carbohydrate metabolic pathways in ruminants, is impaired due to reduced activity of the enzyme methylmalonyl coenzyme A-mutase.[2,5,36] This will cause accumulation of methylmalonic acid in the body, a situation which is made use of diagnostically. Other metabolic processes in the body are also affected.[2,4,36]

The diseases which arise from cobalt deficiency are basically characterized by anorexia, progressive anemia, wasting, emaciation, and death, if left untreated. Affected animals are hypoglucemic, and sheep often show profuse lacrimation. Cobalt deficiency may also occur in a subclinical form, this form probably being the one of greatest economic significance. It is manifested by unthrifteness and impaired growth, and may therefore be very difficult to diagnose. A third form is "ovine white liver disease", which occurs in 2- to 6-month-old lambs, animals again showing signs of ill thrift. The condition may become acute, with almost complete anorexia and loss of weight, and with a fatal outcome. Photosensitivity, indicative of liver damage, may also be seen, as well as disturbances of the central nervous system (ataxia). Although the disease is considered to be related to a marginal supply of cobalt, is it not always possible to demonstrate a true cobalt deficiency. A vitamin B_{12} deficiency nevertheless always exists. The etiology of the condition thus seems to be complex and still not yet completely understood.[42]

There are no truly specific or characteristic clinical symptoms associated with cobalt deficiency, and diagnosis usually has to be based on the determination of cobalt levels in the pasture and/or in the liver from affected animals, or on the estimation of one of the biochemical parameters, vitamin B_{12} in plasma or methylmalonic acid (MMA) in urine or plasma. The determination of MMA in plasma seems to be a reliable method for the demonstration of a subclinical deficiency.[43] The demonstration of formiminoglutamate in urine is also a good indicator of cobalt deficiency in lambs, though only after affected animals have begun to show signs of impaired growth. The method is therefore considered less suitable for revealing subclinical deficiency.[4,36] Dose-response trials may be useful for verification of the diagnosis. Results of chemical analyses of cobalt in liver may be difficult to interpret because of variation in cobalt intake.

Cobalt deficiency can be prevented by adding cobalt to the feed either directly or in a mineral mixture, by the use of "heavy pellets" containing cobalt or by top-dressing of affected pastures with cobalt sulfate. A single top-dressing treatment appears to be effective for periods up to 3 years.[42] Affected animals can also be treated directly with vitamin B_{12}.

B. COPPER — MOLYBDENUM

Due to metabolic interaction, the two elements copper and molybdenum are so interdependent that they must be dealt with together. The interaction is particularly marked in ruminants, and imbalances in the supply of these elements are most common in ruminant animals. Both copper and molybdenum are essential as well as being toxic.

1. Copper Deficiency — Hypocuprosis
a. Introduction and Historical Development

In a global context, copper deficiency conditions are of most significance with regard to imbalances in copper-molybdenum intake in animals. The original studies which demonstrated that copper supplementation had a beneficial effect on grazing cattle are from Florida ("salt sickness") (1931) and Holland (1933) (see Lee[6]). Enzootic ataxia in lambs was described in Western Australia in 1932, the etiology being clarified in 1937.[6] The reason for abnormal wool growth in sheep, associated with loss of crimp and depigmentation of black wool, was also elucidated in Australia.[6] An excess of molybdenum and consequent copper deficiency as a cause of diarrhea in cattle was described in England in 1938, while the foundation for an understanding of the important interrelationship between copper, molybdenum, and sulfur was first laid by Dick while working on chronic copper poisoning in sheep in Southwest Australia in the early 1950s (see Underwood[7]). However, the biochemical basis for this interaction, formation of copper thiomolybdates, has only been clarified in recent years.[44-46] Ender began his pioneer work on copper deficiency in Norway in the late 1930s.[37]

Copper deficiency is a disease in sheep and especially in cattle which is of clinical importance in most European countries such as the U.K. as well as the U.S. and Australia.[2,46-48] Disease may be due either to an absolute insufficiency in copper intake, the condition then being termed *primary* or *genuine* copper deficiency, or may arise through the biological availability of an otherwise adequate supply of copper in pasture vegetation being reduced by interfering factors. In the latter case, a relative insufficiency of copper is induced and *secondary* or *conditioned* copper deficiency occurs. The most important interfering factor under practical farming conditions is molybdenum, which, in the presence of a sufficient amount of sulfur, exerts an antagonistic effect on copper. Secondary copper deficiency is thus usually due to an excess of molybdenum. Excess amounts of molybdenum can also lead to secondary copper deficiency, even when copper intake is higher than the recognized requirements. In such circumstances, it is really a matter of molybdenum intoxication (molybdenosis), although the pathophysiological state is one of copper deficiency. It seems that molybdenum per se is not particularly toxic.[48]

Conversely, a low molybdenum intake may well result in copper poisoning. Again, the latter condition could just as well be described as a molybdenum deficiency, even though disease is actually caused by the accumulation of copper in body tissues. It is thus the ratio between copper and molybdenum levels in feeds or pastures, rather than the absolute amounts, which determines the copper status of the animal.

The extent to which animals are sensitive to imbalances in the copper:molybdenum ratio varies. Ruminants are markedly more susceptible than nonruminants, the reason for this being the metabolic processes which take place in the rumen. However, there are also differences between ruminant species, cattle being more sensitive to copper deficiency than sheep, while sheep, on the other hand, are more prone to copper poisoning than cattle. Of the domestic ruminants, the goat seems to be the least sensitive to imbalances in copper-molybdenum metabolism, especially copper poisoning. No satisfactory explanation of the marked difference between sheep and goats yet seems to have been found.

Of nonruminant species, copper deficiency can arise in pigs, though it does not usually create a problem under practical conditions, as copper is usually added to compound feeds.[2]

b. Mode of Action

Copper comprises part of, or influences, several important enzyme systems or body reactions, such as ceruloplasmin, cytochrome oxidase, lysyl oxidase, -SH oxidation to S-S, and DOPA oxidation (see Mills[3] and Davis and Mertz[46]). Copper thus exerts an effect on important intracellular oxidation processes, and lack of copper will therefore lead to impairment of one or more of these metabolic processes, with the result that pathological changes associated with deficiency may occur in various organs and tissues:

1. Defective melanin synthesis (depigmentation of hair)
2. Defective keratinization of hair
3. Connective tissue defects
4. Demyelinization (ataxia)
5. Anemia
6. Reduced growth
7. Cardiovascular disorders
8. Infertility

The degree and type of clinical and pathological manifestations depend on the degree, duration, and speed of the enzyme inhibition, which in turn are dependent on, inter alia, age, growth rate, nutritional status otherwise, and husbandry practices. Some enzyme systems or biochemical pathways are more sensitive to deficiency than others. This is the reason why some pathological changes arise with moderate deficiency conditions, while changes associated with impairment of less sensitive enzyme systems arise only when a severe deficiency situation exists.[3] Copper deficiency therefore gives rise to complex disease situations, which may vary greatly from animal species to animal species, the clinical picture also varying considerably, according to the prevailing circumstances.

c. Interfering Elements

As mentioned previously, the two principal elements interfering with copper are molybdenum and sulfur. They form thiomolybdates in the rumen, these combining with copper to form copper thiomolybdates, which impair absorption of copper.[44] Copper thiomolybdates may also exert a systemic effect on copper metabolism and thus influence biliary and urinary excretion.[49] Sulfur per se may also interfere with copper metabolism in that sparsely soluble copper sulfide complexes are formed in the rumen.[47]

Normal levels of copper in pasture are 5 to 15 mg/kg DM, providing interfering elements are within the acceptable range.[46-47] The preferable range of copper is 7 to 12 mg/kg.[2] Normal levels of molybdenum are 1 to 2 mg/kg DM.[22] A balanced diet for ruminants should have a copper/molybdenum ratio of 6 to 10.[12] Pastures in which this ratio is less than 3(2) are likely to give rise to copper deficiency, while values between 3 and 5 are suspect.

Of the elements which influence the availability of copper, iron is of importance primarily because it might be supplied directly via soil, which may comprise up to 10% of the dry matter intake under certain grazing conditions.[47] Silage may also contain significant amounts of soil. Other interfering elements are of significance mostly under experimental conditions or as a result of contamination, such as, for example, cadmium contamination from mining spills.[47] Zinc is of practical importance as it can be employed to prevent copper poisoning, though under field conditions it does not seem that raised zinc intake has created problems with regard to copper deficiency.

d. Copper Deficiency Diseases in Sheep

Two major forms of copper deficiency are seen in sheep, namely, enzootic ataxia or "sway back" and "ill-thrift", both forms occurring in lambs. Swayback is seen particularly in infant lambs from ewes which have received an inadequate supply of copper during pregnancy. This is the most common type of copper deficiency in sheep in, inter alia, Great Britain.[50] The disease may occur in a congenital form in which lambs exhibit incoordination or ataxia at birth. They have difficulty in standing or moving about and often die of starvation or secondary infections. Swayback also occurs in a delayed form in which the newborn lambs appear to be normal, only to develop incoordination, especially of the hindquarters (hence, the name) after a few weeks. Swayback is associated with characteristic pathological changes in the central nervous system due to defective myelin formation caused by failure of the enzyme cytochrome oxidase.

In the moderate form, ill-thrift in lambs merely causes reduced growth, the clinical picture being no different from that seen in other chronic conditions yielding unthriftiness. In its more marked form, the texture of the wool is affected, with loss of crimp and elasticity. Depigmentation may also occur. Lambs may often sustain fractures, especially of the ribs. This form is not associated with any characteristic changes in internal organs, and differential diagnosis may therefore pose difficulties.

e. Copper Deficiency Diseases in Cattle

Moderate copper deficiency in cattle also only results in unthriftiness. Diarrhea is commonly seen in secondary deficiency due to a high intake of molybdenum. Changes in coat color in the form of depigmentation of colored areas are typical. Red/brown pigmentation changes to a yellowish dun color, while black becomes brownish/rusty red or grey, and the coat itself becomes rough and staring. Loss of pigment around the eyes ("spectacles") and ears is particularly characteristic. Anemia, skeletal changes (osteoporosis), and hoof defects are also seen in severe cases. Animals suffering from copper deficiency are also more susceptible to infections caused by opportunistic microorganisms. As a result of various ways in which the disorder manifests itself, copper deficiency in cattle has been given names such as:

- Falling disease
- Peat scours (teart)
- Unthriftiness (pine)
- Salt sickness

Copper deficiency is not very common in wild ruminants, although the condition may occur, especially in deer kept in captivity (deer farms, zoos).[51] The relationship between enzootic ataxia in red deer and copper status seems, however, to be unclear.[51,52] Like the situation in sheep,[53] copper levels in cervides may vary over a broad range without any apparent sign of disease.[54]

f. Prevention of Copper Deficiency Diseases

Copper deficiency is controlled by providing additional copper in the diet, by salt licks or mineral mixtures containing copper, by oral or parenteral dosing with copper compounds, or by top-dressing of the pastures with copper sulfate.[2]

2. Copper Poisoning — Hypercuprosis
a. Introduction and Historical Development

Copper poisoning in animals may occur both as an acute and a chronic condition. Acute poisoning is essentially characterized by gastrointestinal symptoms and occurs because of the accidental intake of large quantities of bioavailable copper salts. This form of disease will not be described further.

Chronic copper poisoning is a disease which occurs mostly in sheep,[2,21,55-57] the course of the disease being dominated by an acute terminal hemolytic crisis with icterus and hemoglobinuria after a prolonged symptomless accumulation phase. The term "chronic" is therefore not really appropriate. The disease may also arise in connection with certain feeding regimens in calves and young cattle.[21,58] Copper poisoning may also occur in pigs, though usually only in association with the use of high-copper feeds (125 to 250 mg Cu per kg feed) for growth promotion purposes.[21] Poultry are not very sensitive to copper overload.[22]

The condition of hypercuprosis in sheep was described in Australia ("yellows") in the mid-1920s and in the U.S. in 1934,[59] and it was reported in Germany in connection with the grazing of orchards which had been sprayed with fungicidal copper salts.[13] Bull[59] has given a comprehensive account of the early research carried out on copper poisoning.

Housed sheep fed largely on concentrates (cereals) are particularly sensitive to an excess of copper in the diet, and chronic copper poisoning is well known under such conditions.[2,13,22,60,61] Sometimes this condition is exacerbated by the addition of copper to the compound concentrates.

Spontaneous chronic copper poisoning also occurs, however, under natural grazing conditions under three different sets of circumstances: (1) when the copper levels are abnormally high, (2) when the copper levels are normal but molybdenum levels are low, and (3) when hepatotoxic plants are presented in the pasture plants being grazed. The latter form was first described from Australia and has subsequently been reported in several countries.[2,21,59] In recent years, copper poisoning in sheep has also been seen in association with the practice of fertilizing pasture with manure from pigs fed on high levels of copper as a growth promotant.[13,22]

b. Mode of Action

The liver plays a major role in copper metabolism, and secondary copper poisoning may thus occur as a consequence of liver damage. When copper intake exceeds requirements and excretion capacity, accumulation of copper will take place, especially in the liver. The liver, however, can store considerable amounts of copper without adverse effect. Thus, in chronic copper poisoning, there is a latent or symptomless accumulation phase which may last for months. In fact, under field conditions with a moderate excess of copper, such accumulation may be so gradual that an age-related accumulation is seen. This is clearly reflected by the fact that, under such circumstances, chronic copper poisoning only occurs in adult animals, especially the older ones.[62]

If the copper storage capacity of the liver is exceeded or if the ability of the liver to store copper should be adversely affected in some way or other, significant amounts of copper may be suddenly released into the bloodstream. An acutely critical sitation arises, dominated by intravascular hemolysis, methemoglobinemia, and hemoglobinuria. The crisis may be triggered by stress factors such as a change of feed, sudden exertion, mating, etc.[22] Prior to the onset of the crisis, increased serum levels of liver-specific enzymes can be found (SGOT)), indicating liver damage. Impaired liver function can also be demonstrated by liver function tests.[2]

The acute crisis is due to the greatly increased concentration of blood copper which causes oxidative damage to the erythrocytes.[63,64] The iron in the hemoglobin is oxidized, with formation of methemoglobin as the result. The erythrocytes become fragile and lyse, Heinz bodies being produced. Other organs such as the liver and kidneys are also damaged. Kidney damage is due to the increased copper load, as well as the large amounts of free hemoglobin which have to be eliminated via the renal pathway, and hemoglobin casts may arise in the renal tubules (hemoglobinuric nephrosis).[22] Dark brown urine (hemoglobinuria) is therefore a characteristic feature of a terminal copper-induced crisis. In general, only a small proportion of the animals in an affected flock become ill. The crisis itself usually lasts only 24 to 48 h, death being the almost inevitable outcome.

c. Feed-Related Chronic Copper Poisoning in Sheep

Feed-related chronic copper poisoning in sheep is well known. It occurs both on diets with high levels of concentrates, often with copper added, and on natural pastures.[21,22,55,56] The form of poisoning occurring under natural grazing conditions in Norway was described by Nordstoga in 1962[65] and is mainly seen in the autumn, especially in sheep on mountain pastures. It is, however, also seen in lowland areas and occurs in several inland districts of southeastern and central Norway, as well as in the northern part of the country.[53]

The condition as it occurs in Norway is due to molybdenum levels in pasture and fodder being too low. According to Blood and Radostits,[2] the desired copper level in pasture is 7 to 12 mg/kg DM. Levels above this range have hardly ever been recorded in Norway. Analysis of mountain pasture in central Norway showed mean levels of copper in grasses and leaves of 5.2 and 6.2 mg/kg DM, respectively.[1] Similar copper levels have also been found in previous studies.[37,66] Samples of vegetation with high copper levels have, however, been detected in a copper-poisoned area in northern Norway.[67]

Though varying, levels of molybdenum are generally very low, compared with those found in other countries. According to Osweiler et al.,[22] the normal level of molybdenum in herbage is 1 to 2 mg/kg DM, while Bartíc and Piskač[68] state that pastures in central Europe, under normal conditions, contain 0.2 to 2 mg Mo per kg DM. Blood and Radostits[2] consider levels of molybdenum below 3 mg/kg to be "safe" and define levels between 3 and 7 mg/kg as critically high. They do not indicate any critical lower limit for molybdenum in feed. Osweiler et al.[22] report that dietary levels of molybdenum below 0.5 mg/kg may induce copper poisoning in sheep fed on diets with normal copper levels of 8 to 10 mg/kg. From these data, it is reasonable to consider the lower and upper critical levels of molybdenum in pasture grasses to be 0.5 and 3 mg/kg, respectively. It may also be correct to accept that molybdenum levels below 0.2 mg/kg are indicative of a molybdenum deficiency which may cause copper poisoning in sheep on a normal copper supply. This limit is also given by Lamand.[69]

On this basis, a large proportion of the grazing land in Norway, as so far investigated, must be characterized as being deficient in molybdenum.[1,12] The average value found in a study of vegetation in natural rough pastures in a mountain district in southern Norway was 0.25 mg Mo per kg DM; only 13% of the samples examined containing more than 0.5 mg Mo per kg DM.[1] Considerable variation was seen in the Cu:Mo ratio, and there was a strong negative correlation between molybdenum content and the Cu:Mo ratio. No correlation was, however, found between copper content and the Cu:Mo ratio. This clearly demonstrates the significance of the plant molybdenum content with regard to the risk of copper accumulation in sheep.

Uptake of molybdenum by plants is strongly dependent on soil pH, increasing with increasing pH,[31] while absorption of copper is less influenced by pH. The increase in soil acidity which is currently taking place as a result of the utilization of acidic artificial fertilizers (ammonium nitrate) may therefore be of significance in areas where pasture vegetation contains marginally low levels of molybdenum. Acidification of the environment resulting from acidic precipitation is also of a certain interest in this connection.

d. *Prevention of Chronic Copper Poisoning*

The prevention of copper poisoning is usually carried out by providing supplements of a molybdenum salt and sodium sulfate.[21] Zinc supplementation has also been employed to prevent copper poisoning,[22] this method being claimed to present less of a risk of a copper deficiency crisis with uncontrolled use.

C. FLUORINE

Fluorine is described as an essential trace element because of its prophylactic effect in caries in humans. Otherwise, it is best known as a toxic element, especially in veterinary medicine. This is the reason for the controversy regarding the fluoridization of drinking water to prevent caries. Fluorine can cause both acute and chronic poisoning, the chronic form, fluorosis, being the important one in livestock farming.

1. Fluorosis

Fluorosis occurs when small, but neverthless toxic, amounts of fluorine are continuously ingested for prolonged periods, either in feed or drinking water. The intoxication is characterized by mottling and extensive wear of teeth, as well as osteoporosis.

Due to fluorine's strong affinity for calcium, the element, in the form of calcium fluoride, enters the bone tissue and teeth, where it in due course enters the bone crystals (see Krishnamachari[70]). It is therefore bone and dental structures which constitute the main targets of fluorine and where it may cause structural damage.

While moderate amounts of fluorine are, in fact, beneficial to dental structure, excessive amounts cause dental fluorosis, manifested by dark discoloration of the teeth, bilateral enamel defects (mottling), and uneven excessive wear of the molars. This is the mildest form of fluorosis, and changes are only seen in teeth exposed to fluorine during development. Uneven wear of the teeth is due to the uneven hardness of the teeth, and may result in spikes developing on the teeth which are so long that they penetrate into the pulp cavity of the opposite tooth.

If exposure to an excess of fluorine is both prolonged and severe, osteoporosis (osteo-fluorosis) may also occur. The bones become thickened, and may show periostal encrustations (exostoses) and become brittle. Animals suffering from generalized fluorosis show lameness, are loath to move about, and have difficulties in eating because of painful dental changes. Fractures may also occur. Ruminants are the group most sensitive to fluorosis.

Fluorosis occurs naturally in volcanic regions in many countries, including the U.S., Australia, North and East Africa (Rift Valley), and India, where ground water may be rich in fluorine.[2,21,70-71] Dental fluorosis is an increasing problem in certain regions in developing countries in which deep bores have been drilled to obtain water. This is true in, for example, Kenya, where the use of bore water in certain villages has resulted in an increasing prevalence of dental fluorosis in children.[72] Periodic fluorosis in sheep has been registered in Iceland in connection with contamination of grazing land with volcanic ash.[22,71]

Industrial fluorosis, associated with contamination of pastures by industrial fumes containing fluorine, poses problems in a number of countries. This is a particular problem in connection with fumes from aluminium factories.[71] Industrial fluorosis resulting from pollution from aluminium processing has caused significant problems in the vicinity of aluminium works in western Norway, where topographical conditions have resulted in fumes spreading over large areas. The problem was described by Slagsvold as early as 1934.[73] Fluorosis may also arise in connection with excessive amounts of fluorine in rock phosphate used as a feed supplement.

In fluorosis, increased levels of fluorine accumulate in bones. Apart from clinical and pathological examinations, analysis of the fluorine content of bones therefore constitutes the most useful aid in assessing whether, and to what extent, a fluorosis exists.

Though the prevention of fluorosis is difficult if animals are continuously exposed to natural or industrial sources of fluorine, the situation can be alleviated by ensuring an abundant supply of the macrominerals calcium and phosphorus. Supplements of aluminium salts, such as sulfates or lactates, reduce uptake of fluroine and have been employed in prophylaxis in contaminated areas.[71] It is also possible to reduce the content of fluorine in drinking water by adding freshly slaked lime and allowing the water to settle for several days.[2] Young animals should be kept away from the most heavily contaminated areas. This should help to reduce the severity of the problem.

D. IODINE

Iodine is both an essential and a toxic trace element, though it is its essential function that is of significance in livestock farming. The biological importance of iodine is due to its role as an essential constituent of the thyroid hormones, which play a major role in growth and development in humans and animals. Iodine deficiency causes goiter, which in animals is especially manifested by neonatal mortality and obvious enlargement of the thyroid glands.

1. Iodism

Iodine poisoning does not occur under natural conditions as the toxic dose, although varying considerably between animal species, is very high. Organic iodides have been widely employed in the treatment of various livestock diseases, e.g., footrot in cattle, and poisoning has occurred in this connection.[2,21,22]

2. Goiter

Accounts of goiter in animals appeared at the turn of the present century.[74] The condition was reported from southern Norway as early as 1912.[75]

A simple (or absolute) deficiency of iodine may occur in connection with low levels of iodine in soil, plants, and water in certain geographical regions.[2,74] Such a deficiency is still seen in Norway in certain inland regions, particularly in sheep (lambs). The condition may exist in a subclinical form with increased neonatal mortality as the only clinical sign.

Goiter may also occur in association with the consumption of interfering goiterogenic substances which may induce a secondary iodine deficiency. Major feed plants of the *Brassica* spp., including rape/rapeseed and kale, contain glucosinolates which on hydrolysis form goiterogenic compounds.[2,21] Certain important staple food plants also contain goiterogenic substances, though this aspect will not be discussed further in this article (see Hetzel and Maberly[74]).

The cardinal sign of iodine deficiency is goiter, which is a swelling of the thyroid gland. The major classical manifestation of goiter in animals is as neonatal mortality. Iodine is the major primary etiological factor in the endemic form. Apart from the goiter itself, iodine deficiency in humans has revealed a variety of effects on growth and development, and it has been suggested that the term iodine deficiency disorders be used instead of the term goiter.[74] Assessment of iodine nutritional status is made by analysis of protein-bound iodine in plasma or analysis of iodine excretion in milk or urine. Clinical and pathological changes with enlargement of the thyroid glands, especially in neonatal lambs and calves, are typical in severe cases of iodine deficiency. The disease is controlled by adding iodine to compound feeds and mineral mixtures. Treatment must be undertaken with care, as overdosage will cause toxicity.

E. IRON

Iron, being an essential component of hemoglobin, is a very important trace element, and certainly the one which is most often lacking in both animals and humans.[76] In livestock, however, iron deficiency anemia is mainly a problem in suckling infant animals or young animals on an unvaried diet of milk.[2] Baby pigs are particuarly at risk. The iron content of sow's milk is much too low to satisfy requirements, and when pigs are confined indoors and not allowed to root about in the soil, a deficiency condition is likely to develop. Iron deficiency in baby pigs is therefore usually the rule unless prophylactic measures are taken, the provision of iron supplement, in fact, being standard practice on all pig farms. Iron deficiency also occurs in calves.

Iron deficiency anemia thus has no direct relation to the natural content of iron in soil or feed — it arises as a direct consequence of the husbandry methods employed in modern livestock farming and will therefore not be discussed further in this article; nor will iron poisoning, which usually results from the accidental intake of large doses or an overdosage of iron administered to prevent anemia.

F. MANGANESE

Manganese is a trace element of low toxicity, and manganese poisoning is therefore usually not encountered under natural conditions. Extremely high levels (>500 mg/kg DM) in grass are stated to cause a reduction in growth rate in sheep.[77]

1. Manganese Deficiency

Manganese is an essential component of enzymes involved in the formation of the organic matrix of bone, cartilage, and other connective tissues (mycopolysaccharides). Skeletal defects (chondrodystrophy) develop in cases of severe manganese deficiency, in addition to impaired growth. The most well-known condition is the perosis ("slipped tendon") seen in chickens. The disease can be prevented by adding manganese to the feed. Manganese deficiency in pigs, which also gives rise to skeletal disorders, can be prevented in the same way. Manganese also exerts other essential effects.[2,77]

Manganese deficiency may occur either as a primary condition in certain geographical regions where the soil is poor in manganese,[2] or in a secondary form associated with excess intake of calcium and phosphorus, though the disease does not appear to be of any practical importance in livestock farming. Mineral mixtures or compound feeds usually contain sufficient manganese to prevent a deficiency in the animals to which they are offered. Manganese levels in feed crops are subject to great variation. Extremely high levels of manganese have been measured in native pasture plants, especially leaves (<3,700 mg/kg DM), from a mountain area in southern Norway.[1]

G. SELENIUM

The trace element selenium, which the Swede Berzelius identified in 1817 and which he named after selene, the Greek goddess of the moon, has long been known as a very poisonous substance. Its toxicity is stated to have been established as early as 1842.

1. Selenium Poisoning — Selenosis

Accounts of selenium poisoning in livestock are, however, much older than mentioned above, though the cause of the disease was, of course, unknown. As early as 1295, Marco Polo described a disease in cows which he encountered on his travels in China, which may well have been selenium poisoning. The most characteristic symptom was separation and sloughing of the hooves. Separation of the horny tissue from the corium is a sign of destruction of the horn and is a typical finding in severe cases of selenium poisoning. This effect is thought to be brought about by replacement of sulfur by selenium in sulfur-containing amino acids in the keratin.

Two typical selenium poisoning syndromes occur, a subacute form called "blind staggers" and a chronic form, "alkali disease". Descriptions of the disease derive from Colombia (1560), Mexico (1760), Nebraska (1856), and Wyoming (1893). Regions in which poisoning has subsequently been reported include North and South America, South Africa, Spain, France, Germany, Russia, Bulgaria, Ireland, Morocco, Algeria, and Israel. Soil in the "dangerous" areas is rich in selenium (seleniferous soils) and conditions favorable for certain plants which are able to take up selenium in toxic quantities (seleniferous plants). The disorder is, therefore, in its typical form, a disease of grazing animals, and it is easy to appreciate its practical significance for livestock in the regions in question from the fact that, for example, 15,000 sheep died from selenium poisoning in Wyoming in 1907-1908. These disease outbreaks were, however, not connected with the poisonous properties of selenium until 1933-1935.[78]

The major symptoms seen in the subacute form of selenium poisoning reflect disturbances of the central nervous system i.e., aimless wandering and circling, the assumption of abnormal postures, head pressing, blindness, and pharyngeal paralysis, as well as lameness and stiffness. Chronic poisoning is manifested by general unthriftiness and exhaustion, emaciation, and impaired movement due to changes in the joints and hooves. The coat becomes dry and staring, and hair is lost from the mane and base of the tail. Degenerative changes develop in internal organs such as the liver, kidney, and heart. The breath of affected animals smells of garlic due to the excretion of dimethylselenide, a volatile metabolite (detoxication product) of selenium. The toxic action of selenium is assumed to be exerted by interference with sulfhydryl groups in the cell.

2. Selenium Deficiency
a. Introduction and Historical Development

It is nevertheless not so much the toxic effect of selenium which is of interest in veterinary medicine today, but rather its importance as an essential trace element. Selenium is an element which is deficient in many parts of the world, and has become the focus of attention during the last 30 years. The biological action of selenium is similar to that of vitamin E. Both substances are antioxidants which, each in its own way and at its specific sites, protect the cells of the body against oxidative damage. Deficiency disorders often arise as a result of a combined deficiency of these two nutrients.

The most important disease resulting from such deficiency is nutritional muscular dystrophy (NMD, "white muscle disease", "weisses fleisch") which, in it typical form, is manifested by degenerative changes in the skeletal musculature and often also in the musculature of the heart.

In Scandinavia, accounts of NMD in young cattle and lambs go back as far as the 1880s.[7] Slagsvold and Lund-Larsen[79] gave in 1934 a detailed description of the condition, its occurrence, and pathology which is still worthy of study. In this connection, it is interesting to note that Slagsvold[80] described, as early as 1925, a condition which he termed "cod liver oil poisoning", and which was undoubtedly latent selenium/vitamin E deficiency induced by the intake of unsaturated, rancid fat. Furthermore, it is worthy of note that the work of Slagsvold and Lund-Larsen[79] was published at the time when the role of selenium as a toxic element was being revealed in the U.S., and it took more than 20 years before its role as a deficiency factor became known. Nor was the significance of vitamin E for this disorder ascertained until the 1940s, even though the vitamin had been known as the "fertility vitamin" from the beginning of the 1920s.

In order to follow developments in Scandinavia further, it should be mentioned that Obel published her studies on so-called "toxic liver dystrophy" in pigs in 1953.[81] She found that the disease was nutritional in cause and could be prevented by providing sulfur-containing amino acids. What she didn't know was that such amino acids may contain some selenium on the sulfur site. Moreover, these amino acids are building stones for the peptide glutathione (GSH), which, as we shall see, plays a central role in the body's redox systems. Neverthless, it was a very important step in the right direction when Obel could term the disease "hepatosis dietetica" — the etiology of the disorder had been shown to be nutritional in nature.

It was not until 4 years later, in 1957, that Schwartz and co-workers[82] found that "factor 3" was selenium, the lack of which was of significance in liver necrosis in rats. The two other factors were vitamin E and the sulfur-containing amino acids. After 1957, the etiology of the typical selenium/vitamin E deficiency diseases was rapidly clarified. It is remarkable that no more than 2 years or so elapsed between the discovery of the role of selenium as an essential element and the world-wide introduction of the element as a therapeutic and prophylactic agent in veterinary medicine.

The amount and quality (i.e., degree of unsaturatedness or instability) of fat and fatty acids in the feed constitute what might well be termed a "factor 4" of significance in selenium/vitamin E deficiency conditons. This had already been observed in connection with "cod liver oil poisoning" in calves[80] and later in "yellow fat disease" in mink and pigs offered feed containing a large proportion of fish rich in fat.[83] Furthermore, there is a condition termed "fresh grain poisoning" reported from Sweden[84] in which unsaturated fat in fresh, newly harvested cereal grains contributes to the induction of the deficiency condition. The amount and quality of the fat in the diet are thus recognized as being very important contributory factors in the development of selenium/vitamin E deficiency disorders.

b. Mode of Action

Yet the mode of action of selenium was unknown until 1973. In the same year that the identity of "factor 3" was determined (1957), the enzyme glutathione peroxidase (GSHPx) was isolated, this enzyme being of significance for the reducing effect of glutathione. However, it was not until 1973 that Rotruck et al.[85] and Flohe et al.[86] independently discovered that this enzyme contains selenium. This provided the explanation for selenium's central role in intracellular reduction-oxidation reactions and its complementary role to vitamin E, which acts as an antioxidant in the cell membrane. The seleno-enzyme GSHPx acts by participating in the reduction of GSSG to GSH, which is of importance for the reduction of harmful oxidation products (peroxides) in the cytosol of the cell. The enzyme is present in erthyrocytes and in most of the soft organs and tissues of the body. It is therefore not surprising that selenium deficiency can result in dramatic disease processes.

It has also been shown that selenium influences the metabolism of arachadonic acid and the formation of prostaglandines.[87] Selenium thus seems to be of significance in inflammatory reactions. In addition, it influences the microbicidal activity of polymorphonuclear leukocytes[87] as well as the response of lymphocytes to mitogens,[88] and thereby probably also the immune response in animals to infectious challenges.[89] These modes of action do not yet seem to have been sufficiently elucidated. A second selenium-dependent peroxidase, phospholipid hydroperoxide glutathione peroxidase, has recently been discovered in mammals. This peroxidase is active on membrane-bound hydroperoxides.[90]

Selenium also exerts an antagonistic action against heavy metals such as lead, cadmium, and mercury, though the mode of action in this respect does not appear to have been fully clarified.[91] Selenium is also considered to be an anticarcinogenic agent (see Combs et al.[27]

and Levander[91]). Ongoing and future research will no doubt reveal new aspects of the effects of selenium which may well prove to be of practical interest for animal health.

c. Selenium Deficiency Disorders

Selenium deficiency disorders are manifested by clinical pictures which differ from species to species and occur, in their typical forms, especially in young animals. Hutchinson[87] lists the following selenium deficiency conditions:

1. Skeletal myopathy: cattle, horses, sheep, swine, dogs, chickens
2. Cardiac myopathy: calves, foals, lambs, growing pigs
3. Steatitis: cats, foals
4. Placental retention: cows
5. Unthriftiness: cows, calves, sheep
6. Diarrhea: cattle
7. Decreased egg production and hatchability: chickens
8. Hepatosis dietetica: growing pigs
9. Exudative diathesis: chicks, growing pigs
10. Abortion, stillbirth: cattle
11. Metritis: cattle
12. Cystic ovaries: cattle
13. Mastitis: cattle

It has also been reported recently that a low selenium diet during pregnancy decreases the reproductive performance in sows.[92]

(1) Nutritional Muscular Dystrophy

Though NMD occurs in its typical form in almost all livestock species, it is nevertheless most typical in young, rapidly growing foals, lambs, and calves from dams which receive too little selenium for prolonged periods during pregnancy. In young pigs, muscular dystrophy often occurs together with other severe conditions such as hepatosis dietetica and mulberry heart disease (cardiac myopathy).

The disease occurs in many parts of the world. Typical deficiency regions in which the disease is found are the U.K., U.S., Canada, and New Zealand, together with Scandinavia and Finland.[2] Selenium deficiency also occurs in Central Europe.[93] Forage crops, cereal grains, and corn grown in these areas are poor in selenium.[32] The minimum requirement for selenium is stated to be 0.1 mg/kg DM in the feed, though under field conditions the requirement is undoubtedly far higher. In Norway, for example, selenium levels down to 0.002 mg/kg in feed grain (barley and oats) and down to 0.01 mg/kg in forage crops have been measured, reflecting an extreme selenium deficiency.[94] A review of the selenium status of soil and vegetation in Scandinavia has been given by Mattsson.[95] The geographical distribution of selenium deficiency diseases in Norway corresponds to the distribution of selenium levels in farmland soils.[96]

NMD is associated with degenerative changes (hyalin muscle degeneration) which affect the skeletal musculature to a greater or lesser extent. The cardiac musculature is also often involved. Consequently, animals exhibit difficulty in getting to their feet and moving about ("stiff lamb disease"). They are unable to feed and may succumb to starvation. Animals may also die as a result of respiratory failure if the respiratory musculature is involved or of heart failure due to degenerative lesions in the cardiac musculature.

Damage to the musculature results in a typical increase in the plasma activity of certain enzymes, an increase in the activity of creatine phosphokinase being especially characteristic.[2] This is exploited as an aid in diagnosis. Low selenium levels are also found in blood and organs, as well as low activity of glutathione peroxidase, a situation which is used in the diagnosis of the condition.

The disease occurs in the acute form in infant animals, while the course may be somewhat more protracted in young animals. Subclinical deficiencies may also exist. In calves, and particularly in lambs, the condition is most commonly seen in the spring when animals are put out to pasture. This is often considered to be related to sudden, unaccustomed exertion, though the presence of polyunsaturated fatty acids in lush, fresh spring grass also seems to play an important role in the induction of muscular degeneration in such cases.[97] Vitamin E also plays a decisive role in muscular degeneration, the disease having been described in lambs suffering from vitamin E deficiency in conjuction with an inadequate selenium intake.[98]

(2) Hepatosis Dietetica and Mulberry Heart Disease

Apart from muscular dystrophy, hepatosis dietetica and mulberry heart disease are the most common forms by which selenium deficiency is manifested in pigs. There is often a concomitant lack of vitamin E, and feed offered to the pigs in such cases often contains various amounts of unsaturated fat. It is generally thought that vitamin E plays a significant role, particularly in mulberry heart disease. This is a condition commonly seen in rapidly growing, young pigs in many countries. The geographical distribution of the disease is by and large as for NMD in ruminants, and reflects selenium levels in locally grown feed grains.

In typical cases of mulberry heart disease, characteristic degenerative lesions and hemorrhages arise in the myocardium. The course of the disease is most often acute, with animals dying of heart failure. Most at risk are animals in a rapid growth phase, up to 25 to 30 kg live weight. Moreover, the largest pigs in the litter are those usually affected, problems being greatest in herds showing good overall weight gains. Hepatosis dietetica occurs in somewhat older pigs, and in these cases, typical degenerative lesions and necroses arise in the liver. Selenium deficiency in pigs may therefore present different syndromes, depending on several, as yet not fully understood factors. Age, body condition, and severity of the deficiency all seem to play a significant etiological role. Other characteristics of the feed offered (content of fat, vitamin E) also undoubtedly play a considerable role. Apart from the characteristic lesions described above, changes also affect the blood (anemia),[2] blood vessels (microangiopathy),[99] and bone marrow (dysfunction).[100] Enhanced sensitivity to an excess of iron has also been described in pigs suffering from a deficiency of selenium.[2]

(3) Other Selenium Deficiency Disorders

In chickens, selenium deficiency is usually manifested by exudative diathesis or pancreatic necrosis. Encephalomalacia seems to be mainly associated with vitamin E deficiency, though the chronic subclinical form also seems to be related to selenium deficiency.[101]

Selenium-responsive unthriftiness in calves and lambs is also a typical form of selenium deficiency, occurring especially in New Zealand and the U.S.

In recent years, as previously mentioned, selenium deficiency has been considered to have a relation to general resistance to infectious agents, mastitis, and disorders of the reproductive system of cows (cystic ovaries, metritis, and retained placenta). Much remains to be clarified as far as the significance of the latter conditions are concerned.

d. Prevention of Selenium Deficiency Diseases

As a prophylactic measure against selenium/vitamin E deficiency, both these nutrients are now added to feed. However, due to the fact that selenium is required in only small amounts and is at the same time a very toxic substance, considerable hesitancy was shown in permitting selenium supplementation of feed. Modern compounding and mixing techniques permitting an even dispersal of selenium in the feed have made such addition of selenium a safe prophylactic measure, and the method is now standard procedure in most countries. Selenium is also provided to animals in mineral mixtures, while salt licks and heavy metal or soluble glass boli containing selenium are also used in ruminants.

The indirect application of selenium either by means of selenium-containing fertilizers or top-dressing of pasture with selenium is also a method which has now been introduced in several countries.[91] The method has also been tried in Norway in order to find out how best to apply the selenium and at which concentrations.[102] Such a practice may also result in increased selenium levels in food grains and edible food crops if an increased selenium intake by the population is desirable. This has been an important consideration in Finland, where selenium intake by the population is particularly low.[103] The use of selenium-containing fertilizers has certain environmental implications which make the extent to which this method is likely to be employed in the future uncertain.

H. ZINC

Although zinc is toxic as well as being an essential trace element, its toxicity is slight, and zinc poisoning will hardly occur under natural conditions.[2,13,104]

1. Zinc Deficiency

The element is incorporated in a number of important metalloenzymes which play a significant role in several syntheses in the body. Cells in which the protein and nucleic acid synthesis rate is high, i.e., cells in a rapid growth phase, are particularly sensitive. The effects of zinc deficiency may therefore be complex, though in practical veterinary medicine the pathological consequences are limited mainly to reduced growth rate and parakeratosis (thickening of the skin). An increasing interest is now being shown in zinc deficiency in human medicine,[104] and considerable experimental research is being carried out.

Zinc deficiency occurs especially in pigs and ruminants. The disease in pigs can probably occur in all countries in which intensive pig farming is practiced. The reason for this usually is not too little zinc in the feed, but rather imbalances in other factors. In particular, high calcium levels in the feed mixture may significantly increase the requirements for zinc. High-level copper feeds used for growth promotion in pigs also increase zinc requirements. Apart from impaired growth, the condition is manifested by parakeratosis, sometimes complicated by secondary cutaneous infections.

Although zinc deficiency in cattle has been reported in several European countries,[2,104] the condition does not appear to be as widespread as, for example, selenium and copper deficiency. The clinical findings in cattle are also anorexia and impaired growth. Severe cases show skin lesions in the form of parakeratosis and hyperkeratosis, and sometimes also alopecia. The disease can be prevented by supplying mineral mixtures or concentrates to which sufficient amounts of zinc have usually been added. Requirements in wild ruminants are more likely to be met, as the zinc content of wild forage plants and vegetation is higher than in cultivated pastures (leys) or feed grain.[1]

III. OTHER ESSENTIAL AND NON-ESSENTIAL TRACE ELEMENTS

Diseases caused by deficiencies of chromium, nickel, vanadium, lead, arsenic, silicon, and lithium seem to occur under experimental conditions only. According to scientific criteria, these trace elements are classified as essential trace elements,[16] but so far they have proved to be of little interest in livestock production.

Some of these elements do, however, exert toxic effects at rather low levels and are of interest in veterinary toxicology. In particular, lead and arsenic, and also the nonessential elements cadmium and mercury, have caused poisoning in domestic animals. Vanadium has also induced poisoning in animals. Most incidents occur as accidental cases, with little or no relation to the natural levels of the elements.

Some of these elements are also of great interest as pollutants. Relatively volatile elements like cadmium, lead, and mercury are important in this connection because local and distant sources of air pollution may contribute to an increase in the extent to which animals grazing natural pastures are exposed to these substances.[105] Cadmium exerts a toxic effect, particularly on the kidneys. It is sometimes found in such high levels in wild herbivores, like hares and cervides in heavily affected regions, that possible pathological effects cannot be excluded. Increasing attention has also been focused lately on the leaching of aluminium from the soil as a result of increasing soil acidity due to acid precipitation, which, in turn, is a result of air pollution. All these problems are related mainly to the general question of environmental pollution and are thus considered to fall outside the scope of this article.

IV. MACROELEMENTS

Calcium, phosphorus, and magnesium are the macrominerals of particular significance with regard to deficiency diseases in animals.

A. CALCIUM AND PHOSPHORUS

Among other functions, these two elements are important, together with vitamin D, in the formation of bone, i.e., bone mineralization. In addition to the actual level of the two individual elements, a correct ratio between the two is extremely important. Too high a level of phosphorus may induce a secondary calcium deficiency.

Calcium deficiency is characterized mainly by impaired bone mineralization (rickets, osteomalacia), while lack of phosphorus manifests itself by pica, poor growth, infertility, and, if more severe, osteodystrophy.[2] Calcium and phosphorus, and also vitamin D, are usually added to animal feed as required, and deficiency disorders therefore do not represent a problem in modern livestock farming.

A deficiency of phosphorus is the condition which is most likely to cause a problem under natural grazing conditions. It is widepread in many countries and shows a distinct geographical distribution that is dependent on the phosphorus content of the soil.[2] Deficiency disorders are most common in grazing animals during the dry season, though they may also be a problem in housed cattle fed on hay only.

Phosphorus deficiency can be prevented by providing rock phosphates or bone meal, though the implementation of such prophylacitc measures on natural grazing land (range conditions) can pose difficulties. The addition of phosphate to drinking water is a method which has also been employed. The top-dressing of pasture with superphosphate is an adequate method of correcting a phosphorus deficiency and also has the advantage of increasing the bulk and quality of the pasture.

B. MAGNESIUM

Magnesium is another macroelement, the lack of which can give rise to deficiency disorders. Magnesium deficiency occurs in ruminants, especially cattle, and is manifested by acute convulsions — hypomagnesemic tetany.[2] It occurs in housed animals, especially calves, and also, in its typical form, in animals which have been turned out to graze after winter-housing (''grass tetany'').

Convulsions are due to low blood-magnesium levels (hypomagnesemia), often in combination with depressed calcium levels (hypocalcemia). The occurrence of hypomagnesemic tetany in grazing animals may be associated with patterns of use of fertilizers, a high rate of application of potassium reducing the availability of magnesium.[106] In countries such as Norway, grass tetany may represent a significant problem in certain districts, even when mineral supplements are provided. This may be due to the low and variable availability of magnesium in mineral magnesium sources. The incidence may also vary from year to year for no apparent reasons. Usually, applied methods in the prophylaxis of hypomagnesemic tetany include the addition of magnesium to the feed and top-dressing with magnesium.

REFERENCES

1. **Garmo, T. H., Frøslie, A., and Høie, R.,** Levels of copper, molybdenum, sulphur, zinc, selenium, iron and manganese in native pasture plants from a mountain area in Southern Norway, *Acta Agric. Scand.,* 36, 147, 1986.
2. **Blood, D. C. and Radostits, O.M.,** *Veterinary Medicine. A Textbook of the Diseases of Cattle, Sheep, Pigs, Goats and Horses,* 7th ed., Baillière Tindall, London, 1989, chap. 29, 31.
3. **Mills, C. F.,** The physiological and pathological basis of trace element deficiency disease, in *Trace Elements in Animal Production and Veterinary Practice,* Suttle, N. F., Gunn, R. G., Allen, W. M., Linklater, K. A., and Wiener, G., Eds., British Society of Animal Production, Edinburgh, 1983, chap. 1.1.
4. **Lewis, G. and Anderson, P. H.,** The nature of trace element problems: delineating the field problem, in *Trace Elements in Animal Production and Veterinary Practice,* Suttle, N. F., Gunn, R. G., Allen, W. M., Linklater, K. A., and Wiener, G., Eds., British Society of Animal Production, Edinburgh, 1983, chap. 1.2.
5. **Scottish Agricultural Research Institutes,** *Trace Element Deficiency in Ruminants. Report of a Study Group,* Scottish Agricultural Colleges, Edinburgh, 1982.
6. **Lee, H. J.,** Trace elements in animal production, in *Trace Elements in Soil-Plant-Animal Systems,* Nicholas, D. J. D. and Egan, A. R., Ed., Academic Press, New York, 1975, 39.
7. **Underwood, E. J.,** *Trace Elements in Human and Animal Nutrition,* 4th ed., Academic Press, New York, 1977, chap. 1.
8. **Phillippo, M.,** The role of dose-response trials in predicting trace element deficiency disorders, in *Trace Elements in Animal Production and Veterinary Practice,* Suttle, N. F., Gunn, R. G., Allen, W. M., Linklater, K. A., and Wiener, G., Eds., British Society of Animal Production, Edinburgh, 1983, chap. 3.2.
9. **Agricultural Research Council,** Requirements for minerals, in *The Nutrient Requirements of Pigs,* Commonwealth Agricultural Bureaux, London, 1981, chap. 4.
10. **National Research Council,** *Mineral Tolerance of Domestic Animals,* National Academy of Sciences, Washington, D.C., 1980.
11. **National Research Council,** Nutrient requirements: excesses and deficiencies, in *Nutrient Requirements of Beef Cattle,* 6th rev. ed., National Academy Press, Washington, D. C., 1984, chap. 2.
12. **Frølie, A. and Norheim, G.,** Copper, molybdenum, zinc, and sulphur in Norwegian forages and their possible role in chronic copper poisoning in sheep. *Acta Agric. Scand.,* 33, 97, 1983.
13. **Howell, J. McC.,** Toxicity problems associated with trace elements in domestic animals, in *Trace Elements in Animal Production and Veterinary Practice,* Suttle, N. F., Gunn, R. G., Allen, W. M., Linklater, K. A., and Wiener, G., Eds., British Society of Animal Production, Edinburg, 1983, chap. 5.1.
14. **Mills, C. F.,** Trace element deficiency and excess in animals, *Chem. Br.,* 15, 512, 1979.
15. **Kirchgessner, M., Reichlmayr-Lais, A. M., and Schwarz, F. J.,** Interaction of trace elements in human metabolism, in *Nutrition in Health and Disease and International Development: Symposia From the XII International Congress of Nutrition,* Alan R. Liss, New York, 1981, 189.
16. **Underwood, E. J. and Mertz, W.,** Introduction, in *Trace Elements in Human and Animal Nutrition,* Vol. 1, 5th ed., Mertz, W., Ed., Academic Press, London, 1987, chap. 1.
17. **Alfrey, A. C.,** Aluminium, in *Trace Elements in Human and Animal Nutrition,* Vol. 2, 5th ed., Mertz, W., Ed., Academic Press, London, 1986, chap. 9.
18. **Mertz, W., Ed.,** *Trace Elements in Human and Animal Nutrition,* Vol. 1, 2, 5th ed., Academic Press, London, 1986, 1987.
19. **Suttle, N. F., Gunn, R. G., Allen, W. M., Linklater, K. A., and Wiener, G., Eds.,** *Trace Elements in Animal Production and Veterinary Practice,* British Society of Animal Production, Edinburgh, 1983.
20. **Nicholas, D. J. D. and Egan, A. R., Eds.,** *Trace Elements in Soil-Plant-Animal Systems,* Academic Press, New York, 1975.
21. **Humphreys, D. J.,** *Veterinary Toxicology,* 3rd ed., Bailliere Tindall, London, 1988, chap. 2.
22. **Osweiler, G. D., Carson, T. L., Buck, W. B., and Van Gelder, G. A.,** *Clinical and Diagnostic Veterinary Toxicology,* 3rd ed., Kendall/Hunt, Dubuque, 1985.

23. **Miller, W. J.**, Mineral and trace element nutrition of dairy cattle, in *Dairy Cattle Feeding and Nutrition*, Academic Press, New York, 1979, chap. 5.
24. **Smart, M. E., Gudmundson, J., and Christensen, D. A.**, Trace mineral deficiencies in cattle: a review, *Can. Vet. J.*, 22, 372, 1981.
25. **Bremner, I. and Mills, C. F.**, Absorption, transport and tissue storage of essential trace elements, *Philos. Trans. R. Soc. London Ser. B*, 294, 75, 1981.
26. **Spallholz, J. E., Martin, J. L., and Ganther, H. E., Eds.**, *Selenium in Biology and Medicine*, AVI Publishing, Westport, CT, 1981.
27. **Combs, G. F., Jr., Spallholz, J. E., Levander, O. A., and Oldfield, J. E., Eds.**, *Selenium in Biology and Medicine*, Parts A, B, Van Nostrand Reinhold Company, New York, 1987.
28. **Davies, N. T.**, An appraisal of the newer trace elements, *Philos. Trans. R. Soc. London Ser. B*, 294, 171, 1981.
29. **Young, R. J., Jr.**, *Selenium and Copper Deficiencies in Cattle*, Schering Animal Health, Lenexa, 1988.
30. **Kirchgessner, M., Weigand, E., Schnegg, A., Grassmann, E., Schwarz, F. J., and Roth, H.-P.**, Spurenelemente, in *Biochemie und Physiologie der Erährung*, Vol. 1, Cremer, H.-D., Hötzel, D., and Kühnau, J., Eds., Georg Thieme Verlag, Stuttgart, 1980, 275.
31. **Gupta, U. C. and Lipsett, J.**, Molybdenum in soils, plants, and animals, *Adv. Agric.*, 36, 73, 1981.
32. **Gissel-Nielsen, G., Gupta, U. C., Lamand, M., and Westermarck, T.**, Selenium in soils and plants and its importance in livestock and human nutrition, *Adv. Agric.*, 37, 397, 1984.
33. **Anke, M., Baumann, W., Bräunlich, H., Brückner, Chr., and Groppel, B., Eds.**, *5. Spurenelementsymposium. B, Cr, Co, Cu, F, Fe, Mn, Se, Zn 1986*, Friederich-Schiller-Universität, Jena, 1986.
34. **Anke, M., Baumann, W., Bräunlich, H., Brückner, Chr., and Groppel, B., Eds.**, *5. Spurenelementsymposium. Al, As, Cd, Hg, Ni, Pb, Sn, Tl, Si, V 1986*, Friederich-Schiller-Universität, Jena, 1986.
35. **Anke, M., Brückner, Chr., Gürtler, H., and Grün, M., Eds.**, *Mengen- und Spurenelemente. Arbeitstagung 1987*, Karl-Marx-Universität, Leipzig, 1987, Part 1, 2.
36. **Smith, R. M.**, Cobalt, in *Trace Elements in Human and Animal Nutrition*, Vol. 1, 5th ed., Mertz, W., Ed., Academic Press, London, 1987, chap. 5.
37. **Ender, F.**, Undersøkelser over slikkesykens etiologi i Norge, *Nor. Vet. Tidsskr.*, 54, 3, 1942.
38. **Ender, F.**, Koboltmangelens betydning som sykdomsårsak hos storfe og sau belyst ved terapeutiske forsøk, *Nor. Vet. Tidsskr.*, 58, 118, 1946.
39. **Ender, F. and Tananger, I. W.**, Fortsatte undersøkelser over årsaksforholdene ved mangelsykdommer hos storfe og sau. Koboltmangel som sykdomsårsak belyst ved kjemiske undersøkelser av fôret, *Nor. Vet. Tidsskr.*, 58, 313, 1946.
40. **Ulvund, M. J. and Øverås, J.**, Chronic hepatitis in lambs in Norway, a condition resembling ovine white liver disease in New Zealand, *N. Z. Vet. J.*, 28, 19, 1980.
41. **Schwan, O., Jacobsson, S.-O., Frank, A., Rudby-Martin, L., and Petersson, L. R.**, Cobalt and copper deficiency in Swedish landrace pelt sheep. Application of diagnostics in flock-related deficiency diseases, *J. Vet. Med. A*, 34, 709, 1987.
42. **Ulvund, M. J.**, personal communication, 1988.
43. **McMurray, C. H., Rice, D. A., McLoughlin, M., and Blanchflower, W. J.**, Cobalt deficiency and the potential of using methylmalonic acid as a diagnostic and prognostic indicator, in *Trace Elements in Man and Animals — TEMA 5*, Mills, C. F., Bremner, I., and Chesters, J. K., Eds., Commonwealth Agricultural Bureau, London, 1985, 603.
44. **Mason, J.**, Molybdenum-copper antagonism in ruminants: a review of the biochemical basis, *Ir. Vet. J.*, 35, 221, 1981.
45. **Humphries, W. R., Mills, C. F., Greig, A., Roberts, L., Inglis, D., and Halliday, G. J.**, Use of ammonium tetrathiomolybdate in the treatment of copper poisoning in sheep, *Vet. Rec.*, 119, 596, 1986.
46. **Davis, G. K. and Mertz, W.**, Copper, in *Trace Elements in Human and Animal Nutrition*, Vol. 1, 5th ed., Mertz, W., Ed., Academic Press, London, 1987, chap. 1.
47. **Gay, C. C., Pritchett, L. C., and Madson, W.**, Copper deficiency in cattle: a review, in *Selenium and Copper Deficiencies in Cattle*, Young, R. J., Jr, Ed., Veterinary Medicine Publishing, Lenexa, Kansas, 1988, 10.
48. **Mills, C. F. and Davis, G. K.**, Molybdenum, in *Trace Elements in Human and Animal Nutrition*, Vol. 1, 5th ed., Mertz, W., Ed., Academic Press, London, 1987, chap. 13.
49. **Wang, Z. Y., Poole, D., and Mason, J.**, The biochemical pathogenesis of Mo-induced hypocuprosis in cattle; the effect of increased dietary Mo on copper metabolism systemically, in *Proc. 14th World Congr. Diseases of Cattle*, Vol. 2, Hartigan, P. J. and Monaghan, M. L., Eds., Dublin, 1986, 845.
50. **Whitelaw, A.**, Copper dificiency in lambs, *Vet. Ann.*, 25, 100, 1985.
51. **Barlow, R. M.**, Enzootic ataxia of deer, in *The Comparative Pathology of Zoo Animals*, Montali, R. J. and Migaki, G., Eds., Smithsonian Institution Press, Washington, D. C., 1980, 65.
52. **Wilson, P. R., Orr, M. B., and Key, E. L.**, Enzootic ataxia in red deer, *N. Z. Vet. J.*, 27, 252, 1979.

53. **Frøslie, A.**, Copper in sheep in Norway, in *Geomedical Aspects in Present and Future Resaerch*, Låg, J., Ed., Universitetsforlaget, Oslo, 1980, 183.

54. **Frøslie, A., Holt, G., Høie, R., and Haugen, A.**, Levels of copper, selenium and zinc in liver of Norwegian moose (*Alces alces*), reindeer (*Rangifer tarandus*), roedeer (*Capreolus capreolus*) and hare (*Lepus timidus*) (in Norwegian), *Nor. Landbruksforskn.*, 1, 243, 1987.

55. **Bostwick, J. L.**, Copper toxicosis in sheep, *J. Am. Vet. Med. Assoc.*, 180, 386, 1982.

56. **Hubbs, R. and Oehme, F. W.**, Understanding chronic copper poisoning in sheep, *Bovine Pract.*, 3, 15, 1982.

57. **Søli, N. E.**, Chronic copper poisoning in sheep. A review of the literature, *Nord. Vet. Med.*, 32, 75, 1980.

58. **Baldwin, W. K., Hamar, D. W., Gerlack, M. L., and Lewis, L. D.**, Copper-molybdenum imbalance in range cattle, *Bovine Pract.*, 2, 9, 1981.

59. **Bull, L. B.**, The story of toxaemic jaundice ("yellows") in sheep in Australia: chronic copper poisoning and pyrrolizidine alkaloidosis, *Victorian Vet. Proc.*, 22, 17, 1964.

60. **Todd, J. R.**, Copper, molybdenum and sulphur contents of oats and barley in relation to chronic copper poisoning in housed sheep, *J. Agric. Sci.*, 79, 191, 1972.

61. **Frøslie, A., Norheim, G., and Søli, N. E.**, Levels of copper, molybdenum, zinc and sulphur in concentrates and mineral feeding stuffs in relation to chronic copper poisoning in sheep in Norway, *Acta Agric. Scand.*, 33, 261, 1983.

62. **Frøslie, A.**, unpublished data, 1988.

63. **Søli, N. E. and Frøslie, A.**, Chronic copper poisoning in sheep. I. The relationship of methaemoglobinemia to Heinz body formation and haemolysis during the terminal crisis, *Acta Pharmacol. Toxicol.*, 40, 169, 1977.

64. **Sivertsen, T.**, Copper-induced GSH depletion and methaemoglobin formation *in vitro* in erythrocytes of some domestic animals and man. A comparative study, *Acta Pharmacol. Toxicol.*, 46, 121, 1980.

65. **Nordstoga, K.**, Undersøkelser over en særlig form for kopperforgiftning hos sau, *Proc. 9th Nordic Veterinary Congress*, 1962.

66. **Frøslie, A., Havre, G. N., and Norheim, G.**, Levels of copper and molybdenum in herbage in an area with copper deficiency in sheep and cattle: comparison of data from 1958 and 1983, in *Trace Elements in Man and Animals — TEMA 5*, Mills, C. F., Bremner, I., and Chesters, J. K., Eds., CAB, London, 1985, 855.

67. **Låg, J. and Bølvigen, B.**, Some naturally heavy-metal poisoned areas of interest in prospecting, soil chemistry, and geomedicine, *Nor. Geol. Unders. (Publ.)*, 304, 73, 1974.

68. **Bartíc, M. and Piskač, A.**, *Veterinary Toxicology*, Elsevier, Amsterdam, 1981, 91.

69. **Lamand, M.**, Copper toxicity in sheep, in *Copper in Animal Wastes and Sewage Sludge*, L'Hermite, P. and Dehandtschutter, J., Eds., Reidel, Dordrecht, 1981, 261.

70. **Krishnamachari, K. A. V. R.**, Fluorine, in *Trace Elements in Human and Animal Nutrition*, Vol. 1, 5th ed., Mertz, W., Ed., Academic Press, London, 1987, chap. 11.

71. **Flatla, J. L.**, The fluorine problem in practice — poisoning in ruminants, in *Festskrift til Knut Breirem*, Spildo, L. S., Homb, T., and Hvidsten, H. (Editorial committee), Oslo-As, 1972, 37.

72. **Løkken, P.**, personal communication, 1988.

73. **Slagsvold, L.**, Fluorforgiftning, *Nor. Vet. Tidsskr.*, 46, 2, 1934.

74. **Hetzel, B. S. and Maberly, G. F.**, Iodine, in *Trace Elements in Human and Animal Nutrition*, Vol. 2, 5th ed., Mertz, W., Ed., Academic Press, London, 1986, chap. 2.

75. **Løken, A.**, Struma, *Nor. Vet. Tidsskr.*, 14, 177, 1912.

76. **Morris, E. R.**, Iron, in *Trace Elements in Human and Animal Nutrition*, Vol. 1, 5th ed., Mertz, W., Ed., Academic Press, London, 1987, chap. 4.

77. **Hurley, L. S. and Keen, C. L.**, Manganese, in *Trace Elements in Human and Animal Nutrition*, Vol. 1, 5th ed., Mertz, W., Ed., Academic Press, London, 1987, chap. 6.

78. **Oldfield, J. E.**, The selenium story: some reflections on the "moon metal", *N. Z. Vet. J.*, 22, 85, 1974.

79. **Slagsvold, L. and Lund-Larsen, H.**, Myosit hos lam, kalver og ungfe, *Nor. Vet. Tidsskr.*, 46, 529, 1934.

80. **Slagsvold, L.**, Tranforgiftning hos kalver, *Nor. Vet. Tidsskr.*, 37, 161, 1925.

81. **Obel, A.-L.** Studies on the morphology and etiology of so-called toxic liver dystrophy (hepatosis diaetetica) in swine, *Acta Pathol. Microbiol. Scand.*, (Suppl. 94), 1953.

82. **Schwarz, K. and Foltz, C. M.**, Selenium as an integral part of Factor 3 against dietary necrotic liver degeneration, *J. Am. Chem. Soc.*, 79, 3292, 1957.

83. **Jubb, K. V. F., Kennedy, P. C., and Palmer, N.**, *Pathology of Domestic Animals*, Vol. 2, 3rd ed., Academic Press, Orlando, 1985, 333.

84. **Thafvelin, B., Swahn, O., and Erne, K.**, Foderspannmålens roll vid vissa ämnesomsättningrubbningar hos svin, *Medlemsbl. Sven Vetförb.*, 12, 8, 1960.

85. **Rotruck, J. T., Pope, A. L., Ganther, H. E., Swanson, A. B., Hafeman, D. G., and Hoekstra, W. G.**, Selenium: biochemical role as a component of glutathione peroxidase, *Science*, 179, 588, 1973.

86. **Flohe, L., Günzler, W. A., and Schock, H. H.,** Glutathione peroxidase: a selenoenzyme, *FEBS Lett.,* 32, 132, 1973.
87. **Hutchinson, L.,** The significance of selenium in cattle health and productivity, in *Selenium and Copper Deficiencies in Cattle,* Young, R. J., Jr., Ed., Veterinary Medicine Publishing, Lenexa, Kansas, 1988, 4.
88. **Larsen, H. J., Øvernes, G., and Moksnes, K.,** Effect of selenium on lymphocyte responses to mitogens, *Res. Vet. Sci.,* 45, 11, 1988.
89. **Smith, K. L., Hogan, J. S., and Conrad, H. R.,** Selenium in dairy cattle: its role in disease resistance, in *Selenium and Copper Deficiencies in Cattle,* Young, R. J., Jr., Ed., Veterinary Medicine Publishing, Lenexa, Kansas, 1988, 15.
90. **Ursini, F., Maiorino, M., Roveri, A., and Coassin, M.,** Phospholipid hydroperoxide glutathione peroxidase, in *Selenium in Biology and Medicine,* Abstr. 4th Int. Symp., Tübingen, July 18-21, 1988.
91. **Levander, O. A.,** Selenium, in *Trace Elements in Human and Animal Nutrition,* Vol. 2, 5th ed., Mertz, W., Ed., Academic Press, London, 1986, chap. 3.
92. **Mihailović, M., Gavrilović, B., Veličkovski, S., and Ilic, V.,** Influence of diet low in selenium on reproductive performance in sows, in *Selenium in Biology and Medicine,* Abstr. 4th Int. Symp., Tübingen, July 18-21, 1988.
93. **Hartfiel, W. and Bahners, N.,** Zur Selenversorgung von Wiederkäuer, *VDLUFA-Schriftrehe,* 16, Kongressband 1985, 1986, 511.
94. **Frøslie, A., Karlsen, J. T., and Rygge, J.,** Selenium in animal nutrition in Norway, *Acta Agric. Scand.,* 30, 17, 1980.
95. **Mattsson, A.,** Selenium in animal husbandry in Scandinavia, *Feedstuffs,* 54(9), 21, 1982.
96. **Wu, X. and Låg, J.,** Selenium in Norwegian farmland soils, *Acta Agric. Scand.,* 38, 271, 1988.
97. **Rice, D. A., McMurray, C. H., and Kennedy, S.,** Polyunsaturated fatty acid induced nutritional degenerative myopathy in selenium deficient steers, in *Trace Elements in Man and Animals — TEMA 5,* Mills, C. F., Bremner, I., and Chesters, J. K., Eds., CAB, London, 1985, 232.
98. **Maas, J., Bulgin, M. S., Anderson, B. C., and Frye, T. M.,** Nutritional myodegeneration associated with vitamin E deficiency and normal selenium status in lambs, *J. Am. Vet. Med. Assoc.,* 184, 201, 1984.
99. **Jubb, K. V. F., Kennedy, P. C., and Palmer, N.,** *Pathology of Domestic Animals,* Vol. 3, 3rd ed., Academic Press, Orlando, 1985, 26.
100. **Nafstad, I. and Nafstad, P. H. J.,** An electron microscopic study of blood and bone marrow in vitamin E-deficient pigs, *Path. Vet.,* 5, 520, 1968.
101. **Kristiansen, F.,** personal communication, 1985.
102. **Øvernes, G., Moksnes, K., Mo Økland, E., and Frøslie, A.,** The efficacy of hay from selenized soil as a source of selenium in ruminant nutrition, *Acta Agric. Scand.,* 36, 1, 1986.
103. **Varo, P.,** Effects of selenium fertilization in Finland 1985/86, in *Commercial Fertilizers and Geomedical Problems,* Låg, J., Ed., Norwegian University Press, Oslo, 1987, 35.
104. **Hambidge, K. M., Casey, C. E., and Krebs, N. F.,** Zinc, in *Trace Elements in Human and Animal Nutrition,* Vol. 2, 5th ed., Mertz, W., Ed., Academic Press, London, 1986, chap. 1.
105. **Frøslie, A., Haugen, A., Holt, G., and Norheim, G.,** Levels of cadmium in liver and kidneys from Norwegian cervides, *Bull. Environ. Contam. Toxicol.,* 37, 453, 1986.
106. **Hvidsten, H.,** Potassium fertilization of pasture and magnesium deficiency in ruminants, in *Commercial Fertilizers and Geomedical Problems,* Låg, J., Ed., Norwegian University Press, Oslo, 1987, 77.

Chapter 4

PROBLEMS ON EXCESS OF INORGANIC CHEMICAL COMPOUNDS FOR MANKIND

Jan Alexander, Jetmund Ringstad, and Jan Aaseth

TABLE OF CONTENTS

I. INTRODUCTION

In this chapter, the general problems for mankind due to exposure to an excess of inorganic chemical compounds will be discussed. It is impossible in any way to cover this large area completely. The present chapter is therefore focused on certain aspects of the toxicology of metals, but some comments are also made on mineral particles and inorganic compounds of nonmetalic elements. For further details and information on particular compounds, the reader is referred to the *Handbook on Toxicity of Inorganic Compounds*[1] and *Handbook on the Toxicology of Metals*.[2] These references cover most of this area. Information can also be obtained from general textbooks in toxicology such as *Casarett and Doull's Toxicology*[3] and the Environmental Health Criteria published under the International Program on Chemical Safety by the World Health Organization.

The group of metals consists of some 80 elements, and their compounds range from relatively simple ionic salts to complicated organometallic structures. Most of the metals that cause health problems belong to the trace element category. Many of the trace elements also fulfill essential functions in the human body. Thus, they act in an optimal concentration range, below which deficiency symptoms may occur. When present in excess, essential elements may also become toxic.

Many of the most toxic nonessential elements occur relatively rarely in the earth's crust in comparison with "nontoxic" elements and macroelements essential for life. During evolution, life seems to have adapted to the environment and made use of accessible elements and created protective mechanisms for nonessential abundant elements (e.g., aluminum). This general statement is, however, not always applicable in a changing environment due to human activity and industrialization.

The toxicity to man of several metals has been known for centuries, and they were among the first toxic substances to be described. For example, symptoms of metallic mercury poisoning, including tremor, mercurial erethism, loss of memory, behavioral changes, hallucinations, and constriction of the visual field, were well known in workers in mercury mines in ancient times, as were the lead-induced paralysis and abdominal colic of lead-exposed humans.

There is a wide range of sources of exposure to inorganic chemical compounds. The exposure may occur naturally e.g., by erosion of earth and deposits of metal mineral, but exposure from human activities such as combustion of fossil fuel, mining, smelting, and industrial use of metals are far more important. The general contamination of the human environment and increased human exposure can be illustrated by two examples. The level of lead in human bone and teeth has increased 10- and 30-fold, respectively, from the level in ancient Nubians to adults in our time.[4] Acid precipitation due to air polluted with industrially emitted SO_2 and NOx causing acidification of surface water has increased the solubility and, hence, the mobility of metal compounds. Recent findings from Sweden show that methylmercury in fish from lakes with acidic water was higher than in lakes with neutral water.[5] Raised levels of aluminum in acidic lakes used for drinking water are frequent on the south coast of Norway.[6,7] Thus, acidification may lead to increased human exposure of otherwise less mobile elements via drinking water and food.

II. ROUTES OF EXPOSURE

Exposure via inhalation and ingestion are the most important routes of exposure.[8] Uptake via the skin may be important for some organometallic compounds such as lead, but most often skin exposure leads to local effects, e.g., dermatitis and hypersensitivity.

Some metal compounds may occur as vapors (e.g., mercury), but in general they exist as aerosols and small particles. Exposure to elemental mercury can take place, for example, in mercury mines and chlor-alkali plants. Evaporation of elemental mercury from dental amalgam fillings has recently been recognized as the single most important source of inorganic mercury exposure to the general population. Aerosols and particles may be emitted to the work atmosphere or ambient air by various industrial processes such as melting and refinement of metals as well as production of lead and cadmium-containing storage batteries. Combusion of leaded gasoline by motor vehicles is the most important source of lead in ambient air in areas with heavy traffic. These metal emmitting processes generate different sizes of particles which are handled differently by the human body. Since vapor or gases of metallic compounds are usually insoluble in water, they will reach the alveoli of the lungs, where they easily can penetrate the air-blood barrier. The nose and upper airways are efficient filters for water-soluble gases and particles above 2 to 5 μm; however, a large fraction of particles with an aerodynamic diameter less than 0.5 μm will reach the alveoli. Larger particles are mainly deposited in the tracheobronchial tract and cleared by mucociliar transport.[8]

The other major route of exposure is via the gastrointestinal tract. Food items and drinking water contaminated with various amounts of inorganic compounds constitute the major source for many elements. The fraction of metal that is taken up may vary considerably. For example, the uptake of iron is controlled by the body and is increased during iron deficiency. In adults, about 10% of ingested lead is taken up from the intestine, while less than 0.1% of ingested aluminum enters the body.[9-11] Children seem to take up a greater fraction of several metals, and as much as 40% of ingested lead may cross the intestinal barrier. Other factors may also greatly influence the uptake, such as the chemical form of the element and the presence of chelating agents. While metallic mercury does not pass the intestinal barrier, less than 10% of inorganic mercury salts are absorbed and methylmercury is nearly completely absorbed.[8] The uptake of aluminum is, for example. greatly enhanced by citric acid and maltol.[9-11] The mechanisms by which the metallic compounds are absorbed are, however, only partly understood. Metals may be actively transported via physiological mechanisms (e.g., iron) or just enter by diffusion and other unspecific transport processes. When metal compounds (e.g., iron) are ingested in doses damaging the intestinal mucosal barrier, uncontrolled amounts of the compounds may enter the body, causing systemic poisoning.

Unique routes of absorption of toxic inorganic compounds involve uptake during hemodialysis or infusion of contaminated blood products or parenteral solutions (e.g., albumin contaminated with aluminum). Toxicity due to aluminum in untreated drinking water used for hemodialysis of patients with renal failure has been a great problem. By keeping aluminum in the dialysis fluid at a level less than 10 μg/l, no net transport of aluminum from the dialysis fluid to the bloodstream will occur.[11] Another curious way of exposure is from orthopedic stainless steel implants, and increased levels of nickel in the body fluids of patients with such implants have been reported.[12]

III. DISTRIBUTION, BIOTRANSFORMATION, AND EXCRETION

The major route of distribution is usually the bloodstream. Depending on the chemical form, the metal compound may bind to plasma proteins or enter the red blood cells. Among metals, there are very large differences between the fraction of an absorbed dose that is transported to various organs and the fraction that is excreted. Important determinants for organ uptake are the "diffusible fraction" in plasma, the rate of permeability of cellular membranes and biotransformation, and availability of intracellular ligands for the metal as well as the rate of ligand interchange. While germanium does not bind to plasma proteins, at least 99% of cadmium and mercury in plasma are protein bound. The importance of biotransformation is illustrated by metallic mercury vapor, which is rapidly transported to the brain where it easily enters before it is oxidized to the much less permeable Hg^{2+} form and trapped within the brain cells.[13] Toxic effects are due to this oxidized form.

Methylation and demethylation are of importance for the distribution and toxicity of several metal compounds, including As, Se, and Hg. Demethylation of methyl mercury leads to the less toxic and more easily excreted form Hg^{2+}, while arsenic[14] and selenium[15] are methylated and thereby made less toxic and more excretable. Occasionally, biotransformation reactions may render the compounds more toxic. Thus, the breakage of one carbon-metal bond in tetraethyltin yields a highly toxic metabolite.[16]

Another physiological detoxication mechanism is induction of metallothioneine, a low molecular weight (7000 Da) protein consisting of about 30% cystein. The synthesis of this protein is homeostatically regulated by increased body fluid levels of zinc and low levels of cadmium. The mechanism appears to protect against metal toxicity by intracellular trapping of the metal and is a phylogenetically old way of protection. However, if intracellular metal accummulation exceeds the production and binding capacity of metallothionein, toxicity will appear. This can be seen when cadmium accumulates in the kidney; a certain concentration of cadmium is reached and renal damage becomes apparent.[8,17]

Major excretory pathways are via the urine and feces; saliva, lactation, sweat, and skin are of minor importance. While, for example, manganese, thallium, and methyl mercury are preferably excreted via feces, elements such as selenium, cobalt, and lead are recovered in urine. Fecal metal excretion is derived from bile, pancreatic fluid, intestinal secretions, and shedding of intestinal cells. It has recently been shown that biliary excretion is important for several metals.[18] A common excretory mechanism of metals with a high affinity for sulfhydryl groups, e.g., methyl mercury, inorganic mercury, cadmium, copper, and lead, seems to be translocation from liver to bile bound to the tripeptide glutathione.[18]

Renal excretion is another important route for a number of metals, e.g., lead and selenium. This pathway is of special importance during chelation therapy of patients with acute poisoning, where the low molecular weight metal chelates are readily excreted into the urine. Metals cleared from plasma in the glomerulus may be reabsorbed in the kidney tubule: e.g., cadmium bound to metallothionein is very efficiently reabsorbed in the renal tubule, leading to metal accumulation.[17]

IV. EXCESS EXPOSURE AND TOXIC EFFECTS

The presence of toxic elements in soil or rocks, whether due to natural geochemistry or human activities, including pollution, usually influence human health indirectly via food or drinking water. Although many places in the world rely solely on locally produced food, food consumption in modern industrialized societies it is much more diverse, including food produced in different geographical areas. Drinking water, however, is usually strongly related to local geochemistry. This calls for care when correlations are made between local geochemistry, possible excess exposure, and adverse health effects. Precise dose estimates must be calculated. These are dependent on the amount of locally produced food consumed, its content of harmful elements, and whether a major part of the exposure takes place via drinking water.

Problems of excess intake from drinking water have been encountered for several inorganic compounds, including fluoride in Africa and India; arsenic in certain areas of Argentina, Chile, and Taiwan; selenium in seleniferous areas in the U.S., Venezuela, and China; and nitrate in agricultural areas with heavy use of fertilizers.[14,15,19] Soft and acid drinking water may cause leaching of metals such as lead from lead pipes and copper from pipes made of copper. This may, in certain cases, constitute a considerable source of lead exposure. A few cases of liver cirrhosis in artificially fed newborn children due to excessive copper exposure from contaminated drinking water were recently reported from West Germany.[20]

In general, a closer correlation between local geochemistry and the health of animals living in the area will be seen. Local geochemistry is less important for excess exposure to toxic elements via inhalation.

A. MECHANISMS OF TOXICITY

A wide range of effects due to the excess exposure of elements can be seen in man.[13] Metal compounds may be metabolized in the body to other chemical forms. However, unlike other toxic substances, metals are never broken down and destroyed within the human body. The adverse effects can be local or, when systemic effects are seen, any organ can be affected. Furthermore, metals may also act as allergens, carcinogens, mutagens, or teratogens. Generally, no effects seem to be quite specific for metals or inorganic compounds.

At the molecular level, metal compounds exert their harmful effects by binding to cellular ligands containing oxygen, nitrogen, or sulfur. Depending on the metal, different types of ligands will be preferred. Mercury and other soft metals, so-called sulfur seekers are bond strongly to sulfur-containing ligands, e.g., glutathione and a large number of proteins. Although mercuric compounds bind very strongly to sulfur, the rate of ligand interchange is high. This makes mercury rapidly transportable within the body. In contrast, trivalent chromium forms tend to form rather inert complexes with several ligands; thus, a slower ligand interchange is seen.

Metal toxicity is believed to be mediated through to macromolecules such as proteins with structural, catalytic, or transport functions and DNA. An important mechanism that has been discussed lately is the role of reactive oxygen species in metal toxicity.[21] Particularly transition metal ions, e.g., of iron and copper, can overcome the spin restriction of O_2 and donate a single electron, giving rise to free radical species and chain reactions. Toxicity occurs when such metals are free and reactions become uncontrolled. Several metals, including mercury and cadmium, can also inhibit protection mechanisms against reactive oxygen species e.g., glutathione peroxidase.[15] Other important cytotoxic mechanisms for some metals (e.g., mercury and chromate) are DNA damage and inhibition of cellular respiration (e.g., arsenic and chromate).[22] Although metals can bind to a wide variety of cellular ligands and metals in general can give rise to nearly any lesion, the effects produced seem to be relatively specific for each metal compound at low dose levels. The reason for this is not well understood.

B. ACUTE EFFECTS AND LARGE OUTBREAKS OF DISEASE

The acute adverse health effects can be local, such as allergic dermatitis caused by chromium and nickel compounds. Local irritating effects in the respiratory tract as well as respiratory allergic reactions might also be seen with a large number of compounds. Acute exposure to metal fumes (e.g., from zinc, cadmium, lead, and other metals) generated during welding or welding-like operations may lead to acute alveolitis with systemic symptoms such as fever and muscular pain. Acute toxicity due to massive exposure is well known for metal compounds. The exposure is usually accidental and occurs at the workplace or at home. Particularly children have accidentally ingested arsenic salts, thallium salts, and sublimate, the latter leading to renal failure. Another well-known example is poisoning from the ingestion of seed grain treated with methyl mercury, leading to death or severe neurological symptoms and blindness. The latter effects, however, often develop sometime after exposure and are chronic.[23]

Well-known examples of metal-contaminated soil or sediments leading to serious health problems for a large group of people are known (e.g. Japan). The outbreak of the Itai-itai disease with skeletal dysfunction and renal injury was due to the ingestion of cadmium-containing rice grown on contaminated soil.[17] The Minamata disease, which is caused by methyl mercury, is characterized by overt clinical symptoms from the central and peripheral nervous system and, in severe cases, death occurred. This outbreak was the result of severe industrial contamination of the river and Minamata bay with methyl mercury and other mercury compounds. Methyl mercury, biomethylated in the sediment, accumulated in fish and other seafoods used for human consumption.[23] Although soil in the vicinity of cadmium-emitting industries has been contaminated with cadmium, and certain lakes and areas contain methyl mercury contaminated fish in the Western industrialized world, such large outbreaks have not occurred partly because food consumption is more diverse and does not rely solely on locally produced food. There has also been a recent general awareness of such problems after these outbreaks.

C. LOW-LEVEL EXPOSURE AND NEUROTOXIC EFFECTS

Today, with occupational and environmental standards, such acute, overt symptoms of clincical poisoning are less frequent. There is a growing inquiry regarding more subtle, unspecific, subclinical effects, where cause-and-effect relationships are not obvious. Exposure to low levels of toxic metal may cause adverse health effects after long-term accumulation of the toxic compound in the body or following a number of subclinical injuries. Lead poisoning in children often occurs after long-term ingestion of dirt containing lead-based paint. More subtle symptoms of psychological dysfunction have been seen in children exposed to very small elevations of body fluid levels of lead, either *in utero* or during childhood. It is widely accepted that a major contributing source of such slightly elevated levels in big cities, such as Mexico City, is the combustion of leaded gasoline.

While aluminum previously was considered a nontoxic element, it is now recognized as a potent neurotoxin. Aluminum accummulation in patients undergoing hemodialysis is associated with severe clinical symtoms from the central nervous system and the bone.[10,11] Disputable, however, is the role of aluminum in a number of neurodegenerative diseases, including Alzheimer's disease, Parkinson's disease, and amyotrophic lateral sclerosis.[10,11,24] In all cases, aluminum has been found concentrated in pathological lesions in the central nervous system.[11,26] Some of the diseases are particularly prevalent at Guam in the Pacific Ocean.[25] The soil and rock are very rich in aluminum, but low levels of magnesium also have been found. Recent epidemiological surveys in Norway[7,26] and England[27] have linked Alzheimer's disease to elevated levels of aluminum in drinking water, even when only a small fraction of the ingested aluminum originated from drinking water. The relative risk that has been found is low, between 1 and 2. Hopefully, on-going and future epidemiological studies will resolve these questions about aluminum. If a causal relationship is found, it will be of considerable public health impact.

D. REPRODUCTIVE AND DEVELOPMENTAL TOXICITY

Also of concern are those elements which lead to reproductive and developmental toxicity. In contrast to the generally adverse effects of excessive metal exposure, the effects on reproduction and development are, for most metals, not well described in studies of humans. Inorganic compounds may indirectly affect development via systemic toxicity to the mother, resulting in retardation or abnormal growth of the fetus. It is important to identify those compounds causing developmental effects at levels lower than those which cause systemic toxicity. Due to a lack of data, this is often not possible. It is well known that lead decreases fertility in males by affecting the sperms and the hormone levels.[28] Such effects have not been established for other metals. Animal studies, however, indicate that mercury, cadmium, nickel, and chromium also may affect male reproductivity. Heavy exposure of women to lead is associated with infertility, spontaneous abortions, and fetal death. However, in many cases of metal exposure, it is difficult to differentiate between effects on the female reproductive system, the gamete, the implantation process, or the embryo. Animal studies, however, suggest that the reproductive system can be affected by a number of heavy metals, e.g., lead, mercury, and cadmium.

Developmental effects may occur from exposure *in utero* as well as neonatally. It is accepted that in humans the fetus as well as the developing human organism in the perinatal and early postnatal period are very sensitive to exposure to lead and methyl mercury.[28] Damage is caused in the central and peripheral nervous systems and becomes apparent as retarded mental development and, for methyl mercury, also cerebral palsy and retarded motor development. Studies performed after the Iraq outbreak of methyl mercury poisoning indicate that the prenatal stage of life was roughly three to four times more sensitive to methyl mercury than the adult stage.[28] Recent studies from New Zealand indicate that the fetus might be even more sensitive to methyl mercury exposure.[29]

E. CARDIOVASCULAR DISEASES

Lifestyle factors such as increased intake of saturated fat, obesity, smoking, and low physical activity are major determinants of cardiovascular diseases in humans. However, excess intake of inorganic elements may also play a role in these diseases. The high sodium intake in many parts of the world, 10 to 12 g sodium chloride per day or more, is considered to be of importance for the development of hypertension. This disease is not seen among people living on a diet just covering the daily need of sodium chloride. The effect on blood pressure can, however, only be obtained when the intake is reduced considerably. The World Health Organization has recommended the mean daily intake of the population not to exceed 5 g.

It is well known that several elements at high dose levels may affect the cardiovascular system. This includes cobalt from contaminated beer, lead, cadmium, and mercury. The latter ones often affect the cardiovascular system indirectly via renal failure. A peripheral vascular disorder with gangrene (black foot disease) has been reported among people in Taiwan, while Raynaud's syndrome has been reported to occur frequently in certain parts of Chile, both places with drinking water containing high levels of arsenic.[14]

A possible effect of low-level lead exposure on blood pressure has lately been debated considerably.[30] Although the results are conflicting, several epidemiological studies have shown that the blood-lead level is positively correlated to the blood pressure. The effect, which is not very large, could at least in part be due to confounders. Since mild to moderate hypertension is affecting a large number of people, a causal relationship would be of considerable importance for public health. Animal studies show that low-level lead exposure is associated with elevated blood pressure. At higher dose levels, however, the effect does not seem to increase. Suggested mechanisms of lead-induced blood pressure elevation are interference with the renin-angiotensin system, neuroregulation of the vascular system, or sodium-potassium transport.

F. IMMUNOLOGICAL EFFECTS

In principle, two kinds of effect are seen: stimulation or suppression of the immune system.[31] Hypersensitive or allergic reaction to metal compounds is well known. Chromate and nickel compounds are both potent allergens able to elicit dermatitis by contact. Metal-related occupational allergic lung disease, such as allergic alveolitis and asthma, are seen with compounds of beryllium, cobalt, chromium, nickel, platinum, and vanadium. Lately, it has also been recognized that ingested or injected nickel can elicit general allergic reactions with skin manifestations in patients allergic to nickel. Autoimmunity reactions in the kidney by mercury or gold compounds have been known for some time and are currently being investigated. Immunosuppression with increased susceptibility to infections has been reported for several metals, including lead, mercury, cadmium, and tin, in exposed rodents. Few, if any, studies exist on humans.

G. CARCINOGENIC EFFECTS

Of considerable concern are metal compounds causing cancer. Development of cancer is an incompletely understood and complicated multistage process. Metals could, in principle, influence this process at several stages. It is well known that some metal compounds cause genetic damage, affecting DNA repair as well as the spindle apparatus during cell division.[32]

In addition to skin cancer seen after ingestion of arsenic or contaminated drinking water, respiratory cancer due to arsenic exposure in smelter workers is well known. Even an increased risk of lung cancer among the general population living around a copper smelter has been reported from Sweden.[14] Iatrogenic exposure to arsenic frequently occurred previously when the arsenic-containing Fowler's solution was used for medical treatment. Many of the patients later developed skin changes, including hyperkeratosis as well as skin cancer.[14] The minimal amount of arsenic in well water necessary to induce skin cancer has been estimated to be less than 0.29 mg/l. The relationship generally accepted between arsenic and cancer has been confined to the lungs and skin, but arsenic has also been suspected to induce primary liver cancer.[14]

Metals other than arsenic also cause cancer. These are certain nickel and chromium compounds, and sufficient evidence for carcinogenicity in humans exists. The evidence is based on a large number of epidemiological studies of industrially exposed workers. While industrial chromium and nickel exposure is associated with an increased risk of lung cancer, nickel also induces cancer in the nasal sinuses and larynx. Limited evidence for carcinogenicity, however, exists for cadmium, beryllium, lead, and platinum compounds. For the last group of metals, evidence is also partly based on animal studies.[32]

All metals considered human carcinogens also cause cancer in animals. Mutagenic effects in *in vitro* test systems have been observed for a number of metal compounds, including most of those regarded as carcinogenic. It is generally believed that for genotoxic carcinogenic compounds, no threshold for effect exists. Thus, it is a paradox that a carcinogenic metal such as chromium also is an essential nutrient. Carcinogenicity has been shown for several hexavalent chromium compounds, particularly less soluble particles. It is known that hexavalent chromium may interact and damage DNA by several muchanisms.[33] However, trivalent chromium, which is believed to be the essential form, also binds to DNA, forming interstrand crosslinks and DNA-protein crosslinks. This remains an area of future research.

V. EXPOSURE TO AND EFFECTS OF RADIONUCLIDES

Low-level exposure to radionuclides is usually not associated with acute toxic effects, but, rather, with effects on reproduction, genetic material, and carcinogenesis. These effects are believed to be nonthreshold, stochastic effects.[34]

A naturally occurring source of substantial importance for human health is exposure to radon and radon daughters. A lung disease with high lethality, later recognized as lung cancer, was observed as early as 500 years ago among miners in the Erz Mountains of Central Europe. Radiation from radon was not suspected as a cause before 1920 to 1930 and was generally accepted from 1960. Besides being an occupational health problem for miners, it later became apparent that exposure to radon and radon daughters also constitutes a considerable public health problem.[34,35] The levels of indoor radon is usually higher than ambient air levels. Contamination of indoor air can take place via diffusion of radon gas from the ground into the basement of dwellings, but in some cases ground water with high radon levels may be the primary source. Building materials seem to be a minor source in Norway.[35] Indoor air levels of radon and decay products are highest in basements and decrease upward in the dwellings. The geographical variation of the problem, as well as that between different types of buildings, is considerable. The radiation is α-particles, which work locally and induce lung cancer when inhaled. Combined with cigarette smoking, the risk of lung cancer increases in a synergistic way and the latency time from radon exposure to cancer development becomes shorter.[35-37]

Radionuclides that gain entry into the body via food are naturally occurring, such as ^{40}K and ^{210}Po, or orginate from fallouts from nuclear testing or accidents in nuclear power plants. Among the latter, important elements are (1) the bone-seeker ^{90}Sr, which has a very long biological half-life, (2) ^{131}I, which accumulates in the thyroid gland, and (3) ^{137}Cs, with a physical half-life of 30 years. The latter element was the most important element of long-distance contamination in Norway after the Chernobyl accident in the Soviet Union. ^{137} Cs accumulated in the food chain and the humans mainly exposed were those eating large amounts of reindeer meat and "wild fish". Based on estimates of the collective dose to the Norwegian people and applying ICRP's risk estimates, it has been calculated that the total number of cancer cases due to ionizing radiation after the Chernobyl accident will be about 36 during the next 50 years.[39]

VI. EFFECTS OF MINERAL PARTICLES

Mineral particles mainly cause adverse health effects when inhaled. The form and size of the particles are quite important, as small particles (as mentioned previously) reach the lung tissue. Asbestos is a large group of naturally occurring, filamentous, hydrated metallic mineral silicates.[40] The fibrous form, length, and diameter of these minerals are determinants of the biological activity, which varies considerably among the different asbestos minerals. Most active are fibers with a diameter less than 1.5 μm and a length above 5 to 8 μm. Asbestos produces several types of lesions, including lung fibrosis, pleural plaques, pleural effusion, and malignant mesothelioma.[40] Development of malignant mesothelioma is characterized by a very long latency period between exposure and disease, up to 40 years. Furthermore, asbestos exposure is also associated with an increased incidence of lung cancer, particularly when combined with tobacco, where a multiplicative effect on cancer incidence is seen. Exposure to mineral particles mainly occurs in the work environment. However, the exposure of family members may also take place indirectly via working clothes. Exposure to asbestos may occur when such material has been used in private and/or public buildings. Asbestos exposure and lung disease have also been noted in patients who used steatite mines as a playground during their childhood (unpublished). In some areas, fibrous minerals (e.g., erionite, in Turkey) may even occur naturally, and a general exposure leading to asbestos diseases among the local inhabitants has been seen.[40]

Since asbestos fibers also occur naturally in drinking water in some places and are leached from asbestos-cement drinking-water pipes, the potential health risk has been much debated.[41] Based on animal and epidemiological studies, the current view is that ingestion of asbestos fibers is, in contrast to inhaled asbestos, associated with a very low health risk.

Silicosis, which is characterized by inflammatory nodules and increased levels of connective tissue in the lungs, has been known for centuries to occur in stone cutters and miners. It is caused by free or combined silica, crystaline silica quartz being the most potent agent. Silicates (combined silica) vary considerably with respect to biological activity when inhaled. Heavy exposure to so-called Fullers Earth (clay minerals), talc, and other silicates has been associated with pneumoconiosis, but often the exposure is mixed and it is difficult to establish whether the disease is due to the mineral or the presence of impurities of silicates with high fibrogenic potential.

VII. ASPECTS OF TOXICITY OF INORGANIC COMPOUNDS OF NONMETALLIC ELEMENTS

In a geomedical context, important inorganic nonmetallic elements causing health problems are fluoride, which contaminates drinking water, causing fluorosis which affects bones and teeth, and nitrate, which causes methemoglobinemia and the formation of carcinogenic nitrosamines in the stomach.[19] Other compounds occurring in the industrial work environment are several irritant gases, e.g., NH_3, Cl_2, BR_2, F_2, O_3, NO_2, and $COCl_2$. They cause primary irritation of the respiratory tract. According to their solubility in water, which is highest for NH_3 and lowest for $COCl_2$, as ranked above, they will cause irritation in the upper part (NH_3) or in the lower part (O_3, NO_2, $COCl_2$) of the respiratory tract. Their action is mainly based on their ability to denaturate proteins and cause chemical inflammation. Those compounds reaching the lower parts of the respiratory tract may cause a chemical pneumonia. Gaseous compounds like CO, HCN, and H_2S block respiration or interfere with cellular respiration and oxygen transport.

VIII. ASSESSMENT OF EXPOSURE AND RISK OF ADVERSE HEALTH EFFECTS

Exposure can be estimated by measuring the levels of toxic agent in the exposure medium. These will usually be ambient air, working atmosphere, etc., or food, drinking water, or other ingested materials. It is, however, extremely difficult to calculate the exposure based on such measurements. For a number of metals it is now possible to measure the compounds of interest in body fluids or tissues. The development of modern instrumental analysis such as atomic absorption spectrometry, neutron activation analysis, X-ray fluorescence, and emission spectrometry has made this possible.

For an assessment of the risk of adverse health effects at a given exposure level, dose-effect/dose-response relationships must have been developed. The dose-effect relationship describes the different effects occurring at different dose levels and the dose-response relationship describes the frequencey of a given effect in a population at different dose levels. The first adverse biological effect occurring due to increased metal exposure is usually referred to as the critical effect, the organ as the critical organ, and the dose as the critical dose.[42] The critical effect and organ are usually specific for the different metals and the critical dose may vary among different individuals. The critical effect may not necessarily appear in the organ reaching the highest level of the toxic agent. For lead, bone tissue contains the major part of the body burden, whereas toxic effects are seen in other organ systems such as the central nervous system.

In the process of establishing health criteria for exposure standards, it is important to establish critical effects, critical doses, and critical organs and to develop dose-effect and dose-response relationships for each compound. It is recognized that this is a difficult, but important task. Standards are developed both at national and international levels. In relation to the assessment of the risk of adverse health effects, measurement of the biological level of metals is useful for some metals, e.g., lead, cadmium, and organic and inorganic mercuric compounds. For these metals, metabolic models and dose-effect and dose-response relationships are relatively well described. To predict the level of methyl mercury in the brain, which is the critical organ, it is possible to measure the level in whole blood or hair, while doing measurement in urine is of no use. This is because methyl mercury is not excreted into the urine to any large extent. Regarding exposure to metallic mercury, urinary measurements are valuable. For other metals, however, measurement of biological levels may tell nothing about potential health risks since dose-effect and dose-response realtions and metabolic models have not been developed. It is also important to know about the exposure and metal species. For example, the level of arsenic in urine may say something about inorganic arsenic exposure; however, ingestion of seafood containing organic ''nontoxic'' arsenic compounds will lead to high urinary levels of arsenic. Thus, knowing the metabolic models is a prerequisite for using biological dose indicators.

During recent years, more attention has been focused on the interaction between different metals, other substances as well as host factors. They apparently occur frequently and complicate risk evaluation. Special emphasis has, for example, been put on selenium-metal interactions.[15] There are few examples of data from humans. An example is co-accumulation of very high levels of mercury and selenium in a ratio near one in organs of retired miners. No overt signs of toxicity were noted among these workers. A huge amount of animal data, however, indicate that under certain circumstances selenium can alleviate the toxicity of methylmercury.

IX. CONCLUDING REMARKS

The subject covered in this chapter is enormous and only certain aspects have been touched upon. Human activity apparently changes our environment considerably, with consequences for human exposure to inorganic chemical compounds and to human health. When working with geomedical problems, it is mandatory to have good estimates of human exposure when coupling to health effects is done. Otherwise, one might be fooled by some of the numerous other factors that are determinants of human health and that are unevenly distributed. Simple correlations between disease and local geochemistry can, at best, only be indicative of a relationship, which has to be confirmed by using other methods of investigation.

Important health effects of inorganic elements are probably those associated with low-level exposure, and the developing nervous system seems particularly sensitive to many of the metal compounds, lead and mercury being good examples. Another important health effect is cancer, which is probably influenced by excess exposure to inorganic chemical compounds in our environment as well as a deficiency of certain elements.

REFERENCES

1. **Seiler, H. G., Sigel, H., and Sigel, A., Eds.,** *Handbook on Toxicity of Inorganic Compounds,* Marcel Dekker, New York, 1988.
2. **Friberg, L., Nordberg, G. F., and Vouk, V. B., Eds.,** *Handbook on the Toxicology of Metals,* Vols. 1,2, 2nd ed., Elsevier, Amsterdam, 1986.
3. **Klaassen, C. D., Amdur, M. D., and Doull, J., Eds.,** *Casarett and Doull's Toxicology,* Macmillan, New York, 1986.
4. **Grandjean, P., Nielsen, O. V., and Shapiro, I. M.,** Lead retention in ancient Nubian and contemporary populations, *J. Environ. Pathol. Toxicol.,* 2, 781, 1979.
5. **Jernlöv, A.,** in *Polluted Rain,* Toribara, T. Y., Miller, M. W., and Morrow, P. E., Eds. Plenus Press, New York, 1980, 211.
6. **Gjessing, E. T., Alexander, J., and Rosseland, B. O.,** Acidification and aluminum contamination of drinking water, in *Watershed 89, The Future for Water Quality in Europe,* Vol. I, Wheeler, D., Richardson, M. L., and Bridges, J., Eds., Pergamon Press, Oxford, 1989, 15.
7. **Flaten, T. P.,** An Investigation of the Chemical Composition of Norwegian Drinking Water and its Possible Relationship to the Epidemiology of some Diseases, Thesis no 51, Institutt for Uorganisk Kjemi, Norges Tekniske Høgskole, Trondheim, 1986.
8. **Camner, P., Clarkson, T. W., and Nordberg, G. F.,** Routes of exposure, dose and metabolism of metals, in *Handbook on the Toxicology of Metals,* Vol. 1, 2nd ed., Friberg, L., Nordberg, G. F., and Vouk, V. B., Eds., Elsevier, Amsterdam, 1986.
9. **Slanina, P., Frech, W., Ekström, L.-G., Lööf, L., Slorach, S., and Cedergren, A.,** Dietary citric acid enhanced absorption of aluminum in antacids, *Clin. Chem.,* 32, 539, 1986.
10. **Ganrot, P. O.,** Metabolism and possible health effects of aluminum, *Environ. Health Perspect.,* 65, 363, 1986.
11. Summary report. Workshop on aluminium and health, Oslo, May 2-5, 1988. In Proc. Workshop on Aluminium and Health, Oslo, May, 1988, Alexander, J. and Orme, J., Eds., *Environ. Geochem. Health,* in press.

12. **Hildebrand, H. F., Rumazeille, B., Decoulx, J., Herlant-Peers, M. C., Ostapczuk, P., Stoeppler, M., and Mercier, J. M.,** in *Progress In Nickel Toxicology,* Brown, S. S. and Sunderman, F. W., Eds., Blackwell Scientific, Oxford, 1985, 169.

13. **Clarkson, T. W.,** Effects — general principles underlying the toxic action of metals, in *Handbook on the Toxicology of Metals,* Vol. 1, 2nd ed., Friberg, L., Nordberg, G. F., and Vouk, V. B., Eds., Elsevier, Amsterdam, 1986, 128.

14. **Ishinishi, N., Tsuchihiya, K., Vahter, M., and Fowler, B.,** Arsenic, in *Handbook on the Toxicology of Metals,* Vol. 2, 2nd ed., Friberg, L., Nordberg, G. F., and Vouk, V. B., Eds., Elsevier, Amsterdam, 1986, 43.

15. **Högberg, J. and Alexander, J.,** Selenium, in *Handbook on the Toxicology of Metals,* Vol. 2, 2nd ed., Friberg, L., Nordberg, G. F., and Vouk, V. B., Eds., Elsevier, Amsterdam, 1986, 482.

16. **Cremer, J. E.,** *Biochem. J.,* 68, 685, 1958.

17. **Friberg, L., Kjellström, T., and Nordberg, G. F.,** Cadmium, in *Handbook on the Toxicology of Metals,* Vol. 2, 2nd ed., Friberg, L., Nordberg, G. F., and Vouk, V. B., Eds., Elsevier, Amsterdam, 1986, 130.

18. **Aaseth, J. and Alexander, J.,** Hepatobiliary kinetics, in Trace Elements in Human Health and Disease: Symposium report. Grandjean, P., Andersen, O., Berlin, A., Grant, L., Sors, A., Tarkowski, S. and Victery, W., Eds., Evironmental Health, 26, World Health Organization, Copenhagen, 1987.

19. Working Group on Health Hazards from Nitrates in Drinking-Water, 1984. Health hazards from nitrates in drinking-water, Environmental Health, 1, World Health Organization, Copenhagen, 1985.

20. **Müller-Höcker, J., Meyer, U., Wiebecke, B., Hübner, G., Eife, R., Kellner, M., and Schramel, P.,** Copper storage disease of the liver and chronic dietary copper intoxication in two further German infants mimicking Indian childhood chirosis, *Pathol. Res. Pract.,* 183, 39, 1988.

21. **Gutteridge, J. M. C. and Sunderman, F. W., Jr.,** Free radical formation, in Trace Elements in Human Health and Disease: Symposium report. Grandjean, P., Andersen, O., Berlin, A., Grant, L., Sors, A., Tarkowski, S., and Victery, W., Eds., Evironmental Health, 26, World Health Organization, Copenhagen, 1987.

22. **Ryberg, D. and Alexander, J.,** Inhibitory action of hexavalent chromium (Cr VI) on the mitochondrial respiration and a possible coupling to the Cr VI-reduction, *Biochem. Pharmacol.,* 33, 2461, 1984.

23. **Berlin, M.,** Mercury, in *Handbook on the Toxicology of Metals,* Vol. 2, 2nd ed., Friberg, L., Nordberg, G. F., and Vouk, V. B., Eds., Elsevier, Amsterdam, 1986, 387.

24. Aluminium and Alzheimer's disease. Editorial, *Lancet,* 1, 82, 1989.

25. **Garruto, R. M., Yanagihara, R., and Gajdusek, D. C.,** Models of environmentally induced neurological disease: epidemiology and etiology of amyotrophic lateral sclerosis and Parkinsons-dementia in the Western Pacific, In Proc. Workshop on Aluminium and Health, Oslo, May, 1988, Alexander, J. and Orme, J. Eds., *Environ. Geochem. Health,* in press.

26. **Vogt, T.,** Water Quality and Health — A Study of a Possible Relationship Between Aluminum in Drinking Water and Dementia, Sosiale og Økonomiske Studier 61, Central Bureau of Statistics of Norway, Oslo, 1986.

27. **Martyn, C. N., Osmond, C., Edwardson, J. A., Barker, D. J. P., Harris, E. C., and Lacey, R. F.,** Geographical relation between Alzheimer's disease and aluminum in drinking water, *Lancet,* 1, 59, 1989.

28. **Sager, P. R., Clarkson, T. W., and Nordberg, G. F.,** Reproductive and developmental toxicity of metals, in Friberg, L., Nordberg, G. F. and Vouk, V. B. Eds. *Handbook on the Toxicology of Metals,* Vol. 1, 2nd ed., Friberg, L., Nordberg, G. F., and Vouk, V. B., Eds., Elsevier, Amsterdam, 1986, 391.

29. **Kjellstrom, T., Kennedy, P., Wallis, S., and Mantell, C.,** Physical and mental development of children with prenatal exposure to mercury from fish, Report 3080, National Swedish Environmental Protection Board, Stockholm, 1986.

30. **Victery, W., Ed.,** Symposium on lead-blood pressure relationships, *Environ. Health Perspect.* 78, 3, 1988.

31. **Dean, J. H., Murray, M. J., and Ward, E. C.,** Toxic responses of the immune system, in *Casarett and Doull's Toxicology,* Klaassen, C. D., Amdur, M. D., and Doull, J., Eds., Macmillan, New York, 1986, 245.

32. **Kazantzis, G. and Lilly, L.,** Mutagenic and carcinogenic effects of metals, in *Handbook on the Toxicology of Metals,* Vol. 1, 2nd ed., Friberg, L., Nordberg, G. F., and Vouk, V. B., Eds., Elsevier, Amsterdam, 1986, 319.

33. **Hamilton, J. W. and Wetterhahn, K. E.,** Chromium, in *Handbook on Toxicity of Inorganic Compounds,* Seiler, H. G., Sigel, H., and Sigel, A., Eds., Marcel Dekker, New York, 1988, 239.

34. **Hobbs, C. H. and McClellan, R. O.,** Toxic effects of radiation and radioactive materials, in *Casarett and Doull's Toxicology,* Klaassen. C. D., Amdur, M. D., and Doull, J., Eds., Macmillan, New York, 1986, 669.

35. **Sanner, T., Dybing, E., and Stranden, E.,** Risk of lung cancer by indoor exposure to radon (in Norwegian), Report 1988:3, National Institute of Radiation Hygiene, Oslo, 1988.

36. **Edling, C.,** Lung cancer and smoking in a group of iron ore miners, *Am. J. Ind. Med.,* 3, 191, 1982.

37. **Edling, C., Wingren, G., and Axelson, O.,** Radon daughter exposure in dwellings and lung cancer, in *Proc. 3rd Int. Congr. Indoor Air Quality and Climate,* Vol. 2, Swedish Council for Building Research, Stockholm, 1984, 29.

38. **Pershagen, G., Hrubec, Z., and Svenson, C.,** Passive smoking and lung cancer in Swedish women, *Am. J. Epidemiol.,* 125, 17, 1987.

39. **Sanner, T., Harbitz, O., Trygg, K., Reitan, J. B., Frøslie, A., Aune, T., and Alexander, J.,** Health risk from radionuclides in food. Consequences of the Chernobyl accident, Report 1/87, Advisory Board on Food Toxicology, Directorate of Health, Oslo, 1987 (in Norwegian).

40. World Health Organization, Asbestos and Other Natural Mineral Fibres, Environmental Health Criteria 53, IPCS, WHO, Geneva, 1986.

41. **Millette, J. R., Ed.,** Summary Workshop on Ingested Asbestos, Cincinnati, U.S.A., 1982, *Environ. Health Perspect.,* 53, 1983.

42. Task Group on Metal Toxicity, in *Effects and Dose-Respose Relationships of Toxic Metals,* Nordberg, G. F., Elsevier, Amsterdam, 1976.

Chapter 5

GEOCHEMICAL MAPS AS A BASIS FOR GEOMEDICAL INVESTIGATIONS

B. Bølviken and A. Bjørklund

TABLE OF CONTENTS

I. INTRODUCTION

Geochemical mapping includes (1) sampling of natural material, (2) chemical analysis of the samples, and (3) illustration of the analytical results on maps.

The most frequently used sample types are surface material such as stream sediment (inorganic material from drainage channels) and soil. The sample spacing varies considerably, depending on the purpose of the survey; one sample per 0.002 km² to one sample per 3000 km² have been reported in the literature. Great progress has been made in analytical chemistry during the last part of this century. Through the use of modern methods it is now possible as a routine to determine more than 50 elements in geochemical samples. Computers are used extensively in the compilation of maps and in the interpretation of data.

The most important use of geochemical maps has been in prospecting for mineral deposits. An ore body represents an exceptional concentration of minerals and elements, and can be indicated by anomalous dispersion patterns of major, as well as minor, elements in its surroundings. Since World War II a vast amount of geochemical exploration data have been obtained throughout the world by international agencies, geological surveys, research institutions and private industry. Such data are published in a number of text books, periodicals, and monographs (e.g., see References 6, 7, 27, 44, 52, 58, 63, 64, 77, 79, 82, 90).

During the last few decades one has become more and more aware of the possibility for utilizing geochemical maps in fields other than exploration. Geomedicine is one of these fields. Geochemical atlases have been or are being produced in several countries (e.g., References 3, 15, 16, 21, 30, 56, 93, 100, 102). These and other regional geochemical maps clearly show that the natural contents of chemical elements in surface material vary within wide limits, and depict broad, large-scale distribution patterns. Pollution from anthropogenic sources will modify the natural distribution patterns to a varying degree, depending on local and regional conditions. As a consequence, human and animal population groups are exposed to varying natural and artificial amounts and associations of chemical elements, depending on the location.

The following account is divided into three parts: (1) an overview of methods used in geochemical mapping, (2) a demonstration of selected examples of regional geochemical maps, with emphasis on northern Europe, and (3) some comments on the use of geochemical data in geomedicine.

II. OVERVIEW OF SOME METHODS NORMALLY USED IN GEOCHEMICAL MAPPING

A. SAMPLING

Geochemical samples used for mapping purposes are (1) drainage and (2) terrestrial materials. The most important types of drainage samples are

- Stream sediment
- Lake sediment
- Overbank (levee) sediment
- Stream water
- Lake water
- Stream humic material
- Stream vegetation

Stream sediment is the most widely used sample medium in regional geochemical mapping.[79] It consists mostly of inorganic material (mainly rock fragments and mineral grains) in the stream bed in regular contact with stream water. Small amounts of organic matter and secondary oxides are intermixed, all components being subject to a temporary deposition in the stream bed. A fine fraction of the material (<0.2 mm) is normally used for analysis. The chemical composition of this material is thought to reflect the composition of the bedrock and overburden in the drainage area. A combination of physical, chemical, and biological processes control the dispersion from the sources to the sampling site. Recent literature casts some doubt on the extent to which active stream sediment is representative of whole drainage areas.[73] Stream sediment is vulnerable to contamination by pollution.

Lake sediment is collected from the bottom of the middle or, alternatively, the inlet part of lakes. This material is a mixture of rock fragments, mineral grains, and a substantial but varying component of humic material and secondary oxides. Other features of lake sediment are analogous to those for stream sediment. Lake sediment has been used extensively in North America,[25,54,81] but less so in Europe.[94]

Overbank sediment is produced when major floods occur in a river system and the water discharge exceeds the quantity that can pass through the ordinary stream channel (bankful discharge). Some of the load will then be deposited as overbank sediment on the flood plain at levels above those of the ordinary stream channel. Overbank sediment consists mainly of rock fragments and mineral grains with low contents of humic material and secondary oxides. A fine fraction (<0.06 mm grain size) is used for analysis. The chemical composition of a vertical section of overbank sediment is thought to reflect the average chemical composition of bedrock and overburden in the drainage area. Mechanical movement of inorganic particles suspended in water is the main dispersion mechanism from the source to the place of deposition. The chemical composition of shallow overbank sediment may be modified naturally by *in situ* soil-forming processes and artificially by the effects of anthropogenic pollution.[73]

Stream water is taken by a plastic syringe and filtered through a 0.45 μm plastic filter into plastic bottles.

Lake water is preferably sampled in the autumn just after inversion of the lake.[50] It is normally not filtered before analysis. Both stream and lake waters are acidified in the field with a few drops of ultrapure HNO_3 if cations are to be determined. The chemical composition of pristine natural stream and lake water reflects the amounts of elements released by chemical weathering in the drainage area. The composition varies with time, but this seasonal variation seems generally to be less than the geographical variation.[23a] Stream and lake waters are susceptible to anthropogenic pollution.

Stream humic material is leaf litter and other dead organic matter in various stages of decomposition on the stream banks and bottom intermixed with various amounts of mineral material. This sample material has been used by Brundin and Nairis[18] and Bølviken et al.[21] It is normally sieved to <0.2mm, and in most cases this fraction is ashed before chemical analysis. Humus enriches most of the elements of the periodic table, and the chemical composition of pristine stream humic material depicts general products of a combination of chemical, biological, and physical processes in the drainage area. Stream humic material is susceptible to anthropogenic pollution in the drainage system.

Drainage vegetation of various types is used as a sample medium. Stream moss (aquatic bryophytes) gets most of its nutrition from the stream water through the leaves[17,88] and may enrich elements from the water. This is also the case for drainage-plant roots.[18a] Stream moss and stream-plant roots are washed free of mineral matter and are normally ashed before chemical analysis. These sample materials need less-sensitive analytical methods than stream water. They are both sensitive to anthropogenic pollution in the drainage.

Terrestrial materials for geochemical mapping could be

- Overburden (soil and soil parent material)
- Rocks
- Vegetation

Overburden is sampled from different horizons at various depths. A common soil profile in temperate climates is the podzol profile, which consists of humic material near the surface, then a bleached layer impoverished in dark minerals towards depth, an enrichment layer, and parent material. In tropical and temperate climates the soil parent material often consists of weathering products of the bedrock, but soil parent material may also be eroded from the bedrock by the action of water, glaciers, or wind. Overburden, therefore, represents local to regional bedrock geology and anthropogenic pollution to a varying degree, depending on the circumstances. Soils are the main substratum for plants, and their chemical composition will, to a large extent, be reflected in the food chain. Anthropogenic pollution is more pronounced in surface soils than in soils at depth. The use of overburden in regional geochemical mapping has been applied in many areas (e.g., References 14, 19, 20, 21, 57, 57a, and 89).

Rock samples have been used less in geochemical mapping than have overburden samples, because rocks are more different to sample, and have less areal representation than overburden. However, rocks are less susceptible to pollution than any other type of natural material, and more rock samples can be expected in geochemical mapping. Earlier work has been summarized by Beus and Grigorian[6] and Govett.[45]

Terrestrial moss was introduced by Rühling and Tyler[84,85] as a sampling medium suitable for recording heavy metal deposition. One reason why the composition of natural and anthropogenic atmospheric fallout is reflected in the composition of the moss is that moss has no roots and gets its nutrition through its leaves. The species *Hylocomium Splendens* occurs frequently all over northern Europe and is well suited as a sampling medium. The moss samples are wet-ashed and analyzed with various techniques.[86] Other uses of vegetation sampling in geochemical mapping is reviewed by, e.g., Brooks[17] and Kovalevskii.[58]

B. CHEMICAL ANALYSIS

A high standard is required in the reproducibility of geochemical patterns for geomedicine. If geochemical samples from a survey area are analyzed successively in the order they were collected, then even a very small, often unavoidable analytical bias, such as that caused by the calibration of analytical instruments, may result in significant, but artificial geochemical differences between regions. Such bias can be avoided by analyzing the samples in an arbitrary order. A possible analytical bias between groups or batches of samples will then become a geographically randomly-distributed error, the only effect of which is an enhancement of the background noises.

Many chemical methods are available for the analysis of samples collected in geochemical mapping (e.g., Fletcher[37] and Page et al.[76]). For interpretational purposes, one should distinguish between the total contents of the elements and a partially extracted portion. For many elements in inorganic sample media there is no, or only a slight, correlation between the total and the partially extracted element contents. In geomedicine an easily extractable part is generally of more interest than the total contents because it would normally be a closer approximation of the fraction available for organisms than are the total contents.

The total element contents of geochemical samples are determined directly by methods such as X-ray fluorescence[31] and neutron activation analysis[83] or after chemical attack such as fusion with $Li_2B_4O_7$ or digestion with $HClO_4 - H_2F_2$, followed by the use of various methods of determination, the most important ones being[37]

- Atomic absorption
- Inductively coupled plasma emission spectrometry
- Inductively coupled plasma mass spectrometry
- Ion chromatography

These four methods are also used for direct analysis of waters or various types of extracts of other sample types.

C. PRESENTATION OF GEOCHEMICAL DISTRIBUTION PATTERNS ON MAPS

In geochemistry computers are used extensively in the preparation of maps. Usually the limits of a restricted number of concentration intervals are determined by some statistical method; these intervals are represented on maps by a restricted number of symbols, colors or isoconcentration curves. Due to recent developments in analytical chemistry, there is an increasing demand for mapping techniques which can display geochemical dispersion patterns for the contents of several elements in a great number of samples either individually or in combination, and at any level or contrast of concentration.

Most conventional systems for map construction are ineffective for this purpose, and new techniques have been developed. A convneient method is to illustrate the contents at each sampling site with a dot, the size of which increases with the element concentration according to a continuous curve defined by the user;[11] for examples, see Figures 4 to 9 and 18-19.

For color-surface presentation, various techniques have been applied (e.g., References 30, 102). A moving median method, introduced by Bjørklund and Lummaa[12] and further developed by Bjørklund and Gustavsson,[11] is advantageous because it is not sensitive to outlayers and can disclose areal trends for any type of statistical distribution of concentrations within the moving window.

III. EXAMPLES OF LARGE-SCALE GEOCHEMICAL SURVEYS

Geochemical maps are produced at many scales covering areas ranging from a few tens of square meters to hundreds of thousands of square kilometers. In geomedicine, data from areas as large as states or countries are of greatest interest because health statistics often require a number of large unit areas, each with a population large enough to yield statistically reliable data. A number of country-wide geochemical maps have been published as separate sheets or as geochemical atlases (see list of references). The examples shown here are taken from the first [102] and the last[21] geochemical atlas known to the authors as of 1988. We have also had the opportunity to quote nation-wide data from Finland and Norway that are presently being prepared for publication. The data are all based on the analysis of surface non-consolidated mineral or organic material. Additional examples of geochemical maps based on the chemical composition of waters are taken from Flaten[33,34] (Norwegian drinking water) and from unpublished data from the Nordkalott Project (see Bolviken et al.[23a]) An example of data from chemical analysis of ambient air[80] is also included.

A. ENGLAND AND WALES

The first published geochemical atlas, *The Wolfson Geochemical Atlas of England and Wales,*[102] includes geochemical maps at a scale of 1:2 mill for the contents of 21 single elements and 4 element combinations in stream sediment.

Nearly 50,000 samples of stream sediment were collected within the 125,000 km² survey region. The samples were dried and sieved to a grain size of minus 0.20 mm and analyzed for the total contents of Al, Ba, Ca, Co, Cr, Cu, Fe, Ga, K, Li, Mn, Ni, Pb, Sc, Sn, Sr, and V by direct-reading optical spectrography using DC-arc excitation. The contents of As and Mo were determined colorimetrically after fusion with $KHSO_4$. Hot nitric-acid soluble Cd and Zn were determined by atomic abosrbtion spectrometry. If more than one sample fell within a unit pixel (3 km²), the geometric mean of the analytical results of that pixel was computed. For each element the arithmetic mean of (1) the raw data of the content and (2) the computed geometric means were calculated for 3×3 pixel cells and plotted as color maps. Figure 1 is a reproduction of the Sn map, which was chosen because it is easy to reproduce in black-and-white and shows a simple, but distinct distribution pattern. Cornwall

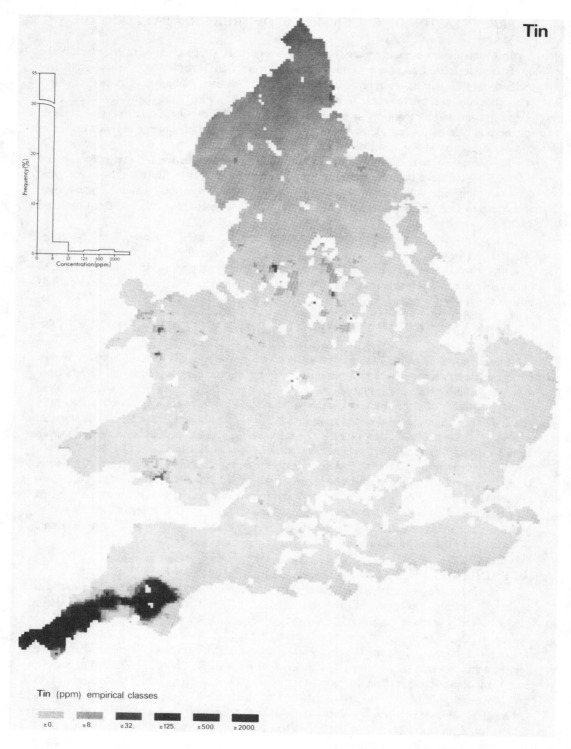

FIGURE 1. Contents of total tin in the minus 0.2 mm fraction of stream sediment from England and Wales. (From Webb et al., The Wolfson Geochemical Atlas of England and Wales, Clarendon Press, 1978, 69. With permission of Clarendon Press, London.)

in southeastern England stands out as a pronounced Sn province. The atlas[102] shows that several other elements occur in high concentrations within this Sn province. Seven hundred years of mining activities have taken place in the region, and it is unclear to which extent these patterns are natural or the results of pollution. In any case, human and animal populations of Cornwall are exposed to more environmental Sn and some other elements than populations from other regions elsewhere in England and Wales. An atlas of cancer motality in Great Britain[38] is available for geomedical comparison.

B. FENNOSCANDINAVIA

The Nordkalott Project (1980-1986) included geological, geophysical, and geochemical mapping in the areas north of the 66° latitude in Finland, Norway, and Sweden and resulted in the compilation of a number of thematic maps presented at the scales of 1:1 mill and 1:4 mill, of which 136 geochemical maps, and a selection of geological and geophysical maps are available in an atlas. [21] The geochemical atlas is based on a multimedia/multielement widely spaced geochemical survey within an area of 250,000 km². Up to six different types of geochemical samples (humic top soil, till, stream water, stream sediment, stream organic matter, and stream moss) were collected at 7267 stations. The last five types of sample were analyzed from 1981 to 1985. The analyses were carried out by several institutions, but all analyses of a certain type were done in the same laboratory regardless of the origin of the samples. The data were recorded on magnetic tape for computerized processing and map drawing as well as for storage and future retrieval.

A selection of seven geochemical maps (Figures 1, 2 and 3 to 9) are shown here as examples of the data. The survey region is sparsely populated, and with certain exceptions (see Figure 7) lightly industrialized and remote from main source-areas of anthropogenic pollution. The geochemical distribution patterns are therefore thought to be predominantly natural, being caused mainly by (1) variations in the chemical/mineralogical composition of the bedrock; (2) recent long-transported airborne elements of marine, volcanic and other derivation;[70,71,75] (3) elements in gases and fluids of deep-seated origin;[8-10,59,66] and (4) environmental conditions such as topography and climate. Dispersion patterns caused by pollution are thought to be the exception (see Figure 7.)

All elements show large variations in content (Tables 1 and 2) and most of them depict broad distribution patterns, including regions and provinces with characteristic element contents and element association. These large-scale patterns are most evident on the maps of the total element contents in the heavy mineral fraction of till and stream sediment (Figures 2 and 4), but are also seen on maps of total- and acid-soluble fractions in various sampling media (Figures 5 to 8). The survey region was divided into 18 sub-areas defined as geochemical provinces. Enrichment factors were calculated for each of 22 elements, defined as the ratio between (1) the average content in the heavy mineral fraction of till within each of the 18 provinces and (2) the average for the whole survey region. The enrichment factors vary between 0.1 and 3.

In many cases the geochemical distribution patterns coincide with known geological units. In other cases, especially for mobile elements, there appears to be no simple relationship between the bedrock geology and the geochemical data (Figures 2 to 9).

Other maps of geochemical data from northern Europe are provided by Rühling et al.[86] who sampled terrestrial moss (*Hylocomium Splendens*) from all the nordic countries and analyzed the samples for As, Cd, Cr, Cu, Fe, Pb, Ni, V, and Zn. This sample medium is thought to reflect the composition of fallout from the atmosphere. Broad dispersion patterns occur for all elements. Some of the distributions are natural, while others, e.g., that of lead (Figure 10), are assumed to be strongly influenced by anthropogenic pollution, since the concentrations are lower in arctic regions than elsewhere.

Possible relation between the geochemical distribution patterns in Fennoscandinavia and the epidemiology of diseases[69] warrant investigation.

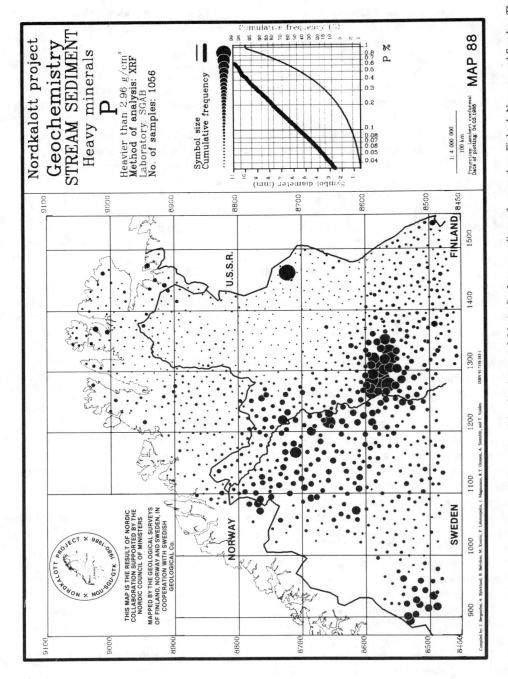

FIGURE 2.   Contents of total phosporous in a heavy mineral fraction (sp.gr. >2.96 g/cm³) of stream sediment from northern Finland, Norway, and Sweden. There is no simple relationship between the P pattern and the bedrock geology. (From Bølviken et al.[21])

FIGURE 3.   Simplified geological map of northern Finland, Norway, and Sweden.

C. FINLAND

In 1971 systematic regional geochemical mapping of Finland was initiated by the Geological Survey. During a 15-year period, an area of 100,000 km² considered promising for prospecting has been mapped geochemically by analysis of till, mineral and organic stream sediment, and lake sediment. Procedures and examples of results have been described by Bjørklund et al.,[13] Gustavsson et al.,[47] Kauranne,[57] Tenhola,[94] and Koljonen et al.[57a]

In 1981 a nation-wide reconnaissance survey was included in the geochemical mapping program. In areas previously sampled at a high density (10 samples per km² at a depth of 1 to 3 m), sets of 5 to 20 existing samples from neighboring sample sites were combined and homogenized into a composite sample representing a sampling station. In previously unsampled areas till samples from a depth of 50 to 80 cm were collected at four subsites and combined into a composite for each station. Composite samples from 1057 stations uniformly scattered over Finland (1 station per 300 km³) were thus obtained. A Geochemical Atlas of Finland based on the determinations of the total contents, as well as the aqua regia soluble part of elements in the 0.06 mm fraction of these samples, is being compiled.

Both the total contents and the acid extractable part of the elements show pronounced regional patterns. For some elements there are significant differences in the distribution patterns for the total and the acid-soluble part (e.g., see Figure 11). A striking feature of the easily mobilized phase of elements in the till is the poor correlation with the lithology of the underlying bedrock. An example is provided by the three very sharp edges (a-a, b-b, and c-c) of the pattern of aqua regia soluble Al in Figure 12, which transect main geological units.

Figure 13 shows results from more dense sampling within a sub-area of Figure 12 as an illustration of the validity of the patterns obtained by the widely spaced sampling.

The radon content in ambient air has been mapped in Finland.[80] Also for this element broad regional patterns appear, regions in the southern part of the country showing Rn concentrations in air up to 8 times that of the normal background in areas elsewhere in

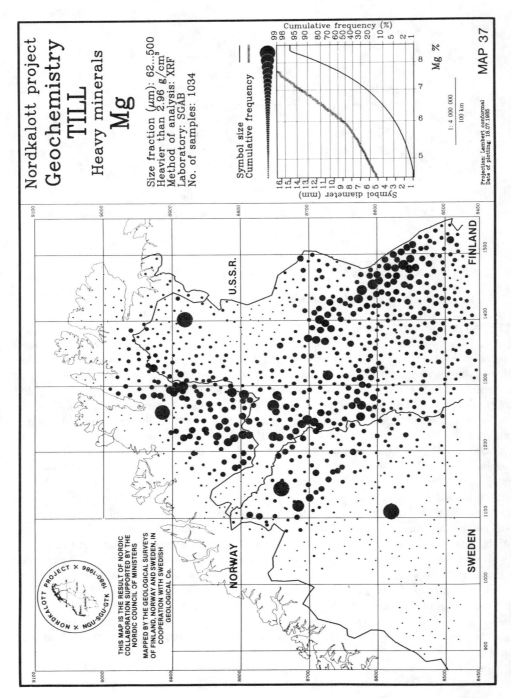

FIGURE 4.   Contents of total magnesium in a heavy mineral fraction (sp. gr. >2.96 g/cm³) of till from northern Finland, Norway, and Sweden. There is no simple relationship between the Mg pattern and the bedrock geology. (See Figure 3.) (From Bølviken et al.[21])

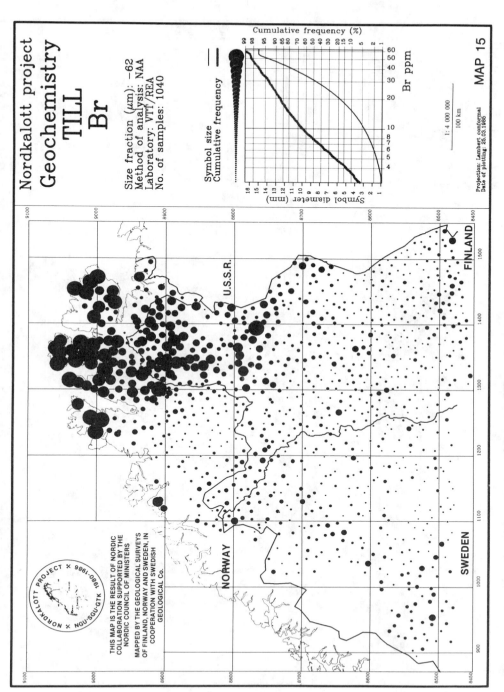

FIGURE 5. Contents of total bromine in the minus 0.06 mm fraction of till from northern Finland, Norway, and Sweden. The bromine province in the northeastern part is thought to reflect properties of the bedrock. The possibility of recent dominating influence from the ocean must be excluded due to an inconsistent pattern in relation to distance from the sea. (From Bølviken et al.[21])

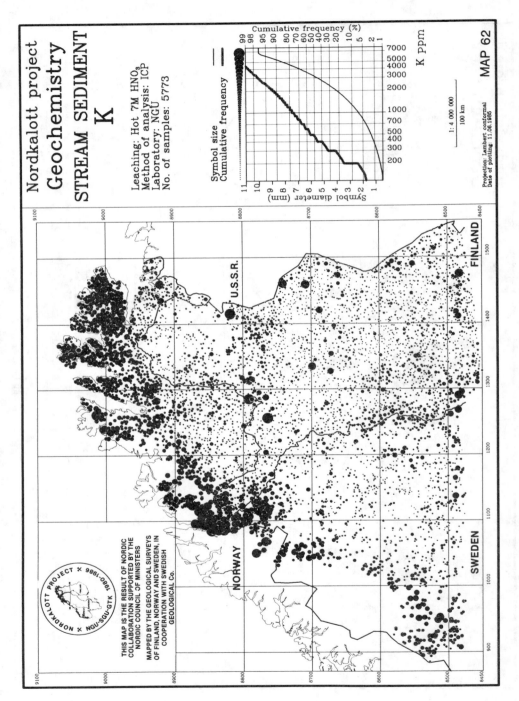

FIGURE 6.   Contents of hot 7N nitric acid soluble potassium in the minus 0.18 mm fraction of stream sediment from northern Finland, Norway, and Sweden. Areas of Caledonian overthrust have higher contents of K than the underlying Precambrian basement. (From Bølviken et al.[21])

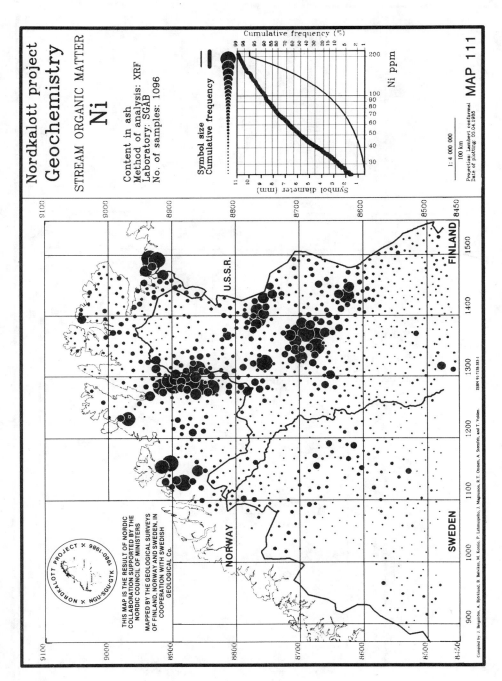

FIGURE 7. Contents of total nickel in the ash of drainage organic matter from northern Finland, Norway, and Sweden. The pattern of high Ni coincide with an eastern greenstone belt, while a western greenstone belt is not associated with high Ni contents. A group of high Ni cocncentrations at 893ON, 1480E is thought to be caused by pollution from the nickel mine at Nikel, a few km further east in the U.S.S.R. (From Bølviken et al.[21])

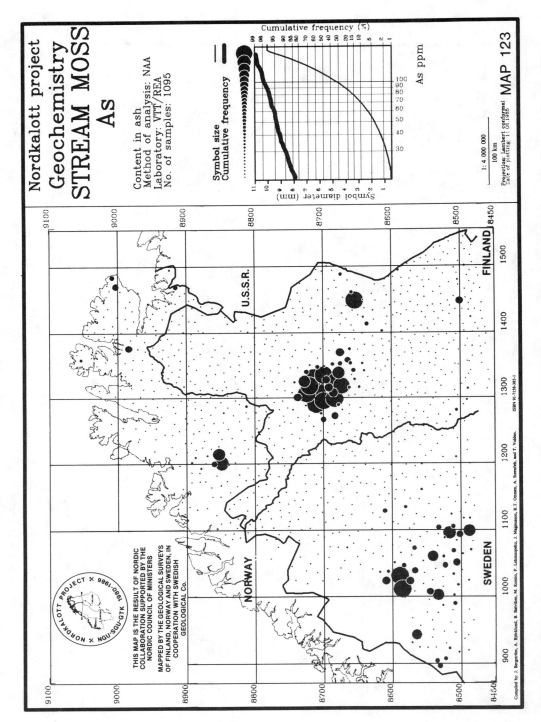

FIGURE 8.   Contents of total arsenic in the ash of aquatic bryophytes from northern Finland, Norway, and Sweden. (From Bølviken et al.[21])

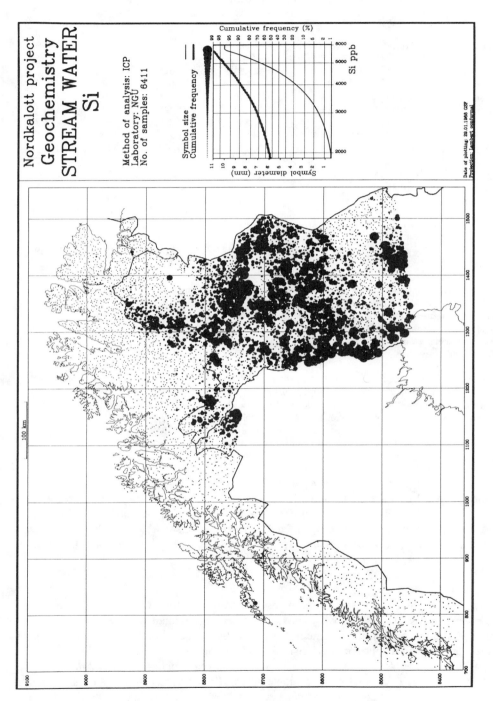

FIGURE 9. Contents of silicon in stream water in northern Finland and Norway. The region of high Si coincides with areas of frequent bogs.[49] The contents of iron in stream water show a similar distribution to that of Si. The patterns of elements such as Ca and Mg are also broad scale but different to those of Si and Fe. These large-scale patterns must have a significant impact on the ecological system in the stream water. (From Bölviken et al.[23a])

TABLE 1
Total Contents of Elements in Combined Samples of Till in Northern Fennoscandia

Element	Min.	Max.	Mean	S.D.	Percentiles 50%	Percentiles 95%
As (ppm)	<1	74	2.4*	3.52*	1.3	8.5
Au (ppb)	<3	436	**	**	**	9.5
Ba(ppm)	<100	1200	636.11*	153.29*	630	900
Br (ppm)	<0.3	98	11.28*	11.12*	7.8	33
Cs (ppm)	<0.6	13	1.85*	1.39*	1.6	4.7
Fe%	<0.5	10.4	4.86*	1.27*	4.8	7
La (ppm)	<1.8	133	36.72*	14.67*	33	61
Na %	<0.01	4.99	2.44*	0.57*	2.5	3.2
Rb (ppm)	<15	182	64.36*	20.41*	61	99
Sb (ppm)	<0.1	115	0.41*	3.58*	0.2	0.9
Sc (ppm)	<0.6	130	19.84*	5.71*	20	28
Sm (ppm)	<0.1	37	6.28*	2.52*	6	10.4
Ta (ppm)	<0.5	39	1.22*	1.25*	1.1	1.8
Th (ppm)	<0.3	45	10.11*	4.06*	9.5	18
W (ppm)	<0.6	9	**	**	**	1.9
U (ppm)	<0.2	11	3.01*	1.28*	2.8	5.2

Notes: Grain Size: <0.062 mm. Analytical Method: Neutron Activation Analysis. N = 1040.

X_i: Content in the i'th sample.

Min.: Minimum value obtained by the indicated method.

Max.: Maximum value obtained by the indicated method.

Mean: Arithmetic mean.

S.D.: Standard deviation.

*   Contents below the detection limit (DL) constituting less than 10% of the values are defined as 2(DL)/3 and included in the calculations of the mean and standard deviation.

** Mean and standard deviation are not calculated because the number of values below the detection limit constitutes more than 10% of the total number of samples.

(The percentiles show cumulative frequencies of element contents beginning with the lowest value.)

After Bölviken et al.[21]

Finland (Figure 14). High contents of Rn in house dwellings are thought to be a cause of lung cancer,[98] especially if the exposure is combined with cigarette smoking.[5,24,28,29]

An atlas of cancer incidence in Finland has been published.[80] Relationships between geochemistry and diseases in Finland have been studied by, e.g., Marjanen,[67] Lahermo et al.,[62] and Piispanen and Nuutinen.[78]

D. NORWAY

Overbank sediment (Ottesen et al.[73]) has been collected from 690 sites uniformly scattered over Norway (320,000 km^2), each site representing drainage areas of 60 to 300 km^2. A vertical section through the sediment was cut with a spade and after excluding the upper polluted 5 to 10 cm, approximately 5 kg of bulk sample was taken from the rest of the section. In the laboratory the samples were dried and sieved to obtain a −0.062 mm fraction, which was analyzed for total element contents by X-ray fluorescence,[31] and acid-soluble elements by inductively coupled plasma spectrometry[103,104] and atomic absorption spectrometry.[61] The analytical results were plotted in black-and-white as single element dots and in colors as moving median maps.[11]

Two of the black-and-white maps are reproduced in Figures 15 and 16. For color maps, the reader is referred to Ottesen et al.[73]

TABLE 2
Contents of Nitric-Acid Soluble (7N, 110°C, 3 h) Elements in Stream
Sediment in Northern Fennoscandia

Element	Min.	Max.	Mean	S.D.	Percentiles 50%	Percentiles 95%
Ag (ppm)	<.8	6	**	**	**	1.3
Al (%)	0.07	5.4	0.79	0.49	0.66	1.7
Ba (ppm)	1	1700	40.77	55.12	27.9	113.5
Ca (%)	0.03	7.12	0.33	0.23	0.3	0.65
Ce (ppm)	3	721	55.47	45.62	43.6	131.
Co (ppm)	0.6	242	6.83	6.99	5.3	16.4
Cr (ppm)	0.3	668	19.04	22.2	14.2	44.1
Cu (ppm)	<10	741	**	**	**	28.3
Fe (%)	0.05	31.9	1.59	2.02	1.1	4
La (ppm)	3.5	317	29.65	20.31	24.6	64.3
Li (ppm)	0.3	62	6.85	6.58	4.7	19.8
Mg (%)	0.01	4.38	0.29	0.24	0.24	0.7
Mn (ppm)	6.5	118000	317.97	1741.1	141	828.2
Mo (ppm)	<1	117	**	**	**	4.4
Ni (ppm)	1	617	9.99	12.87	6.99	26.4
P (ppm)	29	21500	539.45	559.42	442	1200
Sc (ppm)	0.1	24	2.41	1.28	2.21	4.5
Sr (ppm)	1.6	569	14.85	13.81	12.1	32.9
V (ppm)	1	225	23.33	14.77	19.8	50.3
Zn (ppm)	1.2	608	23.27	24.85	15.6	66.6
Zr (ppm)	<5	51	**	**	**	17

Note: Grain size: minus 0.18 mm. Analytical method: inductively coupled argon plasma spectrometry.[104] N = 5773. (See also Table 1 notes.)

After Bölviken et al.[21]

Based on the results one can conclude that:

- All elements depict broad regional patterns with high contrasts, reflecting compositional characteristics of the bedrock and overburden.
- For some elements (e.g., Ba, K, and Na), the total contents and the acid-soluble fraction depict different distribution patterns, while for others (e.g., Mg and Ni) the patterns are similar.
- Overbank sediment can be used as a sample medium to map natural features, even in heavily polluted terrain.

In Norway, nation-wide geochemical maps have also been prepared based on chemical analysis of soil and water. Allen and Steinnes[1] and Steinnes et al.[92] collected humic soil and B_2 samples from the soil profile. The samples were analyzed for a number of chemical elements, several of which reflect natural conditions, while others, such as Pb (Figure 17) are thought mainly to show the effects of long-transported air pollutants.[20] In both cases broad distribution patterns occur.

Analogous results were obtained by Flaten,[33,34] who analyzed drinking water from 386 main waterworks in Norway supplying water to 70% of the entire population. The sampling was done from treated water four times (fall, winter, spring, summer) during 1982-83. The contents of the determined constituents in the water vary substantially, factors between lowest and highest value ranging from 28 for SO_4^{2-} to more than 480 for NO_3^- (Table 3). When plotted on maps most elements depict broad patterns, some of which are thought to

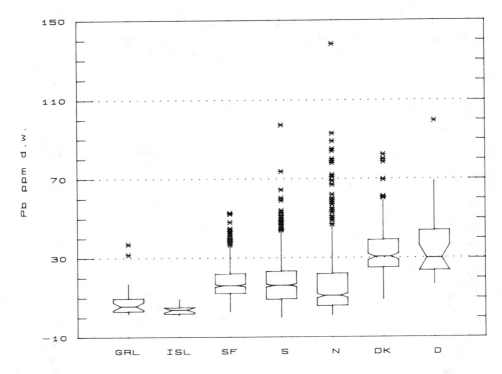

FIGURE 10.   Box-and-whisker plot of the concentration of lead (ppm) in samples of *Hylocomium Splendens* from Greenland (GRL), Iceland (ISL), Finland (SF), Sweden (S), Norway (N), Denmark (DK), and Northern Germany (D). The plot shows the median value, upper and lower quartiles; the whiskers extend to those points within 1.5 times the interquartile range; extreme points beyond this range are plotted as individual adjacent values. (See also Figure 17.) (Data from Rühling et al.[86])

be natural (Figure 18) and others influenced by air-borne pollutants. The high values for Al along the southeastern coast of Norway (Figure 19) is an example of the latter effect. The acid precipitation in the southern parts of Norway[50,74] are thought to cause mobilization of Al in the soil and thereby increased contents of Al in surface waters. High content of Al in drinking water has been suggested as a possible cause of increased rates of Alzheimer's disease in population groups.[35,68]

Possible correlations between geochemistry and epidemiological data in Norway have been studied.[22,23,32,33,41,42,72]

IV. ON THE USE OF GEOCHEMICAL MAPS IN GEOMEDICINE

In some cases similar dispersion patterns are visible on an epidemiological and a geochemical map, indicating that a common causal relationship between the two patterns may possibly exist. Normally, however, a statistical test for covariation is required to compare geochemical and epidemiological data.

In most cases medical ethics demands that epidemiological data are available only by population groups. Before a comparison of epidemiological and geochemical data by statistical methods, both sets must be related to the same geographical units, which normally means that geochemical data must be expressed per municipality or other administrative unit.

Multivariate techniques using a set of disease data as the dependent and various geochemical and other parameters as independent variables are useful in this connection. However, very few studies of this type have appeared in the literature. The following account illustrates some of the problems involved.

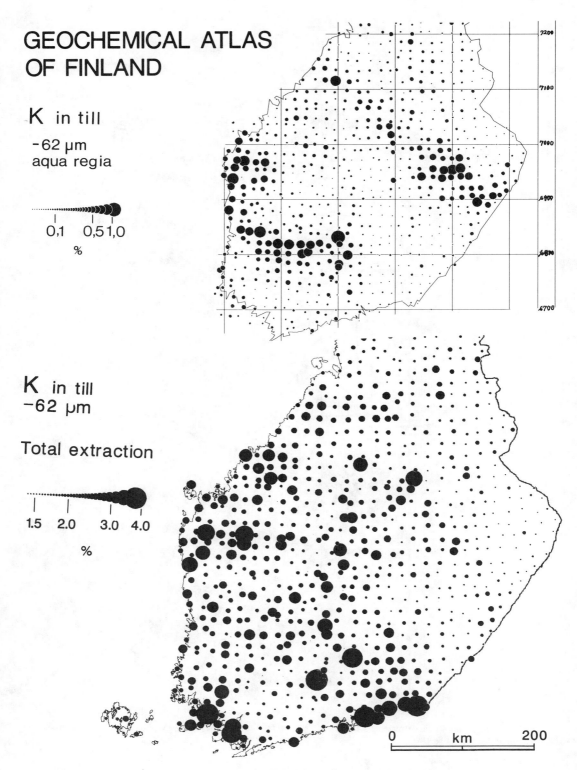

GEOCHEMICAL ATLAS OF FINLAND

K in till

-62 µm
aqua regia

0,1   0,51,0
%

K in till
-62 µm

Total extraction

1.5   2.0   3.0   4.0
%

0   km   200

FIGURE 11.   Aqua-Regia extractable (top part) and total (bottom part) contents of K in the −0.06 mm fraction of till in the southern part of Finland.

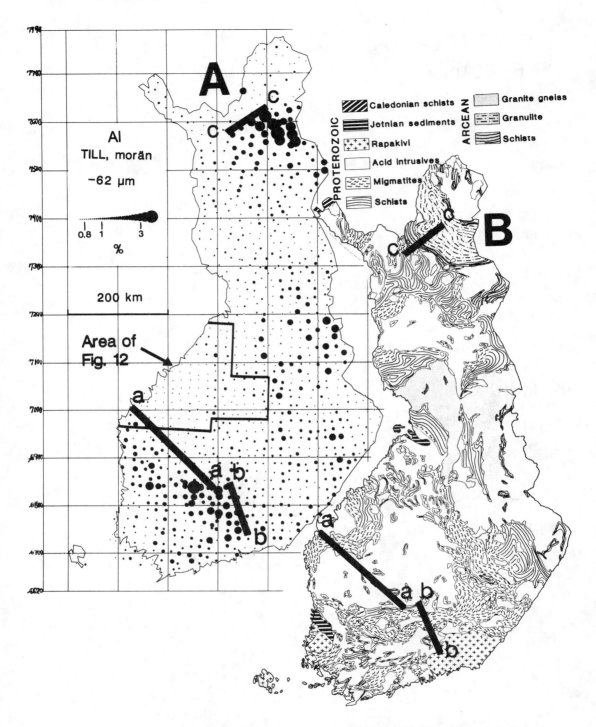

FIGURE 12.   Aqua-Regia extractable Al in the −0.06 mm fraction of till, Finland. (Map of Precambrian bedrock in Finland from Simonen.[91])

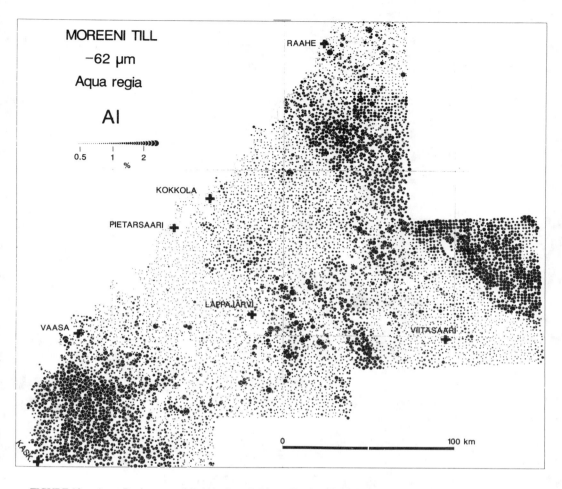

FIGURE 13. Aqua-Regia extractable Al in the − 0.06 mm fraction of till. From the regional geochemical mapping of the Geological Survey of Finland.

Correlation analysis between epidemiological and multi-element geochemical data leads to very large numbers of correlation coefficients. In these large correlation matrices a certain number of the correlation coefficients will be significant by pure chance. If no relevant additional information is available, it is not possible to determine if a certain high-correlation coefficient from a large matrix is significant.

A way of getting around this problem is to divide the data set into subgroups. Some of the high-correlation coeffcients obtained in the whole data set may be maintained in the subgroups, while others will not be reproduced, indicating that their corresponding original values were incidental and therefore spurious (Table 4).

Normally there are also spatial autocorrelations in the geochemical data, resulting in overly optimistic estimates of the significance of covariation between geochemical and epidemiological data if classical statistical methods are employed.

It should be kept in mind that demonstration of a high degree of correlation between epidemiological and geochemical data is no proof of a causal relationship. Unknown, spurious, lurking, or confounded variables may control the epidemiological as well as the geochemical values. The application in geomedicine of comparisons between epidemiological and geochemical data are, therefore, of restricted value in the testing of causal hypotheses. For such purposes, case-control studies are a more powerful tool. However, geomedical

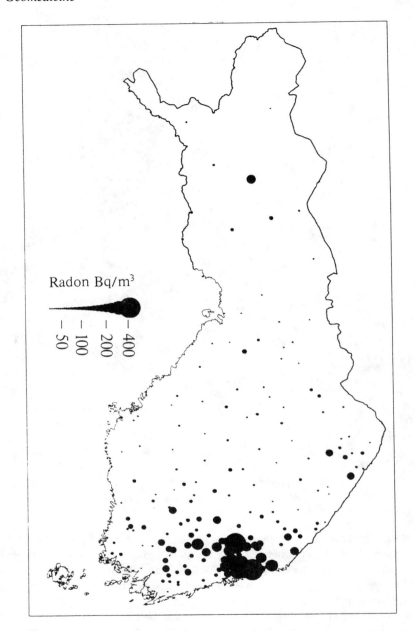

FIGURE 14.   Contents of radon in ambient air, Finland. (From Pukkala et al.[80])

comparisons and correlation studies may disclose earlier unknown epidemiological and geochemical association, from which new causal hypotheses can be generated for testing with other methods.

It is often claimed that varied sources of food would overrule any effects of the local conditions in the relationship between environmental geochemistry and human health. A closer examination reveals:

- In many cases of geomedical comparisons, such large geographical units are used (municipalities, districts, regions, countries) that a substantial part of the food consumed (e.g., milk, meat, certain fruits and vegetables) may have originated within a unit area.[53]

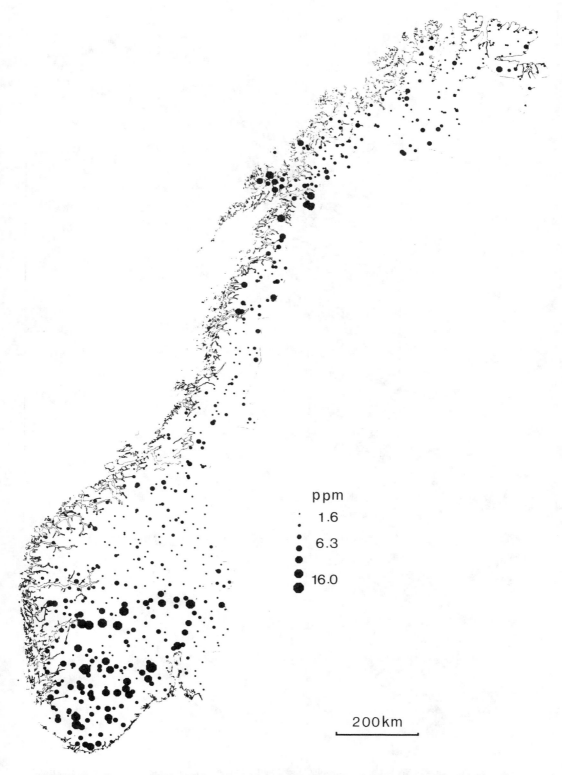

ppm
· 1.6
• 6.3
● 16.0

200km

FIGURE 15.   Contents of 7N hot nitric acid soluble molybdenum in the − 0.06 mm fraction of overbank sediment, from 684 drainage areas (60 to 300 km² in size), Norway.

FIGURE 16.   Contents of 7N hot nitric acid soluble lead in the −0.06 mm fraction of overbank sediments from 684 (60 to 300 km²) drainage areas, Norway. The lead pattern, which is thought mainly to be natural, bears similarities to anthropogenic patterns, indicating that dispersion mechanisms could be similar for natural and anthropogenic material and therefore difficult to distinguish.

10.0 16.0 25.0 39.0 63.0 100.0 160.0 250.0 390.0 >390.0

FIGURE 17.   Contents of acid soluble lead in the −0.06 mm fraction of B_2 layer of soil, Norway.[92,20] The pattern of high lead is thought mainly to reflect effects of atmospheric fallout of long transported pollutants. (See also Fig. 16.) (From Steinnes et al.[92] and Bølviken and Steinnes.[20])

TABLE 3

Some Constituents in Drinking Water From 386 Main Water Works, Norway

Constituent	Arithmetic mean	Range	Ratio highest to lowest
Aluminum	app. 0.06	<0.10-4.1	>40
Barium	app. 0.008	<0.01-0.48	>50
Iron	0.06	<0.01-4.3	>400
Calcium	2.85	0.16-57	350
Magnesium	0.69	<0.07-13.0	>180
Manganese	app. 0.008	<0.01-0.98	>100
Strontium	0.015	0.0015-0.57	380
Bromine	0.011	<0.005-0.44	>90
Fluorine	0.058	0.013-1.20	90
Chlorine	6.6	0.46-117	250
Nitrate (mg N/l)	0.10	<0.01-4.8	>480
Sulfate (mg SO_4/l)	5.2	1.4-39.5	28
Total organic carbon (TOC)	2.4	0.2-10.0	50

Note: All values in mg/l.

After Flaten.[33]

- Some of the mineral uptake of human beings takes place through drinking water,[33] the chemical composition of which in most cases reflects the local geochemical environment.

- Dust, salt, liquids, and gases in the ambient air of local origin may be taken up through the skin or through the respiratory organs. If such materials are temporarily retained in the lungs, they may eventually reach the digestive system by passing up the trachea and down the esophagus.[53,59]

- Direct oral intake of soil particles has been demonstrated not only in grazing animals, but also in human beings, especially in children.[4,95-97]

- The natural background radiation varies on the earth's surface. Various population groups are, therefore, exposed to various levels of natural radiation caused by the decay of elements such as K, Ra, Rn, Th, and U. Radiation could be harmful to health,[98] but in some cases, possibly beneficial.[65,87]

- There might be unforeseeable complicated pathways from natural geochemical environments to animals and human beings. Halvorsen et al.[48] and Andersen and Halvorsen[2] have shown that *Elaphosstrangys rangiferi* (which is a serious parasite on Scandinavian reindeer) requires a Ca-demanding snail as an intermediate host. This points to a possible association between Ca in soils and the occurrence of *Elaphosstrangys rangiferi* in reindeer. Analagous complicated pathways for human beings should not be excluded.

Finnish and Norwegian works show that for the acid-extractable mobile phases of elements there appears in some cases to be a poor correlation between the bedrock lithology as classified on geological maps and the geochemistry of the overburden. This could be due to (1) differences in the susceptibility of minerals and rocks to weathering,[26] (2) differences in the acid solubility of various elements, minerals, and rocks,[43,46] and/or (3) large-scale epigenetic processes in the bedrock created by sources such as meteoric waters,[70,75] crustal fluids, and ascending geogas.[8-10,66] This indicates that geochemical maps of partly extractable fractions of chemical elements can provide important new information for geomedicine in addition to the more traditionally determined total contents.

These features in favor of geomedicine point to an empirical strategy. Possible corre-

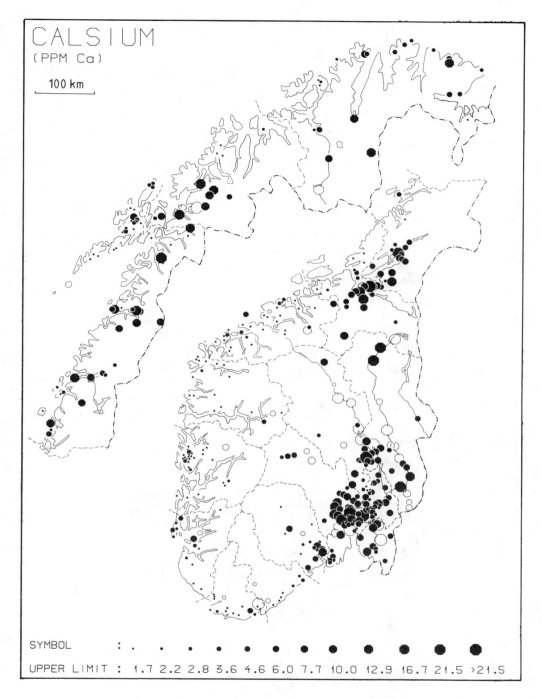

FIGURE 18.   Contents of calcium in Norwegian drinking water, from 384 main water works supplying 70% of the population. The distribution pattern is natural. (From Flaten.[34])

lations between epidemiological and numerous types of geochemical data should be studied systematically. Hypotheses about causes, mechanisms, and pathways have their proper place in the follow-up stage.

It can be concluded that regional geochemical maps provide basic data which will be useful in geomedicine.

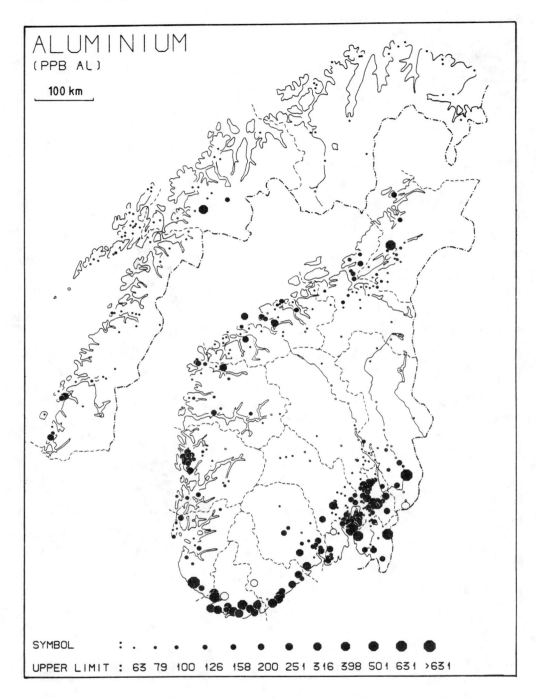

FIGURE 19. Contents of aluminum in Norwegian drinking water, from 384 main water works supplying 70% of the population. The pattern of high Al contents along the southern part of the coast is thought to be an effect of Al mobilization from the soils due to acid rain. (From Flaten.[34])

TABLE 4
Correlation Coefficients Between Geochemistry of Overbank Sediment[73] and Municipal Rates of Cancer Incidence[42] in Norway

Element	Cancer	Sex	N = 73	N = 36	N = 37
Ba	Colon, rectum	M	0.01	−0.05	0.10
		F	−0.47***	−0.45**	−0.50**
Li	Cervix uteri	F	0.31**	0.34*	0.26
K	Female breast	F	−0.30*	−0.40*	−0.07
Ce	Prostate	M	0.21	0.24	0.16
K	Other skin	F	−0.52***	−0.61***	−0.45**
Ce	Thyroid gland	M	0.01	−0.08	0.12
		F	−0.17	−0.35*	0.05

Note: 73 rural municipal aggregates divided into two random groups.

*** $p = 0.001$; ** $p = 0.005$; * $p = 0.01$.

REFERENCES

1. **Allen, R. O. and Steinnes, E.,** Contribution from long-range atmospheric transport to the heavy metal pollution of surface soil, in *Ecological Impact of Acid Precipitation,* Drabløs, D. and Tollan, A., Eds., Oslo-Ås, 102, 1980.
2. **Andersen, J. and Halvorsen, O.,** Species composition, abundance, habitat requirements and regional distribution of terrestrial gastropods in arctic Norway, *Polar Biol.,* 3, 45, 1984.
3. **Appleton, J. D. and Llanos-Llanos, A.,** Geochemical atlas of eastern Bolivia, *British Geological Survey,* Keyworth, Nottingham, 1985.
4. **Barltorp, D., Strehlton, C. D., Thornton, I., and Webb, J. S.,** Absorption of lead from dust and soil, *Postgrad. Med. J.,* 51, 801, 1975.
5. **Bergmann, H., Edling, C., and Axelson, O.,** Indoor Radon daughter concentrations and passive smoking. Indoor air. Radon passive smoking, particulates and housing. *Epidemiology,* Swedish Council for Building Research, 1984, 79.
6. **Beus, A. A. and Grigorian, S. V.,** *Geochemical Exploration Methods for Mineral Deposits,* Applied Publishing, Wilmette, Il, 1977, 287.
7. **Bjørklund, A. J.,** Geochemical Exploration 1983. Selected papers of the 10th International Geochemical Symposium — 3rd Symposium on Methods of Geochemical Prospecting, *J. Geochem. Explor.,* 21, 1, 1984.
8. **Bjørklund, A.,** Relation between structures of the bedrock and geochemistry and morphology of glacial deposits, in *Present Processes and Properties in the Lithosphere; A part of the Swedish ILP,* Mörner, N.-A., Ed., Report 1988, 64, 1988.
9. **Bjørklund, A.,** Regional till geochemistry: Fingerprint of mineralization and control on glacial geology, *J. Geochem. Explor.,* in press, 1989.
10. **Bjørklund, A.,** Fingerprints of ore-forming processes in the geochemistry and in glacial geology, *J. Geochem. Explor.,* 1989.
11. **Bjørklund, A. and Gustavsson, N.,** Visualization of geochemical data on map: New options, *J. Geochem. Explor.,* 29, 89, 1987.
12. **Bjørklund, A. and Lummaa, M.,** Representation of regional, local and residual variability of geochemical data by means of filtering techniques, Proc. 2nd Intern. Symp. Methods of Prospecting Geochemistry, Irkutsk, USSR, 25, 1983.
13. **Bjørklund, A., Gustavsson, N., Kauranne, K., and Tanskanen, H.,** Geokemiallinen kartoitus geologiassa tutkimuslaitoksessa. Summary: Geological mapping at the Geological Survey of Finland, *Vuoriteollisuus-Bergshantering,* 1, 20, 1977.
14. **Bradshaw, P. M. D. and Thomson, I.,** The application of soil sampling to geochemical exploration in nonglaciated regions of the world, in Hood, P., Ed., Geophysics and geochemistry in the search for metallic ores, Geological Survey of Canada, Economic Geology Report 31, 327, 1979.
15. British Geological Survey, Regional geochemical atlas: Great Glen. British Geological Survey, Keyworth, Nottingham, 1987.

16. British Geological Survey, Regional geochemical atlas: Aqryll and Tiree, British Geological Survey, Keyworth, Nottingham, 1988.

17. **Brooks, R. R.**, *Geobotany and Biogeochemistry in Mineral Exploation*, Harper & Row, New York, 1972, 290.

18. **Brundin, N. H. and Nairis, B.**, Alternative sample types in regional geochemical prospecting, *J. Geochem. Explor.*, 1, 7, 1972.

18a. **Brundin, N. H., Ek, J. I., and Selinus, O. C.**, Biogeochemical studies of plants from stream banks in northern Sweden, *J. Geochem. Explor.*, 27, 157, 1987.

19. **Bølviken, B. and Gleeson, C. F.**, Focus on the use of soils for geochemical exploration in glaciated terrain, in Geophysics and geochemistry in the serach for metallic ores. Geological Survey of Canada, Hoods, P., Ed., Economic Geology Report 31, 295, 1979.

20. **Bølviken, B. and Steinnes, E.**, Heavy metal contamination of natural surface soils in Norway from long-range atmospheric transport: further evidence from analysis of different soil horizons, in *International Conference on Heavy Metals in the Environment* Lindberg, S. E. and Hutchinson, T. C., Eds., New Orleans 1987, Vol. 1, 291, 1987.

21. **Bølviken, B., Bergstrøm, J., Bjørklund, A., Kontio, M., Lehmuspelto, P., Lindholm, T., Magnusson, J., Ottesen, R. T., Steenfelt, A., and Volden, T.**, Geochemical Atlas of Northern Fennoscandia, scale 1:4 mill. Geological Survey of Finland, Geological Survey of Norway, Geological Survey of Sweden, 155 maps, 1986, 20.

22. **Bølviken, B., Finne, T. E., Glattre, E., and Olesen, O.**, Geomedical Investigation in Norway, in *Trace Substances in Environmental Health XVIII*, Hemphill, D. D., Ed., Columbia, Missouri, 1984, 534.

23. **Bølviken, B. et al.**, 35 open file reports, Project 1494, Geological Survey of Norway (in Norwegian), 1976-1986.

23a. **Bölviken, B., Bergström, J., Björklund, A., Finne, T. E., Gustavsson, N., Kontio, M., Lehmuspelto, P., Magnusson, J., Ottesen, R. T., Steenfelt, A., Volden, T., and Ødegård, M.**, Stream water geochemistry across northern Fennoscandia (submitted).

24. **Cohen, D., Arai, S. F., and Brian, J. D.**, Smoking impairs long-term dust clearance from the lung, *Science*, 204, 514, 1979.

25. **Coker, W. B., Hornbrooke, E. H. W., and Cameron, E. M.**, Lake sediment geochemistry applied to mineral exploration, in Geophysics and geochemistry in the serach for metallic ores, Hood, P. J., Ed., Geological Survery of Canada, Economic Geology Report 31, 435, 1979.

26. **Davis, J. A. and Hayes, K. F., Eds.**, *Geochemical Processes at Mineral Surfaces*, American Chemical Society, Washington D. C., 1986, 683.

27. **Davy, R. and Mazzucchelli**, Geochemical exploration in arid and deeply weathered terrains, Proceedings of the regional meeting of the Australian Branch of the Association of Exploration Geochemists, *J. Geochem. Explor.*, 22, 1, 1984.

28. **Edling, C., Wingren, G., and Axelson, O.**, Radon daughter exposure in dwellings and lung cancer. Indoor air. Radon, passive smoking, particulates and housing. *Epidemiology*, Swedish Council for Building Research, 1984, 29.

29. **Edward, P. R. and St. Claire Renard, K. G.**, Lung cancer in Swedish iron miners exposed to low doses of radon daughters, *New Engl. J. Med.*, 310, No. 23, 1485, 1984.

30. **Fauth, H., Hindel, R., Siewers, U., and Zinner, J.**, *Geochemischer Atlas Bundesrepublik Deutschland*, Bundesanstalt für Geowissenschaften und Rohstoffe, Hannover, 1985.

31. **Faye, G. C. and Odegård, M.**, Determination of major and trace elements in rocks employing optical emission spectroscopy and X-ray fluorescence, *Norges Geologiske Undersøkelsen*, 322, 35, 1975.

32. **Finne, T. E.**, Comparison of stream sediment data and death rates in Southern Norway, in *Geochemical Research in Relation to Geochemical Registrations*, Låg, J., Ed., Universitetsforlaget, Oslo, 1984, 71.

33. **Flaten, T. P.**, An investigation of the chemical composition of Norwegian drinking water and its possible relationships with the epidemiology of some diseases. Thesis No. 51, Institute of inorganic chemistry, The University of Trondheim-NTH, Trondheim, 278, 1986.

34. **Flaten, T. P.**, Chemical composition of Norwegian drinking water and some health aspects, with emphasis on aluminium and dementia (including Alzheimer's disease). *Norges Geologiske Undersøkelse*, Special publication 2, 86, 1987.

35. **Flaten, T. P.**, Geographical associations between aluminium in drinking water and death rates with dementia (including Alzheimer's disease), Parkinsons's disease and amyotrophic lateral sclerosis in Norway, *Environmental Geochemistry and Health*, in press, 1989.

37. **Fletcher, W. K.**, *Analytical Methods in Geochemical Prospecting: Handbook of Exploration Geochemistry*, Vol. 1, Govett, G. J. S., Ed., Elsevier, Amsterdam, 1981, 255.

38. **Gardner, M. J. et al.**, *Atlas of Cancer Mortality in England and Wales 1968-1973*, J. Wiley & Sons, Chichester, 1978.

41. **Glattre, E., Aslevold, R., and Bay, I. G.**, Norwegian water story, *Lancet*, ii, 1038.

42. **Glattre, E., Finne, T. E., Olesen, O., and Langmark, F.,** *Atlas of Cancer Incidence in Norway 1970-79,* Norwegian Cancer Society — Norwegian Cancer Registry, Oslo. 1985.

43. **Goldschmidt, V. M.,** Om Lerjord mineralenes opplöselighet i syre (On the acid solubility of clay minerals), *Tidsskrift for kjemi og Baergvaesen,* 1, 9, 1924.

44. **Govett, G. J. S.,** *Handbook of Exploration Geochemistry,* Vol. 1-7, Elsevier Scientific, Amsterdam, 1981-88.

45. **Govett, G. J. S.,** Rock geochemistry in mineral exploration, *Handbook of Exploration Geochemistry,* vol. 3, Elsevier, Amsterdam, 1983.

46. **Graff, P. R.,** Analyse og utlutning av anorthositter (Analysis and leaching on anorthosites), Geological Survey of Norway. Open file report 1865, 115, 1981.

47. **Gustavsson, N., Noras, P., and Tanskanen, T.,** Seloste geokemiallisen kartoituksen tutkimusmenetelmistä. Summary: Report on geochemical mapping methods, *Geologisen tutkimuslaitoksen tutkimusraportti,* 39, 20, 1979.

48. **Halvorsen, O., Andersen, J., Skorping, A., and Lorentzen, G.,** Infection in reindeer with the nematode Elaphosstrangys rangiferi Mitskevich in relation to climate and distribution of intermediate hosts, in *Proc. 2nd Int. Reindeer/Caribou Symp.,* Røros 1979. Reimers, E. and Skjenneberg, S., Eds., Direktoratet for vilt og ferskvannsfisk, Trondheim, 449, 1980.

49. **Hamborg, M., Hirvas, H., Lagerbäck, R., Minell, H., Mäkinen, K., Olsen, L., Rodhe, L., Sutinen, R., and Thoresen, M.,** Quaternary deposits, Northern Fennoscandia, Map 1:1 mill, Geological Surveys of Finland, Norway and Sweden, 1987.

50. **Henriksen, A., Lien, L., Traaen, T. S., and Sevaldrud, J.,** *1000 Lake Survey 1986 Norway.* Norwegian State Pollution Control Authority, Norwegian Institute for Water Research, 1987, 33.

52. **Hood, P. J.,** Geophysics and geochemistry in the search for metallic ores, Geological Survey of Canada, Economic Geology Report 31, 1979.

53. **Hopps, H. C.,** The geochemical environment in relationship to health and disease. *Interface,* Vol. 8, (official publication of the Society of Environmental Geochemistry and Health) 1979, 24.

54. **Hornbrook, E. H.,** Lake sediment geochemistry: Canadian applications in the Eighties, in *Proceedings of Exploration 87,* Garland, G. D., Ed., Ontario Geological Survey, Toronto, 1989, 405.

56. Institute of Geological Sciences, *Geochemical Atlas of Great Britain: Orkney Islands,* Institute of Geological Scineces, London, 1978.

57. **Kauranne, K.,** Regional geochemical mapping in Finland, in *Prospecting in Areas of Glaciated Terrain 1975,* Jones, M. J., Ed., Institution of Mining and Metallurgy, London, 1975, 71.

57a. **Koljonen, T., Gustavsson, N., Noras, P., and Tanskanen, H.,** Geochemical atlas of Finland: preliminary aspects, in Geochemical Exploration, Jenness, S. E., Ed.,1987; *J. Geochem. Explor.,* 32, 231, 1989.

58. **Kovalevskii, A. L.,** *Biogeochemical Exploration for Mineral Deposits,* Amerind Publishing, New Delhi, 1979, 136.

59. **Kozlovsky, Y. A.,** The world's deepest well, *Sci. Am.,* 251 (6), 106, 1984.

61. **Kuldvere, A.,** Determination of arsenic in selenium metal or in the presence of high levels of selenium by hydride generation absorption spectrometry, *Analyst,* 113, 277, 1988.

61a. **Låg, J.,** (personal communication) 1989.

62. **Lahermo, P. W., Pukkala, E., Gustavsson, N., Bjørklund, A., and Teppo, L.,** Comparison of geochemical and cancer incidence maps in Finland, *Acta Pharmacol. Toxicol.,* 59, 279, 1986.

63. **Levinson, A. A.,** *Introduction to Exploration Geochemistry,* 2nd ed., Applied Publishing, Wilmette, IL, 1974, 612.

64. **Levinson, A. A.,** Introduction to Exploration Geochemistry, the 1980 Supplement, Applied Publishing, Wilmette, IL, 1980, 615.

65. **Luckey, T. D.,** *Hormesis with Ionizing Radiation,* CRC Press, Boca Raton, FL, 1980, 221.

66. **Malmquist, L. and Kristiansson, K.,** Experimental evidence for an ascending microflow of geogas in the ground, *Earth Planet. Sci., Lett.,* 70, 407, 1984.

67. **Marjanen, H.,** On the relationship between the contents of trace elements in soils and plants and the cancer incidence in Finland, in *Geomedical Aspects in Present and Future Research,* Låg, J., Ed., Universitetsforlaget, Oslo, 1980, 149.

68. **Martin, C. N., Osmond, C., Edwardson, J. A., Barker, D. J. P., Harris, E. C., and Lacey, R. F.,** Geographical relation between Alzheimer's disease and aluminium in drinking water, *Lancet,* 59, Jan. 14, 1989, 1989.

69. **Möller Jensen, O., Carstensen, B., Glattre, E., Malker, B., Pukkala, E., and Tulinius, H.,** *Atlas of Cancer Incidence in the Nordic countries,* The Cancer Societies of Denmark, Finland, Iceland, Norway and Sweden, 1988, 205.

70. **Nriagu, J. O.,** *Sulphur in the Environment,* Part I: The Atmospheric Cycle, John Wiley & Sons, New York, 1978.

71. **Nriagu, J. O. and Pacyna, J. M.,** Quantitative assessment of worldwide contamination of air, water and soils by trace metals, *Nature,* 333, 134, 1988.

72. **Ottesen, R. T.**, Introductory geomedical investigations in Norway, in *Geomedical Aspects in Present and Future Research,* Låg, J., Ed., Universitetsforlaget, Oslo, 1980, 197.

73. **Ottesen, R. T., Bogen, J., Bølviken, B., and Volden, T.**, Overbank sediment: a representative sample medium for regional geochemical mapping, in Geochemical Exploration 1987, Wilhelm, E., Zeegers, E., et al., Eds., *J. Geochem. Explor.,* Vol. 32, 1989.

74. **Overrein, L. N., Seip, H. M., and Tollan, A.**, Acid precipitation — effects on forest and fish, Final report of the SNSF-project, Oslo, 1980.

75. **Pacyna, J. M.**, Atmospheric trace elements from natural and anthropogenic sources, in *Toxic Metals in the Atmosphere. Advances in Environmental Sciences and Technology,* Vol. 17, Nriagu, J. O. and Davidson, C. I., Eds., John Wiley & Sons, New York, 1986, 33.

76. **Page, A. L., Miller, R. H., and Keeney, D. R.**, *Methods of Soil Analysis. Part 2, Chemical and Microbiological Properties,* 2nd ed., American Society of Agronomy, Soil Science Society of America, Madison, WI, 1982, 1159.

77. **Parslow, G. R.**, Geochemical Exploration 1982, Proceedings of the 9th International Geochemical Exploration Symposium, *J. Geochem. Explor.,* 19, 1, 1983.

78. **Piispanen, R. and Nuutinen, M.**, Comparison of health patterns in two geochemically contrasted areas in northern Finland, *Environ. Geochem.,* 10, 74, 1988.

79. **Plant, J. A., Hale, M., and Ridgway, J.**, Developments in regional geochemistry for mineral exploration, *Transactions of the Institution of Mining and Metallurgy* (Sect. B. Applied earth sciences), 97, B116, 1988.

80. **Pukkala, E., Gustavsson, N., and Teppo, L.**, *Atlas of Cancer Incidence in Finland,* Finnish Cancer Registry, 1987, 55.

81. **Rogers, P. J.**, Gold in lake sediments: implications for precious metal exploration in Nova Scotia, in *Prospecting in Areas of Glaciated Terrain 1988,* MacDonald, D. R. and Mills, K. A., Eds., The Canadian Institute of Mining and Metallurgy. The Institute of Mining and Metallurgy, London, 1988, 357.

82. **Rose, A. W., Hawkes, H. E., and Webb, J. S.**, *Geochemistry and Mineral Exploration,* 2nd ed., Academic Press, New York, 1979, 657.

83. **Rosenberg, R., Zilliacus, R., and Kaistila, M.**, Neutron activation analysis of geochemical samples. *Technical Research Centre of Finland, Research Notes,* 225, 15, 1983.

84. **Rühling, Å. and Tyler, G.**, Regional differences in the heavy metal deposition over Scandinavia, *J. Appl. Ecol.,* 8, 497, 1971.

85. **Rühling, Å. and Tyler, G.**, Heavy metal deposition in Scandinavia, *Water, Air, Soil Pollution,* 2, 445, 1973.

86. **Rühling, Å., Rasmussen, L., Pilegaard, K., Mäkinen, A., and Steinnes, E.**, *Survey of Heavy Metal Deposition,* Nordic Council of Ministers, NORD 21, 1987.

87. **Sagan, L. A.**, *Radiation Hormesis, Health Physics,* Special Issue, 52, 517, May, 1987.

88. **Shacklette, H. T.**, The use of aquatic bryophytes in prospecting. Geochemical Exploration 1983. *J. Geochem. Explor.,* 21, 89, 1984.

89. **Shilts, W. W.**, Till geochemistry in Finland and Canada. Geochemical Exploration 1983. *J. Geochem. Explor.,* 21, 95, 1984.

90. **Siegel, F. R.**, *Applied Geochemistry,* John Wiley & Sons, New York, 1974, 353.

91. **Simonen, A.**, The Precambrian in Finland, Geological Survey of Finland, Bull. 304, 58, 1980.

92. **Steinnes, E., Frantzen, F., Johansen, O., Rambak, J. P., and Hanssen, J. E.**, Atmosfaerisk nedfall av tungmetaller i Norge (Fallout of heavy metals in Norway). Norwegian State Pollution Control Authority. Report 334/88, 33, 1988.

93. **Stephenson, B., Ghazali, S. A., and Widjaja, H.**, *Regional Geochemical Atlas Series of Indonesia: 1. Northern Sumatra,* Institute of Geological Sciences, Keyworth, Nottingham, 1982.

94. **Tenhola, M.**, Regional geochemical mapping based on lake sediments in eastern Finland, in Prospecting in areas of glaciated terrain 1988. Mac Donald, D. R. and Mills, K. A., Eds., The Canadian Institute of Mining and Metallurgy. The Institute of Mining and Metallurgy, London, 1988, 305.

95. **Thornton, I.**, *Applied Environmental Geochemistry,* Academic Press, New York, 1983, 501.

96. **Thornton, I.**, Environmental geochemistry and health in the United Kingdom, in *Geochemical Research in Relation to Geochemical Registrations,* Låg, J., Ed., Universitetsforlaget, 1984, 125.

97. **Thornton, I. and Abrahams, P. W.**, Soil ingestion — a major pathway of heavy metals into livestock grazing contaminated land, *Sci. Tot. Environ.,* 28, 287, 1983.

98. **UNEP,** *Radiation, Doses, Effects, Risks,* United Nations Environment Programme, Nairobi, 1985, 64.

99. **Wagner, J. C.**, The pneumoconioses due to mineral dusts, *J. Geol. Soc.,* London, 137, 537, 1980.

100. **Weaver, T., Freeman, S. H., Broxton, D. E., and Bolivar, S. L.**, *The Geochemical Atlas of Alaska,* Los Alamos National Laboratory, Los Alamos, New Mexico, 1983.

102. **Webb, J. S., Thornton, I., Thompson, M., Howarth, R. J., and Lowenstein, P. L.**, *The Wolfson Geochemical Atlas of England and Wales,* Clarendon Press, Oxford, 1978, 69.

103. **Ødegård, M.**, Determination of major elements in geologic materials by ICAP spectroscopy, *Jarrell-Ash Plasma Newsletter,* Vol. 2, No. 1, 4, 1979.

104. **Ødegård, M.**, The use of inductively coupled argon plasma (ICAP) atomic emission spectroscopy in the analysis of stream sediments, *J. Geochem. Explor.,* 14, 119, 1981.

Chapter 6

REGISTRATIONS OF GEOCHEMICAL COMPOSITIONS OF SOILS, PLANTS, AND WATERS AS A BASIS FOR GEOMEDICAL INVESTIGATIONS IN FINLAND

Pertti W. Lahermo

TABLE OF CONTENTS

I. INTRODUCTION

The mechanisms relating the geochemical environment of bedrock, soil, and water with nutrition, health, and disease in the world of plants and animals, including man, are extremely complex and still largely unknown. The pathways of elements and nutrients in the bedrock-soil-air-water-plants-animals-man chain are very varied. Furthermore, elements interfere with each other in a complex way. Hence, it is difficult to draw regional and nationwide geomedical or medical geographical conclusions as there are shortcomings in geochemical information on the sources and distribution of elements, in knowledge of the physiology and biochemistry of plants, and in the epidemiology of humans.

The geochemical composition of soil and water affects plants more than higher animals and man, which obtain nutrients from secondary sources through food, drink, and polluted air. Soil, the main natural source and sink of elements, functions as a dynamic system, responding to and developing along with the impact of the changing environment. Plants reflect the composition of the soil in which they grow in diversified ways; some tend to accumulate particular elements in their roots, stem, or leaves, whereas others are more indifferent to the geochemical soil composition. The physiological barriers of plants, i.e., the mechanism that limits the admission of an excess quantity of elements, vary from one plant species to another and from one part of a plant to another.[11] Trace elements in particular may strongly correlate with soil concentrations. Cattle are still largely fed on local products and are undoubtedly more closely related to "at site" geochemical nutrient sources than man. Animals and man have a homeostatic mechanism that maintains element balances to some degree, irrespective of even large differences in nutrient supply. In advanced western societies, feeds, forages, and diets are composed of an undefined mixture of end members that originate from numerous domestic and foreign sources, lessening or even eliminating the effect of the local geochemical environment. Drinking water is generally more local, but, with a few exceptions, it represents only a small fraction of total element intake. However, the importance of studying the interrelationships between the geochemical environment and human life draws attention to the fact that the shortage of many trace elements in the Finnish diet has been attributed to the soil.[1]

Despite the numerous constraints referred to, the above geomedical research may reveal important nutrition and health interrelationships that derive from the local geochemical environment, particularly among the rural population in developing countries, which is heavily dependent on food produced locally. Because of the lack of regional information, however, the geomedical conditions in developing countries are not discussed in this paper; rather, the emphasis is on the findings of geochemical mapping in Finland, as they constitute a basis of information for geomedical evaluations and may serve as a model for future research, particularly in developing countries.

A vast amount of information about the uptake, metabolism, and elimination of elements within plants, animals, and man has been gleaned from experimental studies. In this paper, however, physiological and medical aspects are put aside, and the focus is on the importance of the geochemical environment as a source of nutrients.

At least 25 elements are considered essential or beneficial to plants, animals, and man; Ca, Mg, Na, K, P, Fe, Mn, Zn, Cu, Cr, Co, Mo, Se, Fe, and I are referred to in the text. Elements or nutrients can be divided into major or macro components, minor or micro components, and into trace components. They can be further divided into essential or beneficial and harmful or toxic elements. Some behave as toxicants at high concentrations (e.g., Zn, Cu, Mo, Se, and F), although they may be essential at lower concentrations. Some again, such as Pb, Cd, Hg, and U, have indisputably deleterious effects, even at comparatively low concentrations.

To complicate the situation still further, plants and animals may respond differently to the same elements. Taking this into account, the elements can be divided into four groups. The first group consists of elements essential to plants but whose necessity to animals and man has not been established (e.g., B). The second group is made up of elements essential to both plants and animals (e.g., Mn, Zn, and Cu). The third group comprises elements essential to animals and man but not to plants (e.g., Se and I), and the fourth group consists of toxic elements with no established biological role in plants or animals (e.g., Cd, Hg, and Pb).

Bedrock and soil compositions are often poor indicators of the availability of certain elements to plants, animals, and man. Of more importance than the total or strong-acid-leached partial concentrations are the exchangeable and biologically reactive forms of elements. A good example is the common Fe deficiency in animals and man, even though iron is one of the most common metals in the soil. The bulk mineral and element contents are, however, important because they provide an estimate of the total element reserves, or "soil capacity factor".[12] Weathering processes release elements to soil, water, and plants, providing an estimate of "the intensity factor" of a much smaller magnitude. Both factors together constitute a measure of the power of soil to supply elements and nutrients. Although occurring in microgram amounts in soil, trace elements (mostly heavy metals) exert an effect on plant and animal growth and welfare far out of proportion to their concentrations.

Some elements behave either synergistically or antagonistically with respect to others by enhancing or retarding uptake, transport, and metabolically beneficial or detrimental effects. An example of a synergism is the addition of readily mobile anions (e.g., nitrates) to soil solutions, leading to enhanced uptake of many captions. In the case of an antagonism, an excess of potassium will depress the availability of Ca and Mg for plants and retard the uptake of radiactive.[137]Cs. Also well known is the fact that a Se surplus depresses Cd toxity. Consequently, it is very difficult to estimate the impact of the upper recommended intake of Cd in humans worldwide as Zn, Cu, Fe, and Se interact with Cd, reducing its toxic effects.

The mobility and consequent availability of elements for plants and animals depend on complex weathering mechanisms and are regulated by factors such as soil wetness, pH-Eh conditions, and biological and bacteriological activities. The solubility of the elements is enhanced by low pH (e.g., Ca and heavy metals). Therefore, with a few exceptions (e.g., Mo), available nutrient concentrations are generally higher in acid than in alkaline soils. Consequently, heavy liming of soil may noticeably reduce concentrations of mobile nutrients. Loosely bound elements in the colloidic humus and clay fraction are mainly responsible for exchangeable nutrients. Also, hydrous Fe, Mn, Al, and Si precipitates can adsorb a variety of important trace elements, providing a more effective souce of nutrients for plants than the primary mineral matter itself. Biological activities in the soil enhance element mobility as organic humus matter and plants form complexes and transport effectively many heavy metals.

Absolute trace element abundances in primary minerals contribute little to the geochemical-biochemical cycling of elements.[9] The adsorbed exchangeable trace element fraction posesses higher mobilities and bioavailabilities and is most important with respect to possible deficiencies or toxicities in plants and, thus, in animals and man.

In this paper, geochemical mapping results are presented as a source of information on the magnitude of element reserves and particularly their regional differences in glacial till (Geochemial Atlas of Finland). A comparison is made with stream sediments (Nordkalott Project) and with less systematic data on cultivated arable land. Geochemical data on till and stream sediments provide suggestive information on element reserves in natural forested lands. So far, only stream sediment studies have been used in regional environmental reconnaissance as the samples are considered to approximate well the erosion products of rocks, overburden, and soil upstream from the sampling point, thus representing the mean element compositions in the catchment area.[2,22] Most Finnish soils derive from igneous and metamorphic rocks, which are generally poor in calcareous components and nutrients. Till and stream sediments represent a natural or only slightly altered environment. Because of their low nutrient content, cultivated soils have received appreciable amounts of elements from the natural and artificial fertilizers — lime, sewage sludges, and pesticides — used to improve soil fertility.

The chemical composition of soil water and ground water is closely related to soil and bedrock composition. Hence, besides the pedogeochemical data presented, hydrogeochemical mapping results are evaluated for their geomedical significance as well.

II. SOURCES OF GEOCHEMICAL BACKGROUND DATA FOR GEOCHEMICAL EVALUATIONS

A. GEOCHEMICAL MAPPING OF SOILS

The Geological Survey of Finland carried out extensive nationwide mapping of glacial till in the course of compiling the Geochemical Atlas of Finland. Four samples of glacial till were collected from the C horizon of the podzol profile per site and mixed to form a composite sample. A total of 1057 samples were used from all over the country. The minus-64 μm fraction, which was sieved and leached for partial content analyses with aqua regia (AR) and digested with HF + B(OH)$_3$ for total content analysis, was analyzed by ICAP spectroscopy. The fine fraction of till was analyzed directly by the INAA method.

Within the framework of the Nordkalott Project, which covers northern Scandinavia above 62° latitude, 5400 to 6400 samples of till, minerogenic and organogenic stream sediments, heavy minerals from till and stream sediments, stream moss, and stream water were collected and analyzed jointly by the Geological Surveys of Finland, Norway, and Sweden.[4] In this project, till (minus-62 μm fraction) was analyzed by direct-reading OES with a tape machine (rough total contents); stream sediments were leached with hot 7 *M* HNO$_3$ and analyzed by ICAP. Organic matter was ashed and analyzed by XRF, and ashed stream moss was analyzed by NAA. Heavy minerals were separated from the 62 to 500-μm fraction (heavier than 2.96 g/cm^3) and analyzed by XRF. Stream water was analyzed by ICAP.

The results of the two geochemical projects were presented as black-dot and colored-surface maps (see Reference 3), whereas some of the Nordkalott data were compiled into composite and predictive ore-potential maps. Some attempts have been made to use the maps in environmental and geomedical research.

The partial leaching method (hot AR) employed is more aggressive than the methods generally used for loosely bound nutrients (acid or neutral NH$_4$Ac(+ HAc) and/or EDTA) in agricultural and silvicultural studies. It is clear that the partial and total element contents of soils differ markedly from each other. Till samples were collected from the slightly altered or unaltered C horizon of the podzol profile (from an average depth of 0.5 m) below the biologically active litter and humus-rich layer and eluviation and illuviation layers, which are the main sources of exchangeable nutrients in the root zone. Till, the main object of geochemical mapping, is predominantly forest soil; the peat lands, clay, silt, and mull soils used for agriculture were not objectives of the geochemical projects mentioned, which were principally targeted at exploration. Nevertheless, the geochemical maps undoubtedly reflect the areal differences in element and nutrient status in northern and eastern Finland, where cultivated till, silt, and peat soils largely obtain their natural elements from bedrock and the minerogenic soils transported and mixed by glacial activity.This is particularly true of the trace elements, most of which are of geogeneous origin.

Cultivated soils and tilled fields are investigated by the Soil Fertility Service Company (Viljavuuspalvelu Oy) and the Agricultural Research Centre and forest soils by the Forest

Research Institute. The less aggressive leaching methods (NH₄Ac, HAc, and EDTA) of the Forest Research Institute are specific for loosely bounds nutrients. Although no systematic mapping of arable land covering the whole country was carried out, there are some data (see Reference 13) that can be used for comparision with the geochemical data.

Comparison of data produced by geochemical mapping projects and different materials (till, stream sediments, their heavy minerals, and stream moss) in relation to, say, cobalt contents reveals some similarities in regional distribution patterns (Figure 1). Although the concentration levels in different materials differ largely from each other, the areal distribution patterns are similar, the highest Co contents being in the mafic and ultramafic bedrock areas. This is because Co is incorporated in ferromagnesian minerals replacing Mg and Fe and goes fairly easily into solution during weathering. These interdependencies justify the use of diverse geochemical materials in evaluating the regional differences between element reserves. The Nordkalott project shows that similar interrelationships prevail for many other elements: Ca, Mg, Na, K, Fe, Mn, P, Zn, Ni, Cr, and As, although many elements do not show covariance between different materials (e.g., Si, Cu, Mo, and U). Till, minerogenic and organogenic stream sediments, and stream moss seem to be the best materials for establishing the regional element patterns; heavy mineral geochemistry does not always result in similar patterns.

Another example is the distribution of chromium in till compared with some data on agricultural soils. Despite the difference in leaching (hot NH₄Ac + HAc + EDTA) and analytical methods (AAS spectroscopy) employed for agricultural soils and in sampling depth (upper root zone), density, and soil materials (clay, silt, mull, and peat), some common features are to be found in the regional Cr patterns. The Cr contents in till and cultivated soils are highest in northern Finland and lowest in the central and southeastern part of the country (Figure 2). The average Cr levels differ largely in both materials (24 and 0.28 ppm, respectively) since the AR method is much more effective in removing elements from minerogenic and organogenic material than NH₄AC + HAc + EDTA leach, which is used for loosely bound nutrients.

B. SOME ELEMENTS OF GEOMEDICAL SIGNIFICANCE

In the following material, the most important elements are discussed briefly in terms of their geochemical mode of occurrence and presumed geomedical significance. The evaluation is made in the order of the abundance of the elements in the AR leach according to Table 1. The first five elements occurring at the percent level can be considered major or macro elements, whereas the rest, at ppm levels, are minor or micro elements or trace elements.

Iron, the most abundant major element in the AR leach (Table 1), occurs in sulfides but also in silicates, replacing Mg and Al. When released during weathering, it is largely retained in soil as oxides, ranging from amorphous to a variety of crystallized hydrous oximinerals, and in clay minerals, substituting for Al. *Manganese,* occurring at ppm levels only (Table 1), is present in silicates and forms precipitates whose stability depends on the pH and Eh conditions.

Both Fe and Mn are essential elements for some plants, but particularly for animals and man, and are therefore often added to fertilizers. However, Mn may be toxic to man when his occupation exposes him to exceptionally high concentrations. The availability of Fe and Mn depends on the solubility of primary sulfides and silicates and secondary Fe and Mn oxihydrates in different Eh and pH conditions, while bacterial activity also contributes to solubilization. Fe and Mn oxides also affect plant nutrition indirectly by adsorbing and accumulating many exchangeable cations, e.g., Zn, Cu, and Ni, in readily available forms. The exchangeable elements show the highest solubility under acid conditions, whereas the lower mobility in alkaline soils may contribute to Fe deficiencies.

The regional Fe and Mn distributions in till are consistent with the mafic and ultramafic

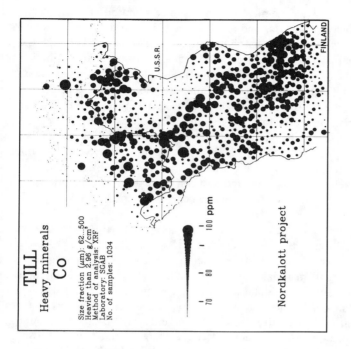

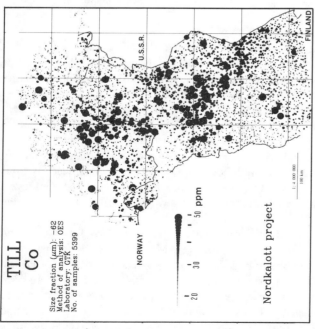

FIGURE 1.   (A to C) The distribution of Co and generalized bedrock composition in northern Finland and Norway (Lapland) according to data from the Nordkalott Project.[4] For materials and analytical methods, see texts on the respective maps.

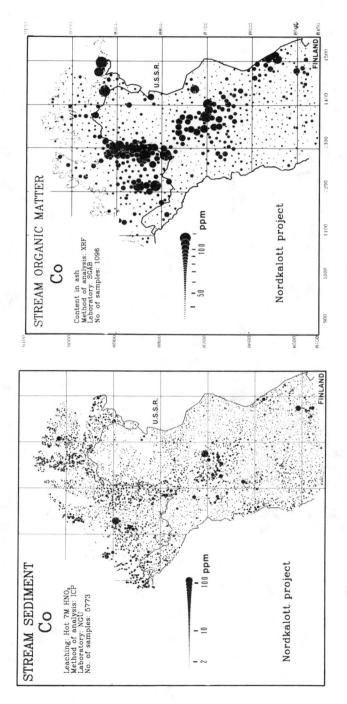

FIGURE 1B.

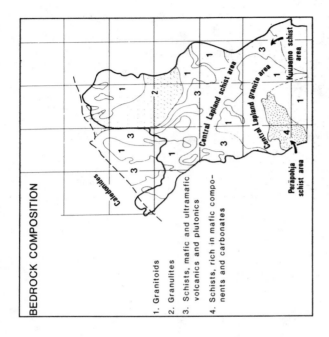

BEDROCK COMPOSITION

1. Granitoids
2. Granulites
3. Schists, mafic and ultramafic volcanics and plutonics
4. Schists, rich in mafic compo-nents and carbonates

Caledonides

Central Lapland schist area

Central Lapland granite area

Kuusamo schist area

Peräpohja schist area

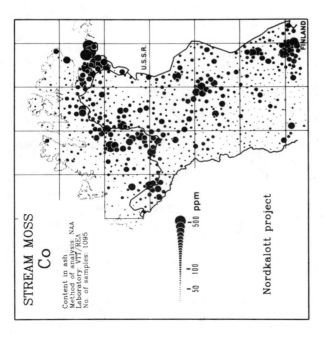

STREAM MOSS
Co

Content in ash
Method of analysis: NAA
Laboratory: VTT/REA
No. of samples: 1095

U.S.S.R.

FINLAND

50   100   500 ppm

Nordkalott project

FIGURE 1C.

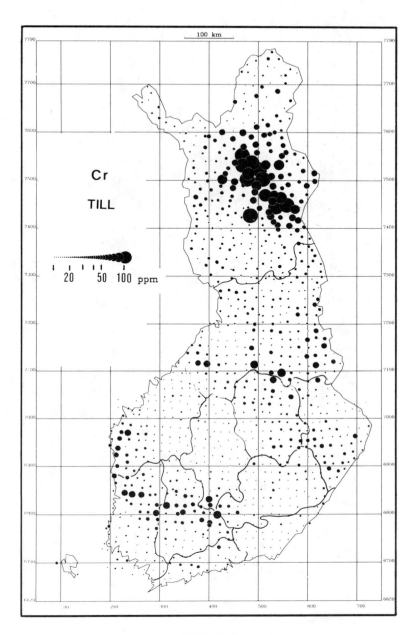

FIGURE 2A.

FIGURE 2. The distribution of Cr in till (1057 samples, -62 μm fraction leached with AR and analyzed by ICAP; data from the Geological Survey of Finland) and cultivated soil (the values are arithmetic means of 2002 samples presented separately in the indicated areas, -2 mm fraction leached with 0.5 N HAc $+$ 0.5 N NH$_4$Ac $+$ 0·02 N EDTA and analyzed by AAS[20]).

bedrock areas in northern Finland, although there are number of areal irregularities (Figure 3). Particularly in stream sediments, the occurrence and precipitation of Fe and Mn depend on pH and Eh conditions and humus content more than on their availability in primary mineral matter. The higher average Fe and Mn contents in northern Finland are presumably due to the abundance of humus matter in acidic soils. The same applies to cultivated soils.

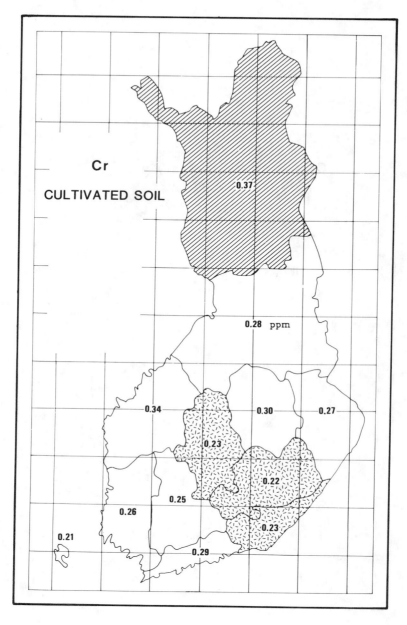

FIGURE 2B.

Aluminium, the most common metal in nature and a primary constituent of silicate minerals, is a largely debated element in terms of its toxic properties to forest stands, fish stocks, and man. Although ubiquitous, its availability to plants depends on its pH-controlled mobility and speciation in soil solutions. The strong acid-soluble Al contents in till show distinct regional differences, a factor that may prove important when evaluating its toxic effects on forests as a result of acidification. As to the effects on the health of higher animals and man, except fish, we do not yet know enough about the pathways, speciation, and possible deleterious impact of Al derived from food and drink.

The principal nutrients, *calcium* and *magnesium,* occur in abundance in silicate and carbonate minerals and show consistent regional distribution patterns on the geochemical

TABLE 1
The Arithmetic Means of Partial and Total Element
Concentrations in the Fine Fraction (-62 μm) of Till

Element	Dimension	ICAP-partial	ICAP-total	NAA
Fe	%	1.7	3.3	3.6
Al		1.3	7.4	—
Mg		0.42	1.0	—
Ca		0.17	1.8	—
K		0.17	2.1	—
Na		0.02	2.2	2.3
P	ppm	622	695	—
Mn		166	516	—
Zn		32	83	87
Cr		28	70	91
Cu		21	24	—
Ni		17	28	37
Co	ppm	7	15	12
Pb		3.2	162	—
As		2.6	9.6	—
Mo		0.2	10.3	1.1
U		20	32	3.4

Note: ICAP-partial = hot HNO_3-HCl (AR) leach; ICAP-total = HF + $B(OH)_3$ fusion, both analyzed by inductively coupled argon-plasma spectrometry (ICAP). NAA = total contents, analyzed by instrumental neutron activation. Composite samples (1057) collected for the Geochemical Atlas of Finland by the Geological Survey.

till, stream sediment, and stream water maps. The highest concentrations are met in areas where mafic, ultramafic, and, in some places, carbonate rocks prevail. The geochemical till and stream sediment maps, and even the heavy minerals in these sediments, seem to reflect relative differences in Ca and Mg reserves in natural soils, regardless of the large concentration differences (five to ten times) between the materials. The cultivated soils richest in Ca and Mg are in southern and southwestern Finland.[20] The clay-rich arable soils typical of the subaquatic coastal areas are naturally richer in Ca and Mg than are the silty and humus-rich soils in the eastern and northern parts of the country. The sporadic occurrence of metamorphic carbonate rocks or, in parts of the southwestern coastal area, the Paleozoic limestone material transported by the continental ice sheets from the Bothnian Sea basin cannot be ruled out either. Effective liming has also undoubtedly increased Ca and Mg concentrations.

Calcium is a necessary element for plants, animals, and man. Because Finnish surficial deposits derived from igneous and metamorphic rocks are poor in Ca, knowledge of regional element differences have both agricultural and silvicultural implications. Since the Ca intake by man in Finland, particularly through dairy products, is among the highest in the world, the element distribution in nature does not play a significant geomedical role. Magnesium is more intricate as its deficiency is suggested to influence the incidence rate of cardiovascular diseases. Although no indisputable epidemiological proofs are available, similar relationships have been proposed in Finland.

Potassium, which is a common element in silicate minerals and clays, behaves differently from alkali earths. On the geochemical till and stream sediment maps, the highest K contents occur variably in granitoid areas and in schist areas that include mica schists and mafic rocks. Consistent areal distribution patterns are not seen in geochemical materials either.

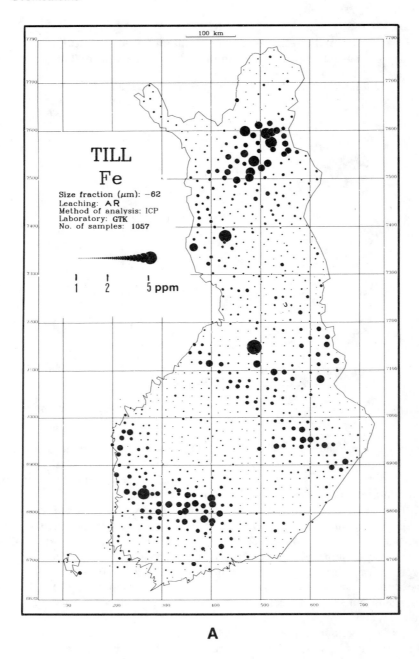

A

FIGURE 3.   Distribution of partial and total concentrations of Fe in till analyzed from AR leach and HF + B(OH)₃ fusion by ICAP. For more information, see texts on the map. (Data from the Geological Survey of Finland.)

The differences are largely due to the adsorptive properties of the element concerned in relation to colloidic clay and humus fractions and to Fe, Mn, and Al hydroxides. Consequently, geochemical maps of potassium do not have a good predictive value in nutrient and geomedical surveys. Neither is there a known relationship between geochemical K distribution and bioavailability and human health.[5]

Sodium is one of the most common elements in silicate minerals and the most abundant

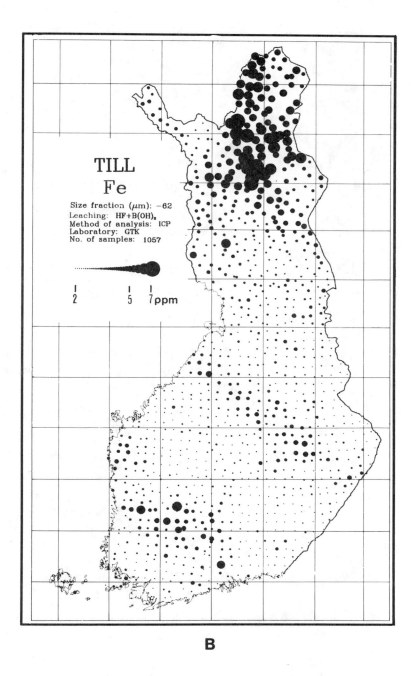

TILL
Fe

Size fraction (μm): −62
Leaching: HF+B(OH)₃
Method of analysis: ICP
Laboratory: GTK
No. of samples: 1057

| | | |
| 2 | 5 | 7 ppm |

B

alkali metal in soil solutions. This element is more conservative than K and is not adsorbed readily. Na is not considered as important a plant nutrient as K, probably partly because it is more ubiquitously available in soil water. The geochemical factor does not play a significant role for higher animals and man either, although Na is the predominant extracellular cation. It is easily adsorbed from water and is a common salt in food.

Phosphorus content in soil is a very important factor for plant growth, but it seems to have hardly any direct effect on man's health. It occurs as a major component only in a few minerals, e.g., apatite, which is considered to be the main source of the element. Phosphorus has a distinct geographical distribution (Figure 4), although the highest concentrations occur

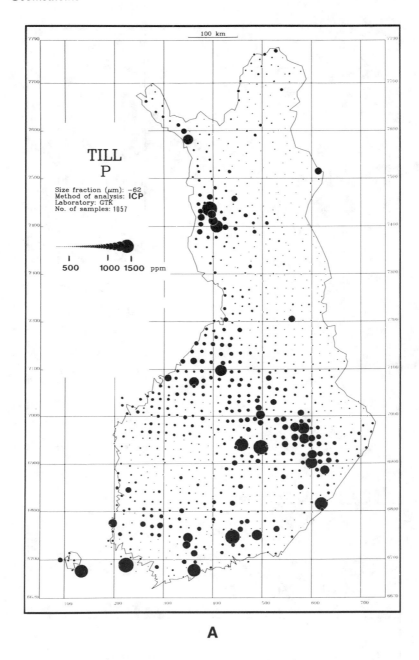

FIGURE 4. The distribution of P according to data from the Geological Survey of Finland and the Nordkalott Project.[4] For materials and analytical methods, see texts on the respective maps.

in areas of widely different bedrock compositions (granites and granodiorites, migmatitic gneisses intermingled with granitic veins, and mica schists). Roughly similar areal trends are seen regardless of the geochemical material, such as till, minerogenic stream sediments and their heavy minerals (mainly apatite), but not the stream organic matter. Although the mean concentrations differ by a factor of two to ten in different geochemical materials, geochemical maps seem to be good sources of information when evaluating the regional

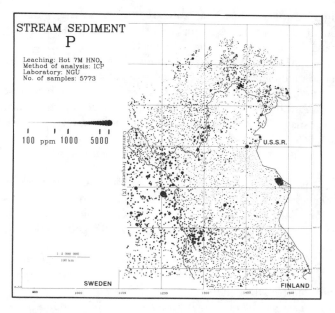

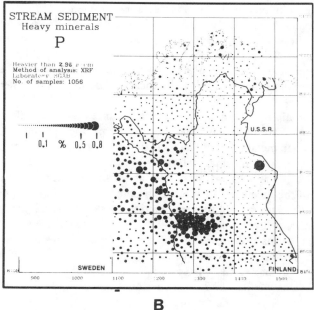

B

differences in P reserves in natural forest lands. The highest P contents in cultivated soils occur in the northern half of the country, where the soils are the most acidic (see Reference 20). The richness in humus may well explain the occurrence of the above-average P contents in these soils, as it does the abundances of Fe and Mn.

Apart from Fe and Mn, the most important heavy metals are *zinc, copper, nickel, chromium,* and *cobalt.* Zn, Cu, and Ni occur in sulfide minerals and also as accessory elements in silicates. Most of the elements concerned go into solution during weathering, but their mobility is reduced by their strong tendency to become concentrated in hydrous Fe-Mn oxides, clay, and the humus fraction. Although ubiquitous in minor and trace concentrations in different geochemical materials, they show roughly similar areal distribution patterns (see Figure 2). The schist belts composed of mafic and ultramafic rocks and the sulfide-ore potential belts appear as anomalous areas for most of the heavy metals. All the above heavy metals seem to be important for metabolic processes in plants and animals, including man, in very low concentrations, although their role has not yet been satisfactorily defined. Another common feature is that they act as toxicants in high concentrations, although no such effects have been recorded in Finland. To complicate the picture still further, their effect may be modified by their mutual interaction. Thus, for example, additional amounts of Ca, Cu, and Cd may inhibit Zn adsorption in animals and man. Furthermore, extensive interaction between Cu, Zn, Fe, Mo, and Pb in various animal systems suggest that imbalance of these elements in the diet may lead to various deficiencies or excesses (see Reference 5).

As a whole, it is not unexpected that plants and ruminants are more susceptible to the effect of trace elements than man. This effect depends strongly on the speciation and the fixation mechanism of the elements concerned in soil material, especially in the humus and clay fraction and Fe and Mn hydroxides. The effect of the geochemical background is not generally very significant.

Molybdenum occurs in igneous and metamorphic rocks, mainly as disseminated sufides, with the highest concentrations in granitoids and various schists. Molybdenum is released during weathering and oxidized into mobile molybdate ions. It tends to be adsorbed on sesquioxides and clay minerals and, as it behaves differently from all other trace metals is most soluble in an alkaline environment (see, e.g., Reference 18). Molybdenum plays a major role in enhancing plant growth, although its role in animal and human metabolism and health is not fully understood. The geochemical maps do not reveal any consistent anomalous geochemical regions. Only in heavy minerals do the anomalous Mo contents seem to be concentrated in granitoid areas, while till and minerogenic stream sediments show higher contents in silicic and subsilicic rock areas without any systematic distribution pattern.

In moderate or high concentrations, *arsenic* is an environmentally toxic element for plants, animals, and man, although it has recently been suggested that in small amounts it is beneficial for human metabolism. In an oxidizing environment, As is released from sulfide minerals, such as arsenopyrite, and transformed into arsenates, which behave like phosphates strongly adsorbed by hydrous Fe oxides. The long-distance atmospheric transport of As is a marked source of the metal, which tends to become enriched in the upper part of the soil profile by humus material. It has also been found that As contents are enriched in overburden composed of primary bedrock.[19] The geochemical maps show the highest concentrations in soils derived from schists rich in mafic and ultramafic components. The anomalous areas represent potential regions for environmental toxication in the event of increasing acidification. So far, no arsenotoxic cases have been reported in this country.

Lead is unanimously considered as an environmental toxicant, whose pathways leading via the environment to man are reminiscent of those of Cd. In rocks, Pb is available in sulfides, but it can also replace K, Ba, Sr, and even Ca in silicate minerals, e.g., feldspars, which may be the primary mineral source.[18] Currently, anthropogenic Pb-containing aerosol derived from automobile exhausts is the primary source of the metal in humus- and sesquioxide-rich top soils, which readily adsorb heavy metals. Elevated Ph contents are detected in soils and plants at a distance of some tens of meters alongside very busy roads. The multi-source origin of Pb is a confusing factor, contributing to the irregular and unsystematic distribution of Pb on geochemical maps. Lead is likely to be fixed by clays and sesquioxides as inactive compounds which, however, may be released by strong environmental acidification. Therefore, there is increasing concern that lead may pose a threat to the health of man in the future.

Cadmium is considered as an environmentally toxic element since it tends to become enriched in the food chain. In natural soils, it derives mainly from Zn and Pb sulfide ore minerals, whereas in cultivated soils it may be a component in fertilizers. Anthropogenic sources locally raise Cd concentrations in soil and plants, although it is of more minerogenic origin than Pb, another environmental toxicant. Long-distance aerial transport and consequent minor Cd fallout cannot be ruled out.

Due to their common source in sulfide minerals and to their chemical similarity, Cd generally correlates with Zn. It behaves like most of the heavy metals in that it tends to become adsorbed in the humus-rich top soil, which may be more effectively leached during acidification. The metabolic pathways are reminiscent of those of lead.[6] Some of the plants adsorb Cd effectively (see Reference 10). Particularly Zn, but also some other elements such as Ca, Cu, and Se, tends to inhibit the Cd uptake of plants. Geochemical mapping does not reveal any conspicuous anomalous areas, and no health hazards induced by elevated Cd contents in soils, plants, animals, or in man have been reported in this country.

Another widely discussed environmental toxicant is *mercury,* which has the ability to become enriched in the food chain, especially in predatory fishes and, at the end of the chain, in man. Mercury is ubiquitous in the environment, including the atmosphere, in very minor concentrations. Because of its relatively high vapor pressure at normal temperature, there is a global Hg cycle through air to soils, water, sediments, and biota enhanced by the anthropogenic input of the element.[7] The main primary sources of Hg are weathering sulfide minerals (their natural sinks), possibly the degassing of the earth's surface, and on a global scale, volcanic activity.

Although absolute Hg concentrations in soils and waters are extremely low, the strong affinity for humic and fulvic acids leads to the accummulation of this toxic metal in organogenic and minerogenic bottom sediments in lakes and reservoirs, and ultimately in predatory fish stock. In this way, humus acts as a sink or vehicle when dissolved and transported in water. The crucial reaction affecting the abundance of available toxic Hg in biota and in the food chains is the formation of methyl mercury, in all likelihood in the topmost layer of bottom sediments. The acidification of lakes, i.e., the lowering of the pH level, promotes Hg fixation to humus in sediments, contrary to the behavior of heavy metals in general.

Some trace elements, e.g., Zn and Se, have been found to have the antagonistic effect of reducing Hg toxicity in biota.[15,16] Although still a matter of concern, the most acute danger for health is over since the use of organomercurial seed dressings and disinfectants in agriculture and antifungal mercurious slimicides in the pulp and paper industry has been reduced or altogether terminated. Since the residence time of Hg in biological end members is, however, quite long, the deleterious effects may continue.

Selenium is one of the most debated environmental elements with implications for animals and man. Although it is beneficial in small amounts, it is toxic in high concentrations. In igneous rocks, it replaces S in sulfides, from which it is released during weathering as highly soluble selenites. In contrast to sulfates, selenites have a strong tendency to become adsorbed on Fe hydroxides and clay minerals. It was found in Norway that soil Se is related to precipitation and thus derives partly from marine and distant natural terrestrial or anthropogenic sources.[17]

The concentrations of Se in soil worldwide vary from very low to high. Consequently, the geographical pattern of Se leads to deficiencies or surpluses and poisonings in animals and sometimes also in man. Because of the very low natural concentrations in soil, Finland is one of the deficient areas. Until the middle of this decade, intake of Se was too low. This is one of the best examples of the importance of geochemical sources of essential elements and of the interrelationships between the geochemical distribution and the animal and human intake of an element. Because Se has been added nationwide to domestic fertilizers since the middle of 1985, the formerly meager Se intake has risen considerably and is now at an adequate level.

Iodine is a classic example of the indisputable geomedical importance of an element for animal and human health. Although I is one of the most widely distributed elements in nature, failure to supply it to the diet in adequate amounts results in a simple endemic goiter. Natural concentrations in Finland are extremely low in soil and waters and derive over-whelmingly from sea spray through rainfall. Iodine is adsorbed readily on clay and humus in soil and sediments.

The natural I intake was very low in Finland until the 1950s, resulting in a high prevalence of goiter. This has now been wiped out as there are many I sources for the diet, such as iodinized common salt and imported food items rich in I, e.g., sea water fish. The present I intake is four to five times higher than previously.[21]

C. HYDROGEOCHEMICAL MAPPING OF GROUND WATER AND SURFACE WATER

Soil water, shallow ground water, and surface waters draining small catchments are geochemically closely related. Hence, it is natural to include water facies in the scope of geomedical evaluations.

The hydrogeochemistry of shallow ground water in rural areas was studied by the Geological Survey of Finland by collecting some 5900 samples from springs and captured springs and from wells dug into overburden and drilled into bedrock. The most typical aquifer is a comparatively thin layer of glacial till yielding meager amounts of water, enough only for private households. An estimated 20 to 22% of the population in rural areas of Finland still rely on their own water supply. Since, for economic reasons, the public water utility networks cannot cover all the most remote farms and dwellings and, further, since the number of isolated summer cottages is increasing, private water wells will retain their importance, including all the constraints on quality they involve.

The National Board of Waters is responsible for supervising the quality of water supplied by public water utilities. Roughly half of the water delivered is ground water drawn from glaciofluvial sand and gravel deposits, which cover only 3 to 4% of the total area of this country. The water is generally of good quality, although the excess amounts of humus (high COD values), Fe, Mn, and F (the last component only in rapakivigranite areas) pose minor problems in some places. The pH level is sometimes also too low, making it necessary to alkalize the water. The quality of water in private wells varies more than the set water quality standards and frequently exceeds the abovementioned quality parameters. NO_3 contents in particular are alarmingly high in many private, deficiently protected and maintained wells.

As to the properties and dissolved components in drinking water considered to have adverse health effects, only NO_3, F, and Rn exceed the set safe limits in some places in Finland. The most detrimental are *nitrates*, which in 12% of the private wells studied exceeded the set limit of 30 mg/l NO_3; in the public water utilities, the corresponding figure was only 1%. The low background concentrations in soils and shallow ground water derive from anthropogenic fallout and from metabolic processes and decay of plants. The high NO_3 contents are solely the result of point-source biological pollution caused by sewage, fertilizers, or other man-induced N-containing compounds.

Fluorides are good examples of chemicals with a significant geomedical impact on animals and man. Fluoride is a typical lithophilic element, the most important sources of which are granitoids, particularly rapakivi-type and related K-rich coarse-grained granites. It can also derive from phosphate fertilizers and from local or regional airborn anthropogenic material from industry and the burning of coal. Fluoride has the affinity to replace hydroxyl groups in hydroxyl-apatite, micas, and amphiboles, which, together with fluorite, are the most important F sources in soils. Liming or abundant natural Ca can impede F mobilization by forming slightly soluble CaF_2. Mobile F can also be partly sorbed by phosphates as

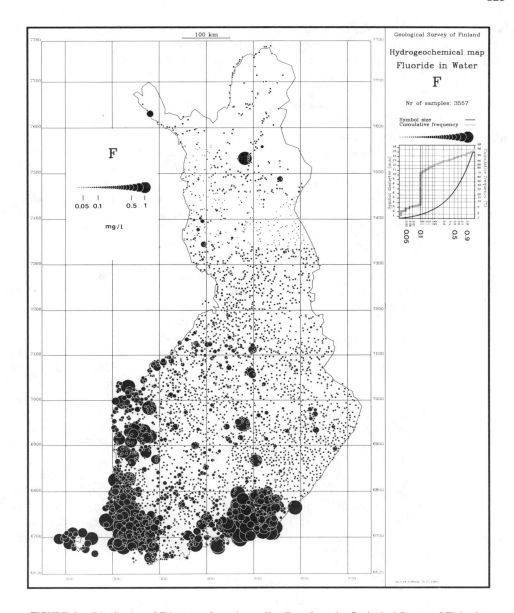

FIGURE 5. Distribution of F in water from dug wells. (Data from the Geological Survey of Finland.)

fluoroapatites. Clays can exchange F and OH reversibly at crystal edges, depending on the occurrence of complexing agents such as Al, Ca, P, Fe, and Si and other factors. Therefore, F tends to become associated and concentrated to a certain extent in clays. Mobility is generally enhanced by a reduction in the pH level.

Centuries ago, symptoms of F toxicity to animals were observed after volcanic eruptions in Iceland.[10] As a result, the participation of plants consumed by domestic animals as sinks of fluorides was also recognized. Although F may not be essential for plant metabolism, some plants (e.g., tea) can accumulate fluorides. Plants are mediating pathways of F from soil and soil water to higher animals and man. Fluoride is, however, an exceptional element in that the bulk of it in a F-anomalous environment is taken up through drinking water.

All types of ground water in the rapakivigranite areas have elevated F contents sometimes exceeding the recommended upper limit of 3 mg/l for drinking water (Figure 5). There are no records, however, of the deleterious health implications of the rare elevated F contents. The average 1 to 2 mg/l F concentrations are found to reduce the incidence of dental caries. On the other hand, in an overwhelming part of the country, F contents in water are below 0.1 mg/l and hence F intake is too low.

Since *radon* is a disintegration product of uranium, there is a strong similarity between the areal distribution of U in ground water sampled from drilled bedrock wells, U in till soils, the total exposure rate of natural radiation measured at the ground surface, and Rn in ambient house air (Figure 6). The Rn contents are most often high in the U-anomalous areas. Although the radiation exposure caused by Rn in water and house air may reach considerably high values and is considered as a health hazard, there is no indisputable evidence of harmful health consequences, e.g., an increase in cancer incidence.

Finnish ground waters are generally very soft and slightly acidic. This is not inevitably beneficial since some scientists consider moderately hard drinking water to be more favorable for health than soft water. Further, acidic CO_2-rich waters are aggressive dissolvers of metal tubing and fittings. This may occasionally increase the concentrations of heavy metals such as Zn, Cu, Ni, Pb, and Cd, which under natural conditions are very low, mostly below the analytical detection limit. The general acidification of the environment also poses a threat to the quality of ground water by increasing the concentrations of heavy metals and dissolved aluminium in waters. So far, however, no indisputable acidification processes in ground water with harmful implications have been observed in Finland, although the Al abundance in some acidic ground waters (below pH 5.5) may exceed 500 to 1000 µg/l. Acid ground water may result from several factors, e.g., oxidation of sulfide minerals, introduction of humus-rich overland flow into wells, and local biogenic pollution. Acidic fallout may have an effect on very shallow ground waters in sand and gravel deposits near the ground surface, and even more on outcropping ground water. The acidification-induced elevated Al concentrations in surface waters have an indisputable deleterious impact on fish populations, whereas there is no proof of the effect on the health of man of aluminous potable water.

Finnish shallow ground waters are frequently rich in dissolved humus components, which in chlorination may form chlorinated organic compounds with potential mutagenic effects. The mutagenesis is now considered one of the most serious quality problems, particularly in chlorinated surface waters, which are among the most humus-rich waters in the world. This is shown by the Nordkalott survey of stream waters, in which the occurrence of Fe and, to a lesser extent, of Mn and Al is associated with the occurrence of dissolved humus in water which, in turn, is really related to the peatlands (Figure 7). Humus is also a vehicle for many heavy metals in water.

III. SUMMARY

Elements and nutrients that are either essential, apparently nonessential, or toxic are ultimately derived from the geochemical environment. As higher animals and man depend directly or indirectly on plants, knowledge of the interrelationships between the geochemical environment and vegetation is vital for geomedical considerations. The bedrock-soil-water-air-plant-animal-man chain involves complicated mechanisms of element migration, which are still largely unknown.

Extensive geochemical mapping of various materials was carried out in northern Fennoscandia and all Finland principally for ore exploration purposes. The most useful materials applied to environmental and geomedical evaluations were glacial till or forest soil and the ground water used by the rural population. The main constraints in the application of pedogeochemical mapping data to geomedicine are the comparatively great sampling depth (C horizon of podsol profile) and aggressive leaching methods (AR) or total dissolution by HF + $B(OH)_3$. These hamper comparisons with material treated with the more specific

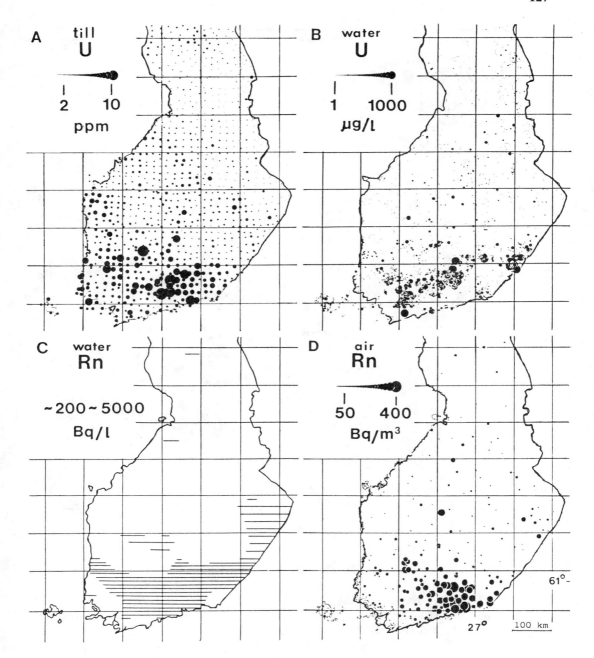

FIGURE 6. Distribution of U (A) in soils and (B) in water from drilled bedrock wells; Rn (C) in water from drilled bedrock wells (data from the indicated areas only, the Geological Survey of Finland), and (D) in ambient indoor air (8149 dwellings in 235 localities; data from the Institute of Radiation Protection). The exposure rate of natural radiation depicts a similar distribution pattern (not shown).

methods (NH_4Ac, HAc, or EDTA) used widely in nutrient studies of forest and cultivated soils. Furthermore, cultivation and animal husbandry are increasingly profiting from appropriately balanced fertilization and the addition of nutrients to fodders, thus diminishing the effect of the geochemical background, let alone the impact on man, who is influenced mainly indirectly at the end of the food chain.

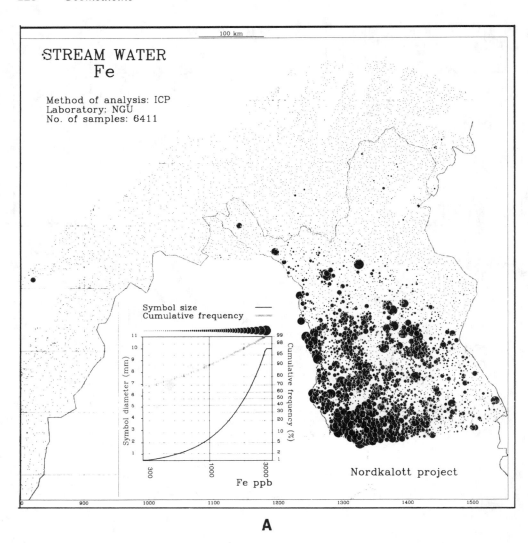

FIGURE 7.   Distribution of Fe in stream water (data from the Nordkalott Project[4]) and peat deposits.[8]

Although direct at-site geomedical connections are few in developed countries, they are very profound in developing countries, where an increasing majority of mankind is living and where the population is mainly fed by local products. Unfortunately, extensive geo-chemical and epidemiological records are seldom available in poor countries where people live virtually at subsistence level. Hence, the behavior of elements and their effect on plants and animals, including man, is considered in this country to the extent the information available allows.

Regardless of the diversified sampling, preparation, and analytical methods employed, different geochemical materials (till, minerogenic and organogenic stream sediments, heavy minerals, stream moss and stream water, and ground water) show similar regional distribution patterns with respect to many elements. Mapping results from till and cultivated soils also show some broad similarities, particularly in the north and east of the country, where fields are largely composed of till, small local silt and clay patches, and shallow peat deposits.

Several tens of elements are considered essential or beneficial for higher plants and

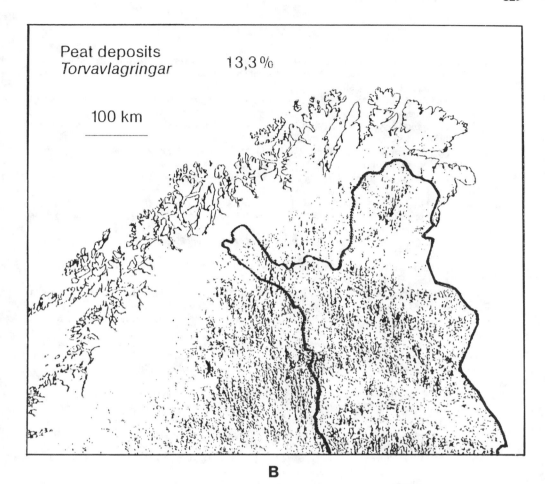

Peat deposits
Torvavlagringar   13,3 %

100 km

B

animals, including man. In this context, the following elements were briefly discussed: Co, Cr, Fe, Mn, Al, Ca, Mg, K, Na, P, Zn, Co, Ni, Mo, As, Pb, Cd, Hg, Se, I, F, U, and Rn. The most indisputable interdependencies between elements in geochemical and biological environments are shown by Se and I, the concentrations of which in Finnish soil and water are extremely low. The effective addition of these elements to fertilizers and the diet have, however, altogether eliminated their deficiencies in animals and man.

There are no elements in Finnish soils known to have toxic implications for the biological environment except for the ubiquitous Al released into soil water and surface water as a result of acidification. A toxic effect on fish stock is evidenced, but possible retardation of tree growth due to Al and its effect on humans through drinking water are still debated. Fluoride is the only component obtained by man in adequate abundance from drinking water in that part of the country where bedrock and soil are composed of rapakivi granites. In the bulk of the country, the F intake is far too low. U and Rn in water pumped from bedrock wells drilled in granites and migmatites intermingled with granitic veins are a possible cause of health hazards, while Rn in ambient house air, mainly seeping from the ground, is the most deleterious component.

REFERENCES

1. **Alestalo, A., Elomaa, E., and Koistinen, O. A.,** Mineral element contents of composts and bark ashes compared with contents in plants, in Mineral Elements 80', Hanasaari Culture Center, Helsinki, Espoo, Finland, December 9 to 11, 1980, 11.
2. **Bowie, S.H.U. and Thornton, I.** 1985. *Environmental Geochemistry. Report to the Royal Society's British National Committee for Problems of the Environment,* D. Reidel, Dordrecht, 1985.
3. **Björklund, A. and Gustavsson, N.,** Visualization of geochemical data of maps: new option, *J. Geochem. Explor.,* 29, 89, 1987.
4. **Bølviken, B., Bergström, J., Björklund, A., Kontio, M., Lehmuspelto, P., Lindholm, T., Magnusson, J., Ottensen, R. T., Steenfelt, A., and Volden, T.,** Geochemical Atlas of Northern Fennoscandia, Scale 1 : 400 000. Geological Surveys of Finland, Norway and Sweden. A project supported by Nordic Councils of Ministers, 1986.
5. **Crounse, R. G., Pories, W. J., Bray, J. T., and Mauger, R. L.,** Geochemistry and man: health and disease. I. Essential elements, in *Applied Environmental Geochemistry,* Thornton, I., Ed., Academic Press, London, 1983, 267.
6. **Fasset, D. W.,** Cadmium, in *Metallic Contaminants and Human Health,* Lee, H. K., Ed., Academic Press, New York, 1972, 97.
7. **Goldwater, L. J. and Clarkson, T. W.,** Mercury, in *Metallic Contaminants and Human Health,* Lee, H. K., Ed., Academic Press, New York, 1972, 17.
8. **Hamborg, M., Hirvas, H., Lagerbäck, R., Minell, H., Mäkinen, K., Olsen, L., Rodhe, L., Sutinen, R., and Thoresen, M.,** Map of Quaternary Geology, sheet 1. Quaternary Deposits, northern Fennoscandia. Compiled at the Geological Surveys of Finland, Norway and Sweden and supported by the Nordic Council of Ministers, 1987.
9. **Hopps, H. C.,** Overview, in *Geochemistry and the Environment,* Vol. 1, *The Relation of Selected Trace Elements to Health and Disease,* National Academy of Science, Washington, D.C., 1974, 3.
10. **Kabata-Pendias, A. and Pendias H.,** *Trace Elements in Soils and Plants,* CRC Press, Boca Raton, FL, 1984.
11. **Kovalevskii, A. L.,** *Biogeochemical Exploration for Mineral Deposits,* Oxonian Press, New Delhi, 1979.
12. **Kubota, J.,** Soils and Plants and the Geochemical Environment, in *Applied Environmental Geochemistry,* Thornton, I., Ed., Academic Press, London, 1983, 103.
13. **Kurki, M.,** On the fertility of Finnish tilled fields in the light of investigations of soil fertility carried out in the years 1955-1980. (In Finnish), Helsinki, 1982.
14. **Lahermo, P. W.,** Hydrogeochemistry in geomedicine, in *Geomedical Research in Relation to Geochemical Registrations,* Låg, J., Ed., Universitetsforlaget, Oslo, 1984, 27.
15. **Leskinen, J., Lindqvist, O. V., Lehto, J., and Koivistoinen, P.,** Selenium and mercury contents in northern pike (Esox lucius, L.) of Finnish man-made and natural lakes, *Water Res. Inst.,* 65, 72, 1986.
16. **Lindeström, L. and Grahn, O.,** Antagonistic effects to mercury in some mine drainage areas, *Ambio,* 11(6), 359, 1982.
17. **Låg, J. and Steinnes, E.,** Regional distribution of selenium and arsenic in humus layers of Norwegian forest soils, *Geoderma,* 20, 3, 1978.
18. **Norrish, K.,** Geochemistry and minerology of trace elements, in *Trace Elements in Soil-Plant-Animal Systems,* edited by Nicholas, D. J. D. and Egan, A. R., Eds., Academic Press, New York, 1975, 55.
19. **Salminen, R., Hartikainen, A., and Lestinen, P.,** Enrichment of As and S in the finest fractions of crushed rock samples — an experimental study simulating the formation of till, *J. Geochem. Explor.,* in press.
20. **Sippola, J. and Tares, T.,** 1978. The soluble content of mineral elements in cultivated Finnish soils, *Acta Agric. Scand. Suppl.,* 20, 11, 1978.
21. **Varo, P., Saari, E., Paaso, A., and Koivistoinen,** Iodine in Finnish foods, *Int. J. Vitam. Nutr. Res.,* 52(1), 80, 1982.
22. **Webb, J. S.,** Regional geochemical reconnaissance in medical geography, *Geol. Soc. Am. Mem.,* 123, 31, 1971.

Chapter 7

MEDICAL REGISTRATION AS A BASIS FOR GEOMEDICAL INVESTIGATIONS

Eystein Glattre

TABLE OF CONTENTS

I. INTRODUCTION

Geomedicine is defined as the science dealing with the influences of ordinary environmental factors on the geographical distribution of health problems in man and animals. This makes geomedicine a branch or subdiscipline of epidemiology[1] and the geomedical investigation an epidemiological study exclusively concerned with the ordinary environment as a disease determinant. The roles played by environmental factors in the etiology and pathogenesis of disease can be studied by relating environment to disease by means of proven epidemiological methods. Access to relevant, adequately observed and recorded health data of a study population is essential for this kind of investigation.

II. FEASIBILITY

A feasibility evaluation consisting of a set of main questions and answers should always be a prior part of a geomedical investigation and starts with evaluation of the hypothesis on which the desire for an investigation is based. An operational, precisely worded hypothesis based on well-defined and accepted concepts is an ideal point of departure for scientific research because it is almost a condensed blueprint of the design of the study. A hypothesis can, however, be insufficient in one or more respects; for example, a hypothesis which claims that environment is important for life is neither precise nor operational and is almost a barrier to the scientific accomplishment of a geomedical investigation. It is, as a rule, possible to strengthen the formulation of an insufficient hypothesis by specifying the environmental factors and morbid aspects in question and their contended, causal relationship in order to facilitate the choice of study design.

A special, geographical incidence pattern for thyroid cancer in Norway was first described by Pedersen in 1956.[2,3] He found higher incidence rates, especially for the papillary carcinoma type, in northern Norway and along the coast and lower rates in southern Norway and inland.

The first national cancer incidence atlas in the world, the Atlas of Cancer Incidence in Norway 1970-1979,[4] was published in 1985. The maps of thyroid cancer seemed to show an incidence pattern not unlike that described by Pedersen for the period 1953-1966, so it became the first subproject for the Norwegian Thyroid Cancer Project (NTCP), which was established in 1985, to investigate whether this pattern persisted into the 1970s. The hypothesis that the incidence pattern for thyroid cancer would persist into the 1970s, is operational and was therefore the point of departure for selecting the study design. A longitudinal study type was chosen in which all new cases of thyroid cancer in the approximately 450 Norwegian municipalities during the period 1970-1979 were observed by sex, histological types and exposure level, i.e., municipality groups aggregated according to latitude, municipal trade type and whether on the coast or inland.

Feasibility also depends upon the choice of study design. Different designs have different features which make them more suitable for some problems than others. There are cross-sectional and longitudinal study types, characterized by whether the ascertainment of exposure and disease relate to one or more points in time. Longitudinal studies are mainly of two types, cohort and case control. In cohort studies, the population is defined on the basis of exposure and the morbidity or mortality of a specified health defect is observed. It is evident that there must be a long observation period for certain categories of health defects, e.g., cancer. The case and control groups are defined on the basis of disease presence and absence, respectively. Cases and controls are observed retrospectively with regard to exposure.

Cohort design provides risk estimates for different levels of exposure. It is time-consuming and therefore economical only when the health problem in question is relatively common. The case-control approach, on the other hand, is preferable when the problem occurs less often. It is less resource demanding and produces answers faster and more economically than the cohort approach. But it has the drawback that population frequencies are not directly deducible. Also, information about exposure to environmental factors is often lacking or defective in ordinary medical records. This makes the acquisition of high-quality exposure data very difficult. It is sometimes possible, though, to estimate exposure indices on the basis of indirect evidence. In cross-sectional studies, observations of exposure and health problem occurrence are made at the same point in time in the defined study population. This design is easier and more economical than the longitudinal type, but it is restricted to permanent or semipermanent conditions.

In the NTCP study mentioned above, verification of the hypothesis showed that the north-south and coast-inland incidence gradients persisted into the 1970s, especially in the case of papillary carcinoma. It also showed that the incidence of papillary carcinoma was highest in fishing municipalities.[5] The cohort study was accomplished with a minimum of manpower and for a low direct cost because it was done retrospectively as an intramural Cancer Registry project.

Available resources and project management are of great significance when assessing the feasibility of a project. In any study there is a certain minimum need for manpower and money below which the study becomes impossible. Good planning and project management must determine this minimum and make sure that it, at least, is available before launching the project. Manpower resources should, if possible, be selected with emphasis on creativity, endurance, and efficiency. It should also be stressed that access to advanced, computational hard- and software is a necessity in all studies involving the use of complicated statistical procedures and data sets with more than a few hundred elements. The time needed to complete a study should always be carefully estimated because it has a major effect on the cost. The period during which an ad hoc project can be productive normally extends not much beyond 5 years. After that, internal problems connected with manpower and motivation tend to increase rapidly.

When geomedical studies are multidisciplinary or multiinstitutional, good management is very important. There must be well-defined leadership — an organization which facilitates execution of decisions and cooperation and which has clear lines of administrative and scientific responsibility. It may be possible on rare occasions to reduce expenses by joining another planned or on-going study. Such ''hijacking'' is possible if the ''hijacked'' study is easily readjustable so as to produce some or all of the desired answers. ''Hijacking'' can, on the other hand, conflict with requirements for good management since it may involve delegating authority without any corresponding, compensational reduction in scientific responsibility.

III. AVAILABILITY

The occurrence of a health problem in a study population can be based either on one's own observations or by utilizing available, relevant information from public or private institutions recording health data for some other purpose. Creating one's own health data from scratch is normally both expensive and time-consuming. It may therefore be crucial for geomedical research that sources of relevant medical information about the study population exist *a priori* and be available for external users.

Reporting of cancer has been compulsory in Norway since the foundation of The National Cancer Registry in 1952. The Registry, which since the beginning has aimed at recording all recognized cases of cancer in the Norwegian population, has now for many years been nearly complete with regard to solid tumors. Reports are required from all institutes of pathology and all hospital departments. In addition, the Registry's material is at intervals matched against the Cause-of-Death Registry in the Central Bureau of Statistics. Death

TABLE 1
Sources of Health Data Referring
to All or Groups of Inhabitants

Adverse drug effects
Births
Causes of death
Census information
Conscripts/veterans
Disability
  Blindness
  Deafness
  Injuries (sequelae)
  Malformations
Employers and employees
Frozen specimens of human tissue
Fertility
Hospital records
Morbidity
  Accidents
  Cancer
  Infectious diseases (notifiable)
    Cholera
    Plague
    Relapsing fever
    (Smallpox)
    Typhus
    Yellow fever
    Others
  Mental illness
  Tuberculosis
  Venereal diseases
  Hospital admissions
Pensioners
Residents (all inhabitants)
Trades
Twins

certificates mentioning malignant disease, but not in accordance with or not found in the database, are subject to further investigation. Throughout the history of the Cancer Registry, the percentage of histologically verified cancers has all the time tended to increase and is now 95 to 99% for the majority of cancer forms.

Geomedical investigators should make themselves familiar with the type and quality of all the main health data sources in their country. The existence and scope of such information systems frequently determine whether a geomedical investigation can proceed from the bare idea to the planning phase. It is essential, though, that data quality is good and that the persons involved are well grounded with regard to the health problem.

Table 1 gives a list of medical information systems that may be available to geomedical investigators as sources of health data about the population being studied. Some of these sources are permanent archives or registers, collecting data continuously or at intervals for administrative or other purposes. Others may be leftover health data files from terminated administrative or scientific projects, and some may be extensive, on-going studies of health problems in the society. Smaller information systems have not been included.

When considering a geomedical hypothesis, it is recommended that the process should start with a preplanning period devoted to searching health data sources that may have information which has a bearing on the hypothesis. This is normally done through libraries, official offices, and statistical institutions. The checklist shown in Table 2 may usefully be applied to this information.[6]

TABLE 2
Important Features of Health
Data Sources: Checklist

Purpose
Information donors
Input
Completeness
Computer equipment
Coding systems
Registered information elements
Output
Accessibility

The purpose of a health data source may serve to explain why it cannot provide the data needed for hypothesis testing. Knowing the *donors* of the information which has been registered often helps us to judge whether the identified source is sufficient or not.

The two most important groups reporting to the Cancer Registry are pathology laboratories and hospitals. In the case of cancer, the former group sends a copy of their reply to the requisitioning physician about the biopsy or autopsy, while the latter must fill out a form by means of which the Registry is notified. The general impression is that the former group is much more expeditious and reliable in their reporting.

This is profitably supplemented by the estimates of *completeness* made by the source institution. The types of *input* data give a general understanding of the quality of the information available from the actual source. Information about *computer equipment*, both hard- and software, and the *systems* used in *classification and coding* make clear what possibilities and limitations there are to data output and communication with the database.

The following list is an output of file elements on Kari Nordmann (fictitious person) recorded in the Cancer Registry:

Person Number	200100 VWX YZ
Death	010579
Site	1940
Date	02-70
Municipality	0301
Histology	8263
(stage)	
Metastases	D
Certainty of diagnosis	3
Basis of diagnosis	78
Age	70
Sex	2
Treatment	11 0000 30
Name	Nordmann, Kari

All Norwegian citizens have a national identification number. That of Kari Nordmann is 200100VWXYZ, where the first six digits represent day, month, and year of birth, VWX is the individual number, and YZ is the so-called check number.

This person died on May 1, 1979. Site 1940 means that the person had a tumor in the thyroid gland — according to our modified version of the International Classification of Diseases, seventh revision of 1955 — in February 1970 (date 02-70). At that time, she resided in Oslo (municipality 0301). 8263 for histology means that the tumor, according to the MOTNAC classification system, was a papillary carcinoma. Stage is not coded for

TABLE 3
(Example)
Checklist Information About the Cancer Registry of Norway

Purpose
Checklist Information About the Cancer Registry of Norway
  To collect information about all cases of cancer in Norway; to counsel the Central Administration with regard
    to the campaign against cancer; to carry out epidemiological cancer research; to cooperate with other research
    and health institutions
Information donors
  All Norwegian hospitals, all pathology departments, and the Central Bureau of Statistics
Input
  Reports on cases of malignant, premalignant, and certain benign tumors; all new cases registered since 1953;
    no deletions
Completeness
  Close to 100% for solid tumors; validity very high
Computer equipment
  ND-500; DDPP, GLIM, BMDP, and specific computer programs
Coding system
  ICD-7 (1955), MOTNAC (1968)
Registered information elements
  Personal identification number, name, municipality of residence, sex, date of diagnosis, clinical diagnosis,
    histological diagnosis, stage, basis for diagnosis, treatment, follow-up data about the cancer, date of death,
    causes of death, autopsy data, hospital or department, admission and discharge data
Output
  Research papers, information requested from external institutions or individuals, reports on incidence, geograph-
    ical variation, trend analyses, and survival
Accessibility
  Data normally available for research purposes on application; also applies to individually identifiable data if
    approval has been granted by the Data Control Office; conditions individually specified for each project

thyroid cancers, but D as a metastasis code means that the neoplasm was infiltrating neighboring tissues at the time of diagnosis. Certainty of diagnosis is 3, which means that malignancy as well as the specified site were regarded as certain by the Cancer Registry coders. Basis of diagnosis is coded 78. According to this, the diagnosis was verified histologically in 1970 and confirmed at autopsy in 1979. The person was 70 years old at the time of diagnosis and female (sex 2). Treatment code as well as the name of the patient is also part of the output.

Kari Nordmann's file also contains information about hospital stays, pathology departments, pathology specimens, and other primary malignancies.

Evidently, the database information elements attract most interest since they reveal whether the source contains the type of information which is required. Sometimes the routine output from the source includes all that is needed. When this is not the case, the source institution must be addressed about data delivery according to the rules regulating external users' access. An example of the use of the checklist (Table 2) is given in Table 3.

It is worth mentioning that not all health data sources can provide an external user with the required set of data unless he or she learns the system from inside or works in close contact with persons having this knowledge.

There are laws in most countries which regulate the access of research workers to person-identifiable health information and also the type of results which can be published from statistical analysis of such data. Knowledge of these laws and regulations is essential in geomedical research.

IV. SUITABILITY

It is the responsibility of the geomedical investigator to make sure (1) that the health

TABLE 4
International Classifications of Disease

1. International Classification of Diseases Vols. 1, 2, 1975 Revision (ICD-9), World Health Organization, Geneva, 1977, 1978.
2. International Classification of Diseases — *Oncology* (ICD-0), World Health Organization, Geneva, 1975, (Encl. International Classification of Diseases, Vol. 1 1975 revision.)
3. International Classification of Diseases, Ninth Revision (basic tabulation list with alphabetical index), World Health Organization, Geneva, 1978.
4. International Classification of Procedures in Medicine, Vol. 1, World Health Organization, Geneva, 1978.
5. International Classification of Impairments, Disabilities and Handicaps, World Health Organization, Geneva, 1980.
6. *International Classification of Health Problems in Primary Care* (ICHPPC-2-Defined), World Organization of National Colleges, Academies and Academic Associations of General Practitioners (WONCA), Oxford University Press, 1983.
7. *International Classification of Primary Care* (ICPC), Lamberts, H. and Wood, M., Eds., World Organization of National Colleges, Academies and Academic Associations of General Practitioners (WONCA), Oxford University Press, 1987.
8. Systematized Nomenclature of Medicine (SNOMED), Vols. 1, 2, College of American Pathologists, Skokie, IL, 1979.

data provided by a source really measure the effect described in the hypothesis and (2) that these data also satisfy quality requirements and possess the characteristics required by the design of the study. Decision problems rarely arise when these criteria are fully satisfied or unsatisfied. It is the many intermediate alternatives which pose the problems. It is, for example, not easy to decide if exposure data measured 5 to 7 years before diagnosis are acceptable in a case-control study on the etiology of cancer. Some of these problems can be circumvented by proper choice of statistical methods; others must be solved by judgement. In general, however, when in doubt, it is better to reject than accept.

Aggregate measures, like variables describing groups of persons rather than the persons themselves, have frequently been used in geomedical research. The use of these aggregate measures involves, however, the possibility of bias, nowadays often termed the ecological fallacy, and serious errors can result if the investigator applies his conclusions directly to persons within or across groups[7] without the necessary reservations. Another type of bias follows from the fact that health data may be subject to rhythmical variations by the day, month, and year, i.e., circadian, circa-lunar, and circannual rhythms.[8] These phenomena must, if present, be carefully considered.

Medical diagnoses constitute important informational elements in the health data sources which are listed in Table 1, where they almost always appear in a classified and coded form.

A list of widely used international classifications of diseases is shown in Table 4; the first five have been issued by the World Health Organization (WHO) in Geneva, two by the World Organization of National Colleges, Academies and Academic Associations of General Practitioners (WONCHA) and SNOMED by the College of American Pathologists. The WHO systems are mainly used in hospitals and for the coding of causes of death in member states, while the WONCHA systems are applied in primary medical care all over the world.

The International Classification of Diseases (ICD), which dates back to 1893, has been the responsibility of WHO since the late 1940s. The sixth revision of ICD was issued in 1948, the seventh in 1955, the eighth in 1965, and the ninth — and for the time being, the last — was issued in 1975. The tenth revision is expected to be adopted by the World Health Assembly in 1993. ICD-7 and ICD-8 are still in use in many source institutions around the world, but ICD-9 is now practically the only one used by the WHO member states in the classification and coding of causes of death.

ICD-9 consists of two volumes. Volume 1 contains the hierarchically designed, ex-

TABLE 5
(Example)
The Hierarchical Structure of ICD-9

Level	Category	Code
Chapter 4	Diseases of the nervous system and sense organs	
Section	Hereditary and degenerative diseases of the central nervous system	330—337
3 Digits	Cerebral degenerations usually manifest in childhood	330
4 Digits	Cerebral lipidoses	330.1
5 Digits	In certain chapters	

haustive disease classification system and also includes directions for use in the introductory chapters and definitions of the causes of death, rules for filling in death certificates, and rules for interpretation, classification, and coding of the causes of death in an appendix. These chapters are therefore recommended reading for geomedical investigators who are going to use ICD-9-coded material, especially data from cause-of-death registers. Volume 2 is an alphabetical index containing not only the terms used in Volume 1, but also many other synonyms and inclusion terms. It is provided with cross-references and references to codes in Volume 1.

Two aspects of ICD-9 are central for understanding how to use this classification system. There are five aggregation levels of diagnosis, as exemplified in Table 5. Most specified is the lowest level, where the disease categories have been provided with five-digit codes. Categories on this level aggregate to broader categories for each upward level, with chapters forming the top level. The five-, four-, and three-digit codes and the ICD-9 short lists are probably the most common codes used by health data sources all over the world.

ICD-9 recommends the principle of multiple coding in order to conserve as much diagnostic information as possible. It also contains a system of double coding. Approximately 150 categories in the international version are double coded, i.e., given one code, marked with a dagger, when classified according to etiology and another, marked with an asterisk, when classified according to manifestation. The possibility of double counting must therefore always be taken into account and eliminated.

From a global perspective, cause-of-death diagnoses are most probably the easiest available disease data which fulfil certain basic quality requirements. The reasons are that (1) cause-of-death registries exist worldwide, (2) they record the causes of death of all deceased in the study population according to recommendations given by WHO, and (3) they can provide all investigators (on conditions defined by their secrecy rules and regulations) with mortality data.

The basic concept is the underlying cause of death, defined as ''(a) the disease or injury which initiated the train of events leading directly to death, or (b) the circumstances of the accident or violence which produced the fatal injury.'' The underlying cause of death initiates, therefore, a sequence of causes terminating in what is called the direct or immediate cause of death. Therefore, the underlying cause of death can be determined by using the inverse cause-and-effect principle to identify retrospectively, cause by cause, the disease or diseases antecedent to the direct cause of death — stopping at the earliest, well-defined nosological unit. In cause-of-death sequences with more than two links, causes between the underlying and direct are termed intermediate (see Figure 1).

Cause-specific mortality statistics always refer, unless it is explicitly stated otherwise, solely to the underlying cause of death.

The death certificate also contains rubric II (see Figure 1) for notification of one or more contributory causes of death. ICD defines the latter type of cause of death as ''any other significant condition which unfavourably influenced the course of events, and thus contrib-

		Duration
I	a) Direct cause of death	
	b) Intermediary cause of death	
	c) Underlying cause of death	
II	Contributory causes of death	

FIGURE 1. International form of medical certificate of cause of death.

uted to the fatal outcome, but which was not related to the disease or condition directly causing death.'' Not infrequently, diseases generally regarded to be nonfatal appear on the death certificate as contributory causes of death. Since an increasing number of cause-of-death registries are recording more than one cause of death per person, it is theoretically possible to study nonfatal disease on the basis of data provided by a cause-of-death registry. There are, however, many pitfalls related to (1) rare disease occurrence, (2) low diagnostic validity,[9] (3) the nebulous and nonoperational definition of contributory causes of death, and (4) the lack of clarity of definitions and rules as to whether the notification of contributory causes of death is strictly compulsory — a fact that undoubtedly causes underreporting.

The frequency with which a nonfatal disease is mentioned on the death certificate gives an estimate of the prevalence of this disease in the deceased population. However, because of (1) through (4) above, many will hold that such estimates are worthless in research. While this may be valid for some diseases, it is not completely right under all circumstances. In a study of the association between aluminium in drinking water and the occurrence of senile and presenile dementia, annual mortality figures of 3.4 to 5.0 per 100 000 were found when only dementias as underlying causes of death were considered. The mortality figures for these dementias, if at all mentioned on the death certificate, were ten times greater, and this ratio tended to be the same for different regions.[10]

In this chapter, geomedicine has been regarded exclusively as a subdiscipline of epidemiology in which health variables always refer to a group or population. Medical registrations or health data have therefore been restricted to observations of this type of variable and do not include observations of lower levels of human and animal biology.

REFERENCES

1. **MacMahon, B. and Pugh, Th. F.,** *Epidemiology. Principles and Methods,* 1st ed., Little, Brown, Boston, 1970.
2. **Pedersen, E.,** Thyreoideacancerens fordeling i Norge, *Nord. Med.,* 56, 1108, 1956.

3. **Pedersen, E. and Hougen, A.,** *Thyroid Cancer in Norway,* UICC Monograph Series, Vol. 12, Hedinger, C., Ed., Heidelberg, 1969, 72.

4. **Glattre, E., Finne, T. E., Olesen, O., and Langmark, F.,** *Atlas of Cancer Incidence in Norway 1970-1979,* Norwegian Cancer Society, Oslo, 1985, 58.

5. **Thoresen, S. Ø., Glattre, E., and Johansen, Aa.,** Thyroidea cancer i Norge 1970-79, *Tidsskr. Nor. Laegeforen.,* 106, 2616, 1986.

6. Registerkatalog, 1st ed., National Institute of Public Health, Oslo, 1988.

7. **Piantadosi, S., Byar, D. P., and Green, S. B.,** The ecological fallacy, *Am. J. Epidem.,* 127, 893, 1988.

8. **Bünning, E.,** *The Physiological Clock:* Circadian Rhythms and Biological Chronometry, 3rd ed., Springer-Verlag, New York, 1973.

9. **Mausner, J. S. and Bahn, A. K.,** *Epidemiology. An Introductory Text,* W. B. Saunders, Philadelphia, 1974, 242.

10. **Vogt, T., Olsen, I., Østdahl, T., Fadum, E., and Glattre, E.,** Water Quality and Health, Central Bureau of Statistics, Oslo-Kongsvinger, 1986.

Chapter 8

POLLUTION AS A GEOMEDICAL FACTOR

H.-W. Scharpenseel and P. Becker-Heidmann

TABLE OF CONTENTS

I. INTRODUCTION: POLLUTION VS. GEOCHEMICAL NORMALITY

Pedogeomedicine, a subdivision of geomedicine, deals with the influence of soil-related, mainly geogenic environmental factors upon the regional distribution of health problems on man and animals. The term "pollution" is derived from the Latin verb *polluere*, soiling, staining, making dirty. Etymologically, pollution therefore implies being polluted by foreign sources of dirt or contaminants in an originally clean environment and ambiance.

Sensu strictu, geomedicine should address environmental factors — related regional health problems existing before anthropogenic influences could pollute, i.e., render the environment "dirty". For pedomedicine we have an ambivalent situation, in as much as soils, the pedosphere, being the product of mutual penetration of lithosphere, atmosphere, hydrosphere and biosphere, can be untouched by human interaction or can be conditioned by anthropic measures such as pollution. In the latter case one would however prefer to speak of contamination and required soil rehabilitation/decontamination rather than of a feature of pedogeomedicine.

Thus, leaning on semantic purism there is only a limited justification for a niche of pollution-induced pedogeomedicine. Taking a broad-minded position and expanding the concept of pedogeomedicine beyond the impact of geochemical/biogeochemical/hydrogeo-chemical influences also to the highly differentiated field of anthropogenic pollution creates a new situation. The breathtaking diversity of the anthropogenic pollution threatens to dwarf the truly pedogeomedicinal problem areas in the scope of the text, conceived to cover primarily pedogeomedical phenomena in the proper sense.

In general, the dilemma of definition and limitation of the term pollution in the light of geomedicine is quite obvious against Greenstein's reminder[15] that the elemental constitution of living organisms reflects more typically that of the whole universe with overwhelming H, O, N, and C contents than that of the planets and of our earth.

It is absolutely imperative to limit the coverage of features of pollution in the context of this book to those who have a link to geochemical sources and strong relevance to soil problems. For the lithosphere, large data collectives, such as those of Vinogradov[62] or Rankama and Sakama[63] for the micro- and macroelements of the earth, reflect "geochemical normality", i.e., "average". Marked deviations in toxicologically relevant elements from these "normal concentrations" in rock analysis would create a "geomedicinal case".

One should keep in mind the classic definition by Paracelsus from Hohenheim (1493—1541) "All things are poison, and nothing is without poison, the *dosis* alone decides that a thing is not poisonous". Acid soils lack in cation exchange capacity (CEC) and buffer capacity, or they are old, strongly weathered substrates with the bases of the sorption complex mostly replaced by protons (hydronium ions) or Al-ions. Thus, they are rarely strongly polluted by toxic elements except Al-ions (ferrolysis soils and Fe-toxicity soils, also Fe-ions).

On the contrary, soils derived from ultrabasic rocks, serpentinites, and ophiolites are often excessive in Cr, Ni, and Co concentrations by approximately 0.1 to 1% and sometimes even >1% (Table 1).[58,64] Other specific mineral associations, also mine cuttings downstream of the mining area, and emergence from the lithosphere into the pedosphere can pollute soils monotypically with metal or nonmetal elements at toxic levels. Reported pedogeomedical consequences of importance, brought about by such elemental excesses or deficiencies are listed in Table 2.

A good example of landscape scrutiny with regard to heavy metals as pedogeomedical background pattern and by substraction method as anthropic pollution was made in the city ecosystem of Greater Hamburg in northern Germany. The geochemical background was measured in three transects of several meters depth by Hintze.[19] A grid of sampling points

TABLE 1
Soils of serpentinites (ultrabasic) in the Philippines

Element	BUGALLON		NATIVIDAD		CAPUDLOSAN	
	H n = 20 %	L n = 19 %	H n = 22 %	L n = 22 %	H n = 5 %	L n = 5 %
Ca	3,7630	3,9668	3,2940	3,2794	1,4302	1,4324
Fe	6,6733	6,7180	6,7324	6,7378	7,4168	6,7106
Mn	0,1196	0,1734	0,1730	0,1414	0,1333	0,0792
K	0,0710	0,0616	0,3628	0,3448	0,9002	0,9336
P	0,0079	0,0069	0,0672	0,0398	0,0604	0,0518
Mg	4,3114	4,5424	1,0267	1,8529	4,8208	6,1374
Zn	0,0051	0,0050	0,0089	0,0089	0,0122	0,0090
Cr	0,1845	0,1816	0,0090	0,0118	0,1387	0,1615
Ni	0,0239	0,0257	0,0036	0,0042	0,0403	0,0538

Note: Bugallon in Pangasinan, Capudlosan in Agusan des Norte, and in basic alluvium, Natividad, Pangasinan. See Mg/Ca ratio as well as Cr and Ni concentrations; H is always the higher catena member (Tropaquent); L is always the lower catena member (Hydraquent).

From Scharpenseel, H. W., Eichwald, E., and Neue, H. U., *Landw. Forsch. Sonderheft*, 38, 224, 1981.

TABLE 2
Pedogeomedical Consequences of Lack or Excess of Elements

Element disorder	Pedogeomedical consequence	Reported location
Excess of Hg	Minamata Disease	Japan
Excess of Cd	Itai-Itai Disease	Japan
Selenosis	Keshan disease	China
Lack or excess of elements (Se ?) also possibly mycotoxins	Kashin-Beck disease, bone and joint degeneration and deformation of children	China
Lack of J	Goiter	Divers
Lack of F	Caries of teeth	Divers
Lack of P	Osteomalacie	Divers
Lack of Se	Nervous diseases	Divers
Lack of Zn	Impeded growth, dwarfism	Divers
Lack of Mg	Nervous diseases, depressions	Divers
Lack of Cr	Impediment for glucose metabolism, one of cofactors of insulin — therefore Cr-Diabetes, cardiovascular diseases	Divers
Lack of Cu + Zn + Se	Arthritis due to overproduction of peroxidase	Divers
Excess of Cd	Destructive effect on kidney and bones	Divers

Cases With Substrate Specificity

Lack of Se and J	Soils of young moraines, mountain soils, related to distance from sea shore	
Lack of Cu, Co, Zn	Histosols, sandy Spodosols, calcareous (sand) soils	
Excess of Cr, Ni	Soils of ultrabasic rocks, serpentines, ophiolites	
Excess of Ni, Cr, Pb, As	Soils near mines (mine cuttings)	
Lack of PO_4, SO_4, and other anions	Histosols due to lack of anion sorption capacity	

ARSENIC , 0.02 494 481

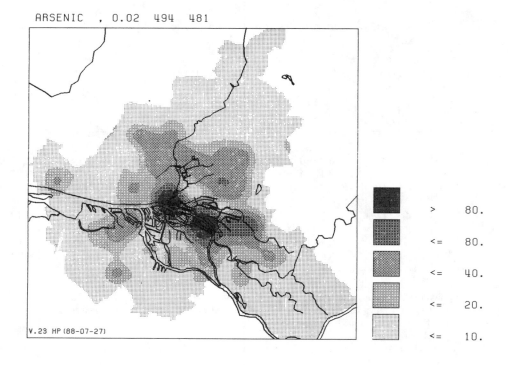

LEAD , 0.02 494 481

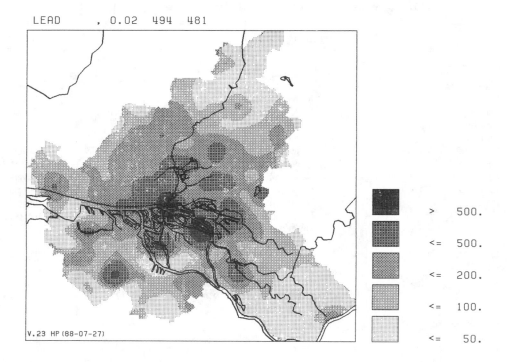

FIGURE 1.   Total arsenic and lead in surface soils (0 to 5 cm) as sampling grid of Hamburg State area. (According to Lux, Hintze, and Piening[69]).

FIGURE 2.   Three-dimensional diagram of zinc concentration in surface soils (0 to 5 cm) of Hamburg State area. (According to Lux, Hintze, and Piening[69]).

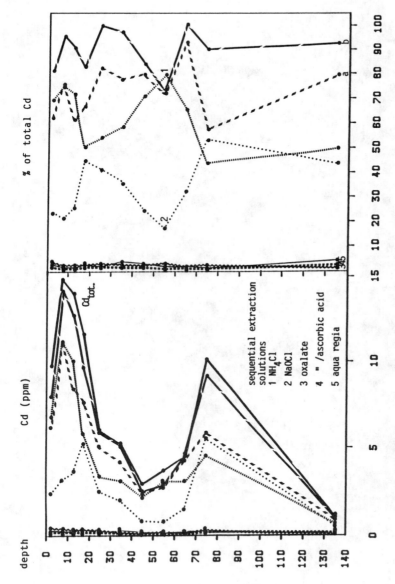

FIGURE 3.   Fractions of solubility of Cd in soil profile "Peuter Elbdeich". Forms of cadmium (in ppm and in percent of total) in a high contaminated sandy Hamburg soil after sequential extraction procedure (exchangeable: 1; organic: 2; Fe-Mn oxide: 3; crystalline oxide: 4; residual: 5) and single extractions ($CaCl_2$: a, and EDTA: b). (Adapted from Dues, G., *Hamburger Bodenkundliche Arbeiten,* 9, 1, 1987.)

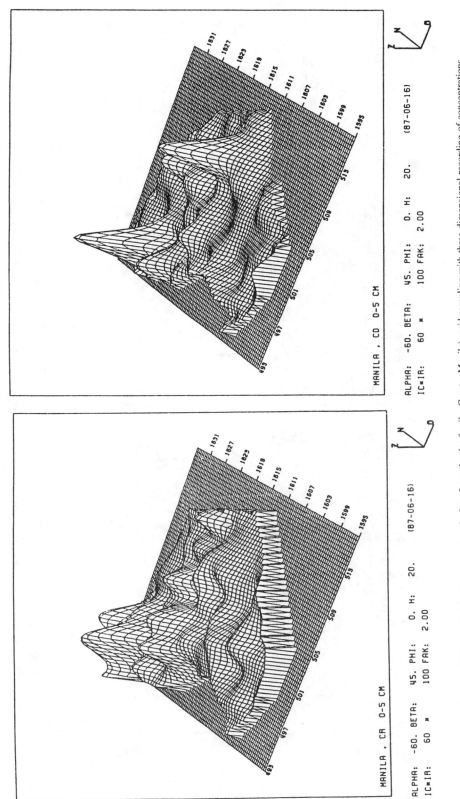

FIGURE 4. Cr and Cd pollution in city ecosystem (in 0 to 5 cm depth of soils, Greater Manila), grid sampling with three-dimensional recording of concentrations. (Left) Cr, (right) Cd. (Adapted from Pfeiffer et al., *Hamburger Bodenkundliche Arbeiten*, 11, 264, 1988.)

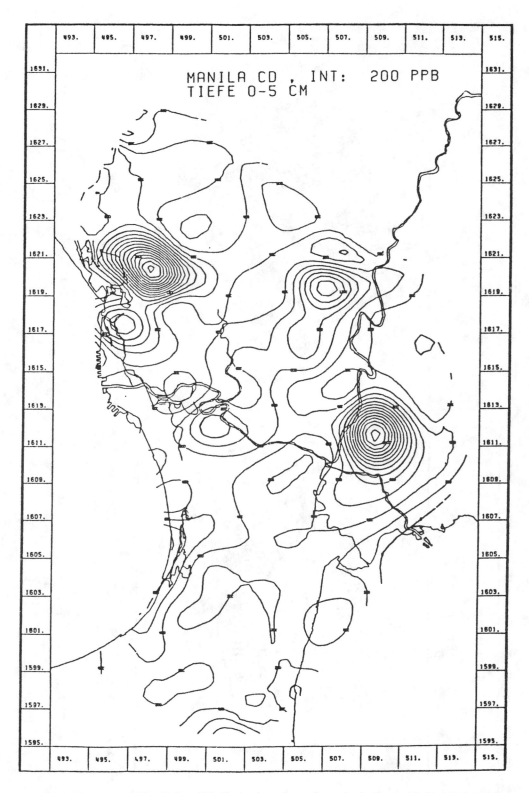

FIGURE 5.   Isopletes of Cd pollution. (Distribution in surface soil samples in Greater Manila.) (Adapted from Pfeiffer et al., *Hamburger Bodenkundliche Arbeiten*, 11, 264, 1988.)

for soil and plant samples over the whole mapping area (State of Hamburg), reflecting geochemical background plus anthropic pollution was analyzed by Lux,[33] and Dues[10] completed the overview by adding the ecological component with measurements into the profile depth and availability to the trophical planes by sequential analysis of solubility fractions (see Figures 1 to 3). A similar toxic element project in the tropical environment was carried out in "Greater Manila" and adjacent race fields, lake sediments, and the colluvial ultrabasic riceland of Pangasinan in the Philippines.[37] (See Figures 4 and 5.) In addition to toxic concentrations of critical elements in basic rock weathering products, such as basalts in Hawaii or in some areas of the Indian Deccan plateau with high Mn contents (leading to iron deficiency due to the higher E_h value of Mn compared to Fe) or in weathered ophiolitic earth-mantle material with extremely high Cr, Ni, and other heavy metal concentrations, periodical volcanic emissions must be mentioned as a source of multiple pollution. An example may be Mt. Merapi, a living volcano near the city of Yogyakarta, Java, which is heavily contaminating the surrounding landscape with, e.g., S, Zn, Se, Pb, As, Hg, Cd.[31] According to Georgii[13] at 40°N, the ordinary SO_2 concentration in the air is ca. 3 ppb, but in the tropical belt it reaches 0.1 to 0.2 ppb only because of less industrial pollution and due to the sink brought about by air humidity and precipitation. Around radial SO_2 sources of industrial origin concentrations up to 100 ppb are possible and observed.[18] In total, the SO_2 contribution by volcanoes is rated minor compared to that of conventional power plants and chimney exhausts,[31] but in wind direction near the volcanoes it can be very high and is often associated with heavy metals and other toxic elements.[6,48]

While in the relatively dry temperate climate zone it may take (according to Rhode[38]) 30 to 70 h before the S compounds will be precipitated, in the tropical belt with >80% relative humidity this happens almost immediately. At Mt. Merapi, which is our example, according to R. G. Sieffermann,[49] it rains above an altitude of 600 m almost every day, and the emission as well as its precipitation is estimated by the volcanologic observatory of Mt. Merapi[1] as being of the order of 30 to 100 t of SO_2 per day, during enhanced volcanic activity of >200 t per day. Sieffermann[49] estimates for the Greater Merapi area an SO_2 concentration in the rain water of 23 to 36 ppm and an input of 60 to 90 g of SO_2 per square meter per year.

It is a phenomenon often ignored, that despite the polluting action of living volcanoes (See Figure 6) — in our example by several hundred kg $SO_2 \cdot ha^{-1} \cdot y^{-1}$ — the vegetation around the cones is often overwhelmingly lush, with, for example, no signs of acid rain damage to trees (often inferred to be a consequence of high SO_2 concentrations) despite the enormous amounts of exhaled SO_2. Little is known regarding the action of the strong anion fixation capacity of the Andisols during their allophane/immogolite phase as a buffer and a sink of such pollutants. A release of such fixed high rates of anions should then produce problems in the decoupling stage of the anions, including humic acids, and later, when the allophanes turn into halloisites and the Andisol changes slowly into its zonal steady state. Little is known about these release mechanisms.

Sometimes rock varnishes[9] are quite revealing for air and rainfall-caused contamination, pollution, or for reaction mechanisms, when observing their element or isotope ratio differences to those of the bedrock. Similarly, element or isotope ratio differences between soil, corresponding saprolitic zone, and bedrock can be indicative of litho- or biospheric contamination and pollution.

II. POLLUTION OF PEDOMEDICAL CONSEQUENCE DEPENDENT ON GLOBAL POOLS AND SOIL COMPARTMENTS

A. INORGANIC MATERIALS

The mobility and fixation of inorganic cations is in a first line dependent on the permanent and variable charge of the sorption complex. HAC (high activity clay) soils of mainly

FIGURE 6.   Taal Volcano, Luzon Island with major activity of emission.

permanent charge micaceous and smectitic clay minerals have over a wide pH range a large negatively charged surface (smectites of ca.800 $m^2 \cdot g^{-1}$) with correspondingly high CEC (cation exchange capacity). On the other hand, we sometimes find scarcely any CEC at all from the part of the clay fraction at the soil pH (Figure 7), in LAC (low activity clay) soils of the tropics with mainly candite clay minerals of small surface area (<100 $m^2 \cdot g^{-1}$) and variable charge[57] at lower pH-level. In these cases the organic matter compartment whose ZPC (zero point of charge) lies at pH 3 to 3.5 may be the only major one with negative charge at soil pH and mainly responsible for sorption, fixation, or release of polluting cations. Anionic pollutants have, according to the negative permanent charge surplus in HAC soils, lower probability for retention due to CEC>AEC discrepancy. In LAC soils, fixation at protonized clay and oxide edges is remarkable. We are reminded of the great PO_4 fixation problem in Ultisols and Oxisols. The "Chinese way" of keeping P plant-available in South Chinese Ultisols by spreading the P fertilizer first over compost piles to give the PO_4 an organic ligand for better solubility, reveals that anionic pollutants may also be incorporated in such soils at larger scale, if they are rich in organic matter, and still be readily released again (for example, in Humults >12 kg C/m^2, 1 m depth or Humoxes with >16 kg C/m^2, 1 m depth). Otherwise, excessively negatively charged histic horizons in Histosols are known to be almost inert to anion sorption. PO_4 is here repelled from the sorption points and percolates in these peat soils almost freely in the drainage water.

A special situation is existent in the 8 to 12,000 year-long allophane/immogolite phase in Andisols. The terminal Gibbsitic structures of the allophane are capable of exceptionally high anion fixation rates of, e.g., phosphate and humic acids. Anionic pollutants can be expected to have optimal conditions for extended fixation and residence time in Andisols.

Major inorganic polluting materials are listed in Table 3. Some typical inorganic pollutants of specific pedogeomedical relevance are members of the uranium, thorium, and actinium natural radioactivity decay chains, especially U, which is at "normal concentration

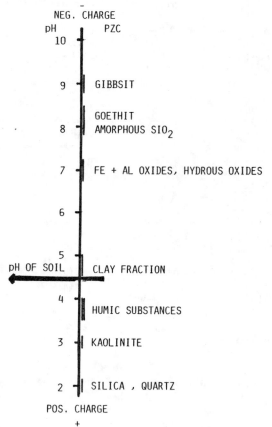

HAC - SOILS

60 % SMECTITES = 60 MEQ

1.5 % HUMUS = ca 4 MEQ

LAC - SOILS

25 % LOW ACTIVITY CLAY = 3 MEQ

1.5 % HUMUS = ca 4 MEQ

$$\Delta \, pH = pH_{1KCL} - pH_{H_2O}$$

(Δ pH negative = CEC dominant ; Δ pH positive = AEC dominant)

NEG. CHARGE −

pH   PZC

10

9 — GIBBSIT

GOETHIT

8 — AMORPHOUS SIO_2

7 — FE + AL OXIDES, HYDROUS OXIDES

6

5

pH OF SOIL   CLAY FRACTION

4

HUMIC SUBSTANCES

3 — KAOLINITE

2 — SILICA , QUARTZ

POS. CHARGE +

FIGURE 7.   Soil compartments of different PZC (point of zero charge) versus soil pH of LAC-soils; humus as CEC-matrix.

level'' of ca.1ppm available in crystalline rocks and can be in mineral soil and especially in soil organic matter (preferably at pH 4 to 5) enriched to concentrations up to several hundred ppm.[4,40,52] It is well known that natural radioactivity in crystalline building stones can double or multiply the 1.7 millisievert / 170 millirem (1.1 from natural sources of radiation, 0.6 from sources of civilization) normal annual radiation dose of the average man in central Europe. Other radioactive pollutants with pedogeomedical impact are treated in a special chapter of this volume.

The fixation of inorganic pollutants is widely dependent on soil pH, which is stabilized by the buffer capacity of the substrate. Ulrich[58] distinguishes 4 main buffer zones and systems in soils (Table 4). Since 1% free $CaCO_3$ represents in a 10 cm topsoil layer a buffer capacity of 300 kmol H^+ per ha, acid inputs, e.g., in the form of SO_4 or $(NO)_x$ of 1 to 2 kmol·y^{-1} per ha can be tolerated without representing a threat for imminent acid damages in soil with only slight amounts of free lime (see also Ulrich and Matzner[59]).

TABLE 3
Major Inorganic Pollutants — Levels of Orientation for Tolerable Total Contents of Contaminating Elements in Cultivated Soils

		Total content in air dry soil (mg/kg)		
		Frequently	**Special highly contaminated soils**	**Tolerable level**
As	Arsen	0.1—20	<8.000	20
B	Bor	5—20	<1.000	25
Be	Beryllium	0.1—5	<2.300	10
Br	Brom	1—10	<600	10
Cd	Cadmium	0.01—1	<200	3
Co	Cobalt	1—10	<800	50
Cr	Chrom	2—50	<20.000	100
Cu	Kupfer	1—20	<22.000	100
F	Fluor	50—200	<8.000	200
Ga	Gallium	0.1—10	<300	10
Hg	Quecksilber	0.01—1	<500	2
Mo	Molybdän	0.2—5	<200	5
Ni	Nickel	2—50	<10.000	50
Pb	Blei	0.1—20	<4.000	100
Sb	Antimon	0.01—0.5	?	5
Se	Selen	0.01—5	<1.200	10
Sn	Zinn	1—20	<800	50
Tl	Thallium	0.01—0.5	<40	1
Ti	Titan	10—5000	<20.000	5.000
U	Uran	0.01—1	<1.000	50
V	Vanadium	10—100	<1.000	50
Zn	Zink	3—50	<20.000	300
Zr	Zirkon	1—300	<6.000	300

According to Kloke.[66]

TABLE 4
Buffer Regions and Capacities of Soils in Dependency of their pH-Value

pH Value	Buffer region	Buffer capacity (10 cm topsoil)
>6.2	Carbonate	300 kmol H^+/% $CaCO_3$
6.2—5.0	Silicate	25 kmol H^+/% silicate
5.0—4.2	Cation exchange	7.5 kmol H^+/% clay
<4.2	Aluminum	150 kmol H^+/% clay

According to Ulrich.[67]

An inorganic element of high actuality and pedogeomedical consequences under deficiency as well as excess pollution is Selenium. B. A. Gamboa-Lewis[12] as well as Låg and Steinnes[26,27] and Lag[28,29] laid the groundwork for correct assessment (see also Table 2). Most other inorganic pollutants, especially the heavy metals, are highly influenced in their distribution by non-pedogeomedical anthropogenic measures. Their maximum permissible levels (Table 5) are more temporary fixations due to a lacking sound toxicological base of assessment. This leads to the major remaining problem of converting total element analysis into an eco-toxicologically meaningful result. Assessment of non-effect levels and plant availability still comprises the whole spectrum of competing methods, i.e., physiological

TABLE 5
Maximum Permissible Levels Indicating Need of Closer Investigation or Rehabilitation/Excavation

Component/level	Soil (mg/kg dry matter)			Groundwater (ug/l)		
	A	B	C	A	B	C
Metals						
Cr	100	250	800	20	50	200
Co	20	50	300	20	50	200
Ni	50	100	500	20	50	200
Cu	50	100	500	20	50	200
Zn	200	500	3000	50	200	800
As	20	30	50	10	30	100
Mo	10	40	200	5	20	100
Cd	1	5	20	1	2.5	10
Sn	20	50	300	10	30	150
Ba	200	400	2000	50	100	500
Hg	0.5	2	10	0.2	0.5	2
Pb	50	150	600	20	50	200
Inorganic contaminations						
NH_4 (as N)	—	—	—	200	1000	3000
F (total)	200	400	2000	300	1200	4000
CN (total-free)	1	10	100	5	30	100
CN (total-complex)	5	50	500	10	50	200
S (total)	2	20	200	10	100	300
Br (total)	20	50	300	100	500	2000
PO_4 (as P)	—	—	—	50	200	700
Aromatic compounds						
Benzene	0.01	0.05	5	0.2	1	5
Ethyl-benzene	0.05	5	50	0.5	20	60
Toluene	0.05	3	30	0.5	15	50
Xylene	0.05	5	50	0.5	20	60
Phenols	0.02	1	10	0.5	15	50
Aromates (tot.)	0.1	7	70	1	30	100
Polycyclic Hydrocarbons						
Naphtalene	0.1	5	50	0.2	7	30
Anthracene	0.1	10	100	0.1	2	10
Phenanthrene	0.1	10	100	0.1	2	10
Fluoranthrene	0.1	10	100	0.02	1	5
Pyrene	0.1	10	100	0.02	1	5
3,4-Benzopyrene	0.05	1	10	0.01	0.2	1
Polycyclic hydrocarbons (tot.)	1	20	200	0.2	10	40
Chlorinated Hydrocarbons						
Aliphatic Cl-hydrocarbons (individually)	0.1	5	50	1	10	50
Aliphatic Cl-hydrocarbons (total)	0.1	7	70	1	15	70
Chlorbenzenes (individually)	0.05	1	10	0.02	0.5	2
Chlorbenzenes (total)	0.05	2	20	0.02	1	5
Chlorphenols (individually)	0.01	0.5	5	0.01	0.3	1.5
Chlorphenols (total)	0.01	1	10	0.01	0.5	2
Chlorhydrocarbons, (total)	0.05	1	10	0.01	0.2	1
PCBs (tot.)	0.05	1	10	0.01	0.2	1
EOCL (tot.)	0.1	8	80	1	15	70
Pesticides						
Org. chlor- (individually)	0.1	0.5	5	0.05	0.2	1
Org. chlor- (total)	0.1	1	10	0.1	0.5	2
Pesticides (total)	0.1	2	20	0.1	1	5

TABLE 5 (continued)
Maximum Permissible Levels Indicating Need of Closer Investigation or
Rehabilitation/Excavation

Component/level	Soil (mg/kg dry matter)			Groundwater (ug/l)		
	A	B	C	A	B	C
Other contaminants/pollutants						
Tetrahydrofuran	0.1	4	40	0.5	20	60
Pyridine	0.1	2	20	0.5	10	30
Tetrahydrothiophene	0.1	5	50	0.5	20	60
Cyclohexanon	0.1	6	60	0.5	15	50
Styrol	0.1	5	50	0.5	20	60
Benzene	20	100	800	10	40	150
Mineraloil	100	1000	5000	20	200	600

Note: Indicative levels: (A), category of reference; (B), category requiring closer investigation; (C), category requiring soil rehabilitation/excavation.

According to Leidraad Bodemsaneering.[68]

ones like field and pot experiments, microbial cultures, Azolla cultures as well as non-physiological buffer, diluted acid, chelate extracts,[32] and ion exchanger batches, as well as labeling procedures such as A- and L-value, Freundlich- or Langmuir ad- and desorption isoterms[46] and more recent approaches like EUF (electro ultrafiltration) and sequential extraction processes.[10] Also, the use of transfer factors (ppm in plant / ppm in soil) remains ambiguous and self-deceptive without proper correction for the main involved modifiers, texture, humus content, and pH as a minimum. The translation of analytical into ecological results still requires improvement.

Evidently the heavy metals are mainly fixed by the soil organic matter compartment in most soils. Stevenson[50] observed the formation of organic metal complexes by systematic titration curves, Scharpenseel[39] by radiometric precipitation curves, Brümmer and Herms[5] by equilibrium curves.

An overview on bonding of heavy metals by clay fractions and soil oxides can be taken from reports by Schlichting and Elgala,[45] Mayer,[34] Herms and Brümmer,[17] and Tiller, Gerth and Brümmer.[53,54]

B. NON-IONIC PRINCIPLES, ESPECIALLY ORGANICS

Due to the extremely large number of possibly polluting organic compounds compared with the already-known inorganic agents, the polluting status of organics is not yet as clear as that of most inorganics. A large list of some 120 volatile, semivolatile, and non-volatile potentially hazardous compounds was issued at the CODATA workshop in Montreal (1986). For practical reasons, efforts to identify the most prominent and dangerous compounds have led to OECD, EC and national lists of priorized compounds, which were selectively tested in different countries for plant uptake, ad- and desorption as well as their persistency in soils, their mean residence time (MRT) in organisms at different trophic planes and regarding the tableau of possible damages (See Table 6).[2,11,43]

Since most of the organic pollutants are anthropogenic and not at all pedogeogenic, an exhaustive description of the effects (and of only the major representative compounds) would reach beyond the scope of the pedogeomedical mandate to which this text is limited. To illustrate by an example of environmental relevance the closeness of pedogeomedical and biogeochemical/hydrogeochemical problem fields in the evaluation of organic pollutant dynamics, we may refer to the behavior of PCBs in sea water[16] and soils.[24,41] Harvey and Steinhauer scrutinized marine food chains without finding any confirmation of the commonly

TABLE 6
Possibly Damaging Organics, Physico-Chemical Properties and General Data

I: physico-chemical properties   II: general data

no.	chemical	abbr.	vapor pressure 20-25°C [mbar]	water solubility [g·l⁻¹]	distribution c(n-octanol)/c(H₂O)	c(soil)/c(H₂O) for 2%C	hydrolysis k [sec⁻¹]	dissociation $K_A|K_B$	K(OH-radicals) [l mol⁻¹ sec⁻¹]	UV-absorption λ>290nm	MAK BRD [mg/m³]	class of risk BRD	Input in the environment [t a⁻¹]
1	hexachlorobenzene	[HCB]	$2.8\cdot10^{-5}$	$5\cdot10^{-6}$	$1.5\cdot10^{6}$	78	0		$2.4\cdot10^{9}$	+			>2000
2	1,1'-(2,2,2-trichloroethylidene) bis (4-chlorobenzene)	[DDT]	$2.3\cdot10^{-7}$	$5.5\cdot10^{-6}$	$1.55\cdot10^{6}$	$2\text{-}8\cdot10^{3}$	$1.9\cdot10^{-9}$ (pH 3,9)			+	1		$6\cdot10^{4}$
3	1,1'-(2,2-dichloroethylidene) bis (4-chlorobenzene)	[DDD]		$5\cdot10^{-6}$	$1.05\cdot10^{6}$	900*				−			$<10^{4}$
4	1,1'-(2,2-dichloroethylene) bis (4-chlorobenzene)	[DDE]		$1.0\cdot10^{-5}$ $1.3\cdot10^{-6}$	$5.8\cdot10^{5}$	660*	$1.4\cdot10^{-10}$						
5	di-(2-ethylhexyl)-phthalat	[DOP]	$1.9\cdot10^{-7}$	$9\cdot10^{-4}$	9500	92	$\approx8\cdot10^{-6}$ (pH 1) $\approx8\cdot10^{-7}$ (pH 10)			+	10		$<1.5\cdot10^{5}$
6	urea		$1.6\cdot10^{-5}$	$1.2\cdot10^{3}$	10^{3}	$4\cdot10^{2}$	$>3.5\cdot10^{-5}$	$K_B=1.5\cdot10^{-14}$	$<7\cdot10^{5}(H_2O)$	0	−		$2.7\cdot10^{7}$
7	perylene		$\sim7\cdot10^{-9}$	$5\cdot10^{-7}$	$6.3\cdot10^{6*}$	$5.7\cdot10^{3}$				++			
8	fluoranthene		$1.3\cdot10^{-2}$	$2\cdot10^{-4}$	$1.6\cdot10^{5}$	67.2*							
9	benzene		97.1	1.79	135	1.7	0		$8.5\cdot10^{8}$	0	26(TRK)	A 1	$10^{5}\text{-}10^{6}$
10	toluene		35.9	0.515	490	2.8*	0		$3.6\cdot10^{9}$	+	750	A 1	$>10^{6}$
11	phenol		0.83	82	29	0.54		$K_A=10^{-10}$	$5\cdot10^{9}(H_2O)$	+	19	A 3	$7.5\cdot10^{4}$
12	4-nitrophenol			16	81	0.42*		$K_A=7.1\cdot10^{-8}$		+			
13	2-nitrophenol		0.29	2.1	61.7	1.3*		$K_A=5.9\cdot10^{-8}$		+			
14	pentachlorophenol	[PCP]	10^{-3}	$1.4\cdot10^{-2}$	$1\cdot10^{5}$	18	0			+	0.5		
15	aniline		1.15	36.6	9.5	0.27		$K_B=5\cdot10^{-10}$	$5.3\cdot10^{9}(H_2O)$	+	19	A 3	10^{4}
16	4-chloroaniline		$3\cdot10^{-2}$	3.0	68			$K_B=10^{-10}$		+	−		
17	Na-dodecylbenzene sulfonic acid	[LAS]	$<10^{-7}$	400						(0)			$5\cdot10^{5}$
18	(2,4-dichlorophenoxy) acetic acid	[2.4D]	$<10^{-7}$	0.9	37	0.4		$K_A=1.7\cdot10^{-3}$		+	10		10^{5}
19	(2,4,5-trichloro-phenoxy) acetic acid	[2.4.5T]	$<10^{-7*}$	0.24	$\approx10^{3}$	1-4		$K_A=1.3\cdot10^{-3}$		+	10		$5\cdot10^{4}$
20	6-chloro-N-ethyl-N'(1-methylethyl)-1,3,5-triazine-2,4-diamine	[atrazine]	$4\cdot10^{-7}$	$3.3\text{-}7\cdot10^{-2}$	$4.8\cdot10^{2}$	3	$7.6\cdot10^{5}$ (pH 7)	$K_B=4.8\cdot10^{-13}$		+	−		$9\cdot10^{4}$
21	HgCl₂		$1.3\cdot10^{-4}$	69	(1,82)					0	0,01(org.)		
22	methanol		$1.5\cdot10^{2}$	∞	0.22			$K_A=10^{-17}$	$5.3\cdot10^{7}$	0	260	B	$>10^{6}$
23	ethylacetate		$1.1\cdot10^{2}$	77	5.4		$1.1\cdot10^{-11}$		$1.2\cdot10^{9}$	0	1400	A 1	$4.5\cdot10^{5}$
24	trichloroethylene		97	1,1	195	2	(0)		$1.4\cdot10^{9}$	0	260		10^{6}
25	ethylene		$>4\cdot10^{4}$	0,11	13.5				$3.8\cdot10^{8}$	0			$7\cdot10^{5}$
26	1,2,4-trichlorobenzene		0.5	$3\cdot10^{-2}$	1500					+	40		

* estimated

+: yes
0: no

From Scheele, B.[43]

advocated food chain magnification of the pollutants' (PCB,DDT) incorporated concentrations. On the other hand the limited time period of PCB usage, between its introduction and ban in the mid-seventies, opened possibilities to study by its concentration or lack of presence in sea water the vertical and horizontal advection and diffusion of the sea water as well as the movement of the particulate organic carbon (POC), to which PCB is especially attracted in sea water.

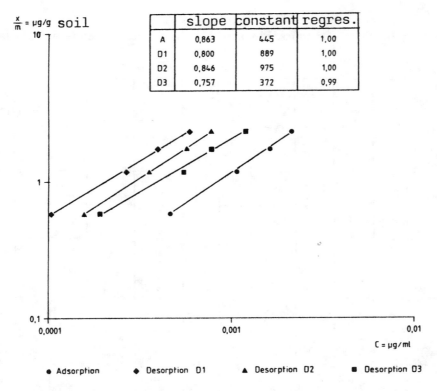

FIGURE 8.   Freundlich plot for adsorption-desorption for ^{14}C-PCB tested by a gleyed Luvisol (Ap: 1.6%; pH 5.7; soil:solution = 1:20). (Adapted from Krogmann et al.[24])

Scharpenseel and Krogmann with their co-workers studied the behavior and concentration balance of ^{14}C-labeled PCB in small biotops as well as in field, pot, and column tests and by ad/desorption isoterms, measurements of persistency, and also their biotic, abiotic and photochemical decomposition — all in a cross-section of central European soil types. From these and other similar results, it became evident that for assessment of water soluble organics in soils, ad/desorption studies and Freundlich K_{ads} plus K_{des} coefficients may suffice, while water-insoluble organics and only very slightly soluble ones (in ppb concentrations) may require generally pot and possibly field experiments beyond the scope of pure lab tests (see Figures 8 and 9).

PCB-intake can lead to swollen liver, jaundice, chloracne, alteration in the peripheral nerve system, suppression of the immune system, adverse influence on the metabolism and power of vision; perhaps it also exerts fetotoxicity.[25] Thus, in a course of water, insoluble PCBs sustained existence in soils for a long time can cause a pseudo-pedogeomedical situation.

Undoubtedly there exist similar situations as a consequence of the >300 mil t organic compounds as a whole[47] and among them Ca.1 mil t of pesticides[22] as worldwide annual production and input in our agrarian and industrial ecosystems, much of it in the hydro- and pedosphere. While hydro- and atmosphere are governed by mixing processes and regeneration cycles, the pedosphere is resentfully adsorbing and storing the products of human misbehaving at least partly. The magnitude of the steadily aggravating threat for the soil related biosphere has so far been widely disregarded, slowly creating the kind of above-mentioned pseudo-pedogeomedically relevant situation.

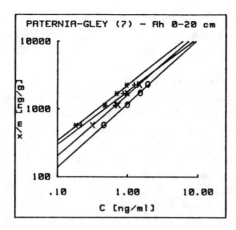

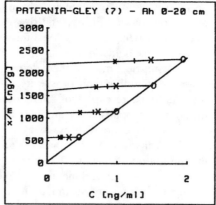

K_{Ads} = 1195, K_{D3} = 2254, 2.4 % C, 1.65 o/oo Fe_o, pH 5.0

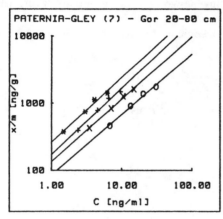

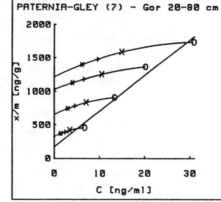

K_{Ads} = 84, K_{D3} = 256, 0.11 % C, 0.35 o/oo Fe_o, pH 6.5

FIGURE 9. Adsorption and desorption isoterms and corresponding K-values according to Freundlich for PCB. o = adsorption; x = 1.desorption; + = 2.desorption; * = 3.desorption; Soil: aeric Haplaquept. (Adapted from Krogmann, H.[25])

III. POLLUTION BY SEWAGE, SLUDGE, GARBAGE, INDUSTRIAL WASTES, MANURE, COMPOST

It can be ascertained that the whole spectrum of these pollutants is anthropogenic. The polluting consequences, though, have a deep impact on the pedosphere and partly even the lithosphere; however, the medicinal aspect of the problem is not primarily of a pedo-geomedical nature.

In facing the effects on soils of these waste products of the developed civilizations, we must remember that, contrary to air and water which are commonly owned goods, soil is not only a product of nature, but also a private property asset. Its vulnerability is high to violations of ecological standards.

Preindustrial society and particularly the "early industrial founders" knew land dev-astation due to excessive consumption of wood and charcoal as well as environmental pollution by smaller pioneer enterprises, the "milky smoke" or metal smelters loaded with poisonous arsenic (As_2O_3), tannin and chromium as well as sulfides from hide curing, sulfite wastewaters from paper milling, and dye metals from glass manufacture. They knew the problems of the associated professional diseases, e.g., infections from recycling of old rags collected from households, etc.[56]

Industrial countries, most of them densely populated (>2 men/ha) suffer especially from the mechanics of building their prosperity, which requires importation of large quantities of bulky raw materials, re-export of smaller-size commodities of high value, and having to cope with unproportionally large amounts of wastes. In addition, the economically well-to-do populations (often called ''throw-away societies'') are producing ever-increasing volumes of garbage. Since they prefer proteinaceous adequate food of animal origin to bulky vegetarian nutritives, the import of feed-stuff, especially grains and correspondingly the production of ''imported sewage, sludge and manure'' also contributes to aggravation of the waste disposal problem. Due to rising nutrient inputs in all agronomic and husbandry systems, the eutrophication of lakes, rivers, harbor basins (and, recently, even the sea) produces constantly increasing subhydrous organic matter and reduced muds which require dredging, excavation, and space for disposal. Poisonous spores of butylitis may kill water birds, and other biotypes are even dangerous for human beings swimming or working in the muddy layer, which gives the matter additional drama and urgency.

Vetter[61] identifies the media with non-pedogeomedical pollutants of this genre which affect agriculture, horticulture, forestry, and fishery:

> Air (possibly polluted)
> Water
>   Drinking water for domestic animals
>   Flooded water
>   Irrigation water
>   Sprinkling water
>   River water
> Wastes
>   Sewage water
>   Sludge
>   Garbage
>   Industrial waste, including lime products
>   Muds
>   Harbor basin muds
>   Excavated, polluted soil
> Commercial feeds
> Paints and solvents

Notwithstanding the fact that the enrichments are almost exclusively anthropogenic and turn slowly pedogeogenic/pedogeomedical only after sustained periods of persistence in the pedosphere, a pollution discussion necessitates assessment of tolerable concentration limits and requires identifying the thresholds of necessary monitoring of measures of soil sanitation. Among the national proposals the Dutch system (1983) regarding such reference levels is quite distinct and comprises a wide array of organic and inorganic pollutants, which are also the active polluting principles in the above-discussed media (Table 5).

IV. POLLUTION BY ELEMENTAL POOL TRANSFERS

In summary, the occurrence of pedogeomedical pollution that excludes human activities, is rather casual. In comparison, long-term repetitive anthropogenic pollution that creates,

TABLE 7
Carbon and Nutrient Balance, Effect of Pool Transfers

Pools and pool changes

Total organic matter in sediments 95% at clays and shales	3.9×10^{15} T
Total pedosphere (soil 1 m deep)	2×10^{14} T
Total soil organic matter	1.5—2.0×10^{12} T
Annual C-turnover by photosynthesis and recycling	115 BIL T
Approximate annual N turnover by recycling	1 BIL T

Pool changes per annum

Into the atmosphere	Ca. 5.5 bil t C from coal, gas, oil
	Ca. 1.5 bil t C from forest clearing
	Ca. 100 mil t C as methane, emanating from rice fields
	Ca. 100 mil t C as methane from ruminants
	Ca. 25 mil t C as methane from vehicles
Into the pedosphere	
From atmosphere	Ca. 75 mil t N (diazotrophs, rhizobium and blue-green algae)
	Ca. 75 mil t N (from mineral fertilizer)
From lithosphere	Up to to 40 mil t P (from mined phosphates and steel converters)
	Up to 40 mil t K (from diapirs and salt deposits)

Pool changes into atmosphere: greenhouse effect, ozone layer
Pool changes into pedosphere: increasing eutrophication

over time, a pseudo-pedogeomedical situation of equal impact, could be shown to occur more frequently. One of those basically anthropogenic but steady-state pollution trends, often ignored but soon to become extremely relevant, is the sum of gigantic pool transfers of biotically most powerful elements, such as C, N, P, K, and S. They have occurred increasingly for a hundred years, with a critical phase since the Second World War. Låg[30] refers to this effect in a treatise on commercial fertilizers and geomedical problems.

Table 7 lists the approximate C, N, P, and K pool transfers against the background of C and N pool sizes. 5.5 billion tons of C from fossil fuel combustion + 1.5 billion tons of C from slash and burn (together ca.7 billion tons of C, or 1% of the total atmospheric C) are annually transfered into the atmosphere. In addition to ca.250 million tons of C as CH_4 from rice fields, ruminants, swamps, and combustion engines, ca.115 billion tons are turned over annually by photosynthesis and inversely by recycling. By the latter, more than 1 billion tons of N are released, ca.6 times the total mineral fertilizer N plus diazotrophic rhizobium related fixed N, which by pool transfer from the atmosphere are imported to the pedosphere. Additionally, an import of up to 40 million tons of P and K originates by pool transfer annually from the lithosphere to the pedosphere. While the rapid increase of CO_2 and CH_4 concentrations undoubtedly contributes to the greenhouse effect, the nutrient pool transfer (which with increasing world population on a shrinking area of fertile, cultivated soils will inevitably further increase) may lead to total eutrophication. The effect of the latter is visible already in the moist depressions and lacustrine soils. At the interface of atmosphere and pedosphere, the additional CO_2 as well as nutrient imports have led to increases of organic matter production. Bertram[3] in his work, based on isotope ratio testing, concludes that the soil has become a sink for ca. 60 gigatons of additional C in the past few decades.

V. SOIL PURIFICATION AND REHABILITATION

Pollution in its pedogeomedical as well as in its anthropic origin requires rehabilitation measures if maximum permissible levels of the pollutants are exceeded. The Dutch system, for example, gives concentration levels (see Table 5) beyond which the soil has to be excavated and purified via solvent extraction, combustion, or *in situ* methods. Correspondingly, in the Netherlands there are privately run facilities for soil purification by dilute acid extraction (Vanadiumsweg 5, 3812 PX Amersfoort, Netherlands, which serve mainly inorganic, especially heavy metal, decontamination) as well as by combustion (Ecotechniek bv, Beveluxlaan 9, 3527 HS Utrecht) primarily to rehabilitate soils polluted with organics. In Berlin cooperation between private enterprises and the Technical University has recently led to development and rigid testing till market maturity of different extraction, combustion and biological procedures.[65]

According to Mischgofsky and Kabos,[36] among the three principal techniques of soil purification/rehabilitation, e.g., soil excavation, soil enclosure (quarantine), and *in situ* curative treatment, the latter with its higher options of specificity and refinement may gain ground with widening experience.

In this connection we may be reminded that, according to Kovda and Glazovsky,[13] as either a concentrating or diluting principle, the pollution path by dust transport represents from 100 to 5000 million $t \cdot y^{-1}$, based on an ordinary Aeolian deflation module of 50 to 300 $t \cdot (km^2)^{-1} \cdot y^{-1}$.

Rehabilitation of soil and landscape stricken by long-term geomedical inflictions of polluting agents is particularly difficult, demanding, and costly. With improvement of *in situ* curative treatments in this difficult field of soil rehabilitation, new affordable methods of treatment appear possible.

REFERENCES

1. **Bahar, I.,** Volcanologic survey of Indonesia, in Sieffermann, G., *Application of Research Programme on Environmental Protection to EC,* ORSTOM, Sekip K 3, Yogyakarta, 1987.
2. **Barth, D. S.,** A position paper for the CODATA workshop; directions for internationally compatible environmental data. Measurement methods and standards for soil environment. CODATA Workshop Montreal, McGill University, 1986, 151.
3. **Bertram, H. G.,** Zur Rolle des Bodens im globalen Kohlenstoffzyklus, *Naturforschende Gesellschaft zu Emden von 1814,* 8, Ser. 3-D3, Emden, 1986.
4. **Bloomfield, C. and Kelso, W. L.,** The mobilization and fixation of molybdenum, vanadium and uranium by decomposing plant matter, *J. Soil Sci.,* 24, 368, 1973.
5. **Brümmer, G. and Herms, U.,** Influence of soil reaction and organic matter on the solubility of heavy metals in soils, *Effects of Accumulation of Air Pollutants in Forest Ecosystems,* D. Reidel Publ., Dordrecht, 1983, 233.
6. **Buat-Menard, P. and Arnold, M.,** The heavy metal chemistry of atmosphere particulate matter emitted by Mount Etna volcano, *Geophys. Res. Lett.,* 5, 4, 245, 1978.
7. **Chen Xi, Chen Xu, Yan G., Wen Z., Chen, J., and Ge, K.,** Relation of selenium deficiency to the occurrence of Keshan disease, *Selenium in Biology and Medicine,* Spallholz, J. E., et al., Eds., AVI, Westport, CT, 1981, 171.
8. **Crounse, R. G., Pories, W. J., Bray, I. T., and Mauger, R. L.,** Geochemistry and Man: Health and disease, in *Applied Environmental Geochemistry,* Thornton, T. J., Ed., Academic Press, London, 1983, 267.
9. **Dorn, R. J. and Niro, M. J.,** Stable carbon isotope ratios of rock varnish organic matter: a new paleoenvironmental indicator, *Science,* 227, 85, 1472, 1985.
10. **Dues, G.,** Untersuchungen zu den Bindungsformen und ökologisch wirksamen Fraktionen ausgewählter toxischer Schwermetalle in ihrer Tiefenverteilung in Hamburger Böden, *Hamburger Bodenkundliche Arbeiten,* 9, 1, 1987.
11. **Führ, F., Biehl, H. M., and Thielert, W., Eds.,** *Methoden zur ökotoxikologischen Bewertung von Chemikalien,* Zentralbibliothek, KFA Jülich, FRG, 1983, 4.
12. **Gamboa-Lewis, B. A.,** Selenium in biological systems and pathways for its volatilization in higher plants, *Environ. Biogeochem.,* Vol. 1, Chap. 26, Ann Arbor Science, Ann Arbor, MI, 1976.
13. **Georgii, H. W.,** Contribution to the atmosphere sulfur budget, *J. Geophys. Res.,* 75, 2365, 1970.

14. **Goldschmidt, V. M.,** *Geochemistry,* Clarendon Press, Oxford, 1954, 730.
15. **Greenstein, J. L.,** Stellar evolution and the origin of the chemical elements, *Am. Sci.,* 49, 449, 1961.
16. **Harvey, G. R. and Steinhauer, W. G.,** Biogeochemistry of PCB and DDT in the North Atlantic, *Environmental Biogeochemistry,* Vol. 1, Nriagu, J. O., Ed., Ann Arbor Science, Ann Arbor, MI, 1976, chap. 15.
17. **Herms, U. and Brümmer, G.,** Einflussgrössen der Schwermetall-Löslichkeit und-bindung in Böden, *Z. Pflanzenernähr., Bodenkunde,* 147, 3, 400, 1984.
18. **Hidy, G. M.,** Removal processes of gaseous and particulate pollutants in, *Chemistry of the Lower Atmosphere,* Rasool, S. I., Ed., Plenum Press, New York, 1973, 121.
19. **Hintze, B.,** Geochemie umweltrelevanter Schwermetalle in den vorindustriellen Schlickablagerungen des Elbe-Unterlaufs, *Hamburger Bodenk. Arbeiten,* 2, 230, 1985.
20. **Jian-An, T.,** The Keshan disease in China. A study of ecological chemico-geography, *Natl. Geogr. J. India,* 28, 15, 1982.
21. **Kitagishi, K. and Yamane, I.,** *Heavy Metal Pollution in Soils of Japan,* Japan Scientific Society Press, Tokyo, 1981.
22. **Korte, W.,** Chemikalien im Ökotest, Umschau, 80, 643, 1980.
23. **Kovda, V. A. and Glazovskiy, N. F.,** Human activities and soil cover of the earth, Adv. in Soil Science, USSR-Acad. Sci., Soviet Pedologists to the XIIIth Intl. Congr. Soil Science, "Nauka", Moscow, 1986, 9.
24. **Krogmann, H., Maass, V., and Scharpenseel, H. W.,** Radiometrische Untersuchungen zu Sorptionsverhalten und Abbau von ^{14}C-PCB sowie ^{14}C-Pichloram in verschiedenen Böden, *Z. Pflanzenernähr. Bodenkunde* 148, 248, 1985.
25. **Krogmann, H.,** Methoden zur ökotoxikologischen Bewertung von Umweltchemikalien. Laborversuche zur Adsorption/Desorption von PCB und Pichloram, *Hamburger Bodenkundliche Arbeiten,* 1986.
26. **Låg, J. and Steinnes, E.,** Soil Selenium in relation to precipitation, *Ambio,* 3, 6, 237, 1974.
27. **Låg, J. and Steinnes, E.,** Regional distribution of selenium and arsenic in humus layers of Norwegian forest soils, *Geoderma,* 20, 3, 1978.
28. **Låg, J.,** *Geomedical Aspects in Present and Future Research,* Universitetsforlaget, Oslo, 1980, 11.
29. **Låg, J.,** A comparison of selenium deficiency in Scandinavia and China, *Jordundersøkelsens Saertrykk,* 323, 1984.
30. **Låg, J.,** *Commercial Fertilizers and Geomedical Problems,* Norwegian University Press, 1987, 9.
31. **Le Guern,** Les débits de CO_2 et SO_2 volcaniques dan l'atmosphère, *Bull. Volcanol.,* 45-3, 197, 1982.
32. **Lindsay, W. L. and Norvell, W. A.,** Development of a DTPA soil test for zinc, iron, manganese and copper, *Am. J. Soil Sci. Soc.,* 42, 421, 1978.
33. **Lux, W.,** Schwermetallgehalte und Isoplethen in Böden, subhydrischen Ablagerungen und Pflanzen im Südosten Hamburgs, Beurteilung eines Immissionsgebietes, *Hamburger Bodenkundliche Arbeiten,* 5, 250, 1986.
34. **Mayer, R.,** Natürliche und anthropogene Komponenten des Schwermetallhaushalts von Waldökosystemen, 70, *Göttinger Bodenkundliche Berichte* 1981.
35. **Mengel, K. and Kirkby, E. A.,** *Principles of Plant Nutrition,* 3rd ed., Int. Potash Institute, Bern, Switzerland, 1976.
36. **Mischgofsky, F. H. and Kabos, R.,** in *Altlastensanierung 88, Conference Hamburg,* Wolf, K., van der Brink, W. J., Colon, F. J., Eds., Kluwer Academic Publisher, Dordrecht, 1988, 539.
37. **Pfeiffer, E. M., Freytag, J., Scharpenseel, H. W., and Miehlich, G.,** Trace elements and heavy metals in soils and plants of the SE Asian metropolis Metro-Manila and of some rice cultivation provinces in Luzon, Philippines, *Hamburger Bodenkundliche Arbeiten,* 11, 264, 1988.
38. **Rhode, H.,** A study of the sulfur budget for the atmosphere over northern Europe, *Tellus,* 24, 128, 1972.
39. **Scharpenseel, H. W.,** Tracer investigations on synthesis and radiometric combination of organo-mineral complexes, *Trans. Comm. II and IV, Int. Soil Sci. Soc.,* Aberdeen, 1966.
40. **Scharpenseel, H. W., Pietig, F. and Kruse, E.,** Uranium contents of hydromorphic soils and soil fractions derived from accumulation sites, *Environmental Biogeochemistry,* Vol. 2, Ann Arbor Science, Ann Arbor, MI, 1976, chap. 38.
41. **Scharpenseel, H. W., Theng, B. K. G., and Stephan, S.,** Polychlorinated biphenyls (^{14}C) in soils: Adsorption, infiltration, translocation and decomposition, *Environmental Biogeochemistry and Geomicrobiology,* Vol. 2, Krumbein, W. E., Ed., Ann Arbor Science, Ann Arbor, MI, 1978, chap. 50.
42. **Scharpenseel, H. W., Eichwald, E., and Neue, H. U.,** Transfer von Chrom und Nickel am Beispiel ophiolithischer Reisböden der Philippinen, *Landw. Forsch. Sonderheft,* 38, 224, 1981.
43. **Scheele, B.,** Reference chemicals as aids in evaluating a research programme, selection aims and criteria, *Chemosphere,* Pergamon Press, Oxford, 1980, 9, 293.
44. **Scheffer, F. and Schachtschabel, P.,** Lehrbuch der Bodenkunde, 10th ed., F. Enke Verlag, Stuttgart, FRG, 1979.
45. **Schlichting, E. and Elgala, A. M.,** Schwermetallverteilung und Tongehalte in Böden, *Z. Pflanzenernähr., Bodenkunde,* 6, 563, 1975.

46. **Schlichting, E., Monn, L., Kleudgen, H. K., and Spohn, G.,** Bindung, Aufnahme und Bewegung von Schwermetallen in bzw. aus Böden. Methoden zur ökotoxikologischen Bewertung von Chemikalien, Böden, Jül-Spez. 224, *Zentralbibliothek KFA Jülich,* FRG, 2, 43, 1975.

47. **Schmidt-Bleek, F.,** Umweltchemikalien. Das Problem der Umweltchemikalien vor der Verabschiedung eines Chemikaliengesetzes in der Bundesrepublik Deutschland, *Chemosphere,* 9, 583, 1979.

48. **Scott, W. D., and Hobbs, P. V.,** The formation of sulphate in water droplets, *J. Atm. Sci.,* 24, 54, 1967.

49. **Sieffermann, R. G.,** Personal communication in connection with mutual project application, 1987.

50. **Stevenson, F. J.,** Binding of metal ions by humic acids, *Environmental Biogeochemistry,* Vol. 2, Nriagu, J. O., Ed., Ann Arbor Science, Ann Arbor, MI, 1976, chap. 33.

51. **Sticher, H.,** Chrom-und Nickeldynamik in Serpentinböden, *Mitt. Dtsch. Bodenk. Ges.,* 27, 239, 1978.

52. **Szalay, A. and Samsoni, Z.,** Investigations on the leaching of uranium from crushed magmatic rocks, *Inst. Nucl. Res. Hung. Sci.,* Debrecen, 1970.

53. **Tiller, K. G., Gerth, J., and Brümmer, G.,** The sorption of Cd, Zn and Ni by soil clay fractions: Procedures for partition of bound forms and their interpretation, *Geoderma,* 34, 1, 1984.

54. **Tiller, K. G., Gerth, J., and Brümmer, G.,** The relation affinities of Cd, Ni and Zn for different soil clay fractions and goethite, *Geoderma,* 34, 17, 1984.

55. Toetsingstabel for de broordeling van de concentration niveaus van diverse verontreinigingen in de bodem. Leidraad Bodemsanering. Ministere van Volkshuivesting, Ruimtelijke Ordening en Milieubeheer (VROM), Directoraat-Generaal vor de Milieuhygien, Staatsuitgevery, s'Gravenhage, 1983.

56. **Troitzsch, H. and Bayerl, G.,** *Technikgeschichte,* Verein Deutscher Ingenieure Verlag, Düseldorf, FRG, 48(3)177, 1981.

57. **Uehara, G. and Gilman, G.,** The mineralogy, chemistry and physics of tropical soils with variable charge clays, *Westview Tropical Agriculture Ser.,* 4, Boulder, CO, 1981.

58. **Ulrich, B.,** 1984, as cited in Sauerbeck, D., Funktionen, Güte und Belastbarkeit des Bodens aus agrikulturchemischer Sicht, *Sachverständigenrat für Umweltfragen, Kohlhammer,* 10, 181, 1985.

59. **Ulrich, B. and Matzner, E.,** Abiotische Folgewirkungen der weiträumigen Ausbreitung von Luftverunreinigungen, *Luftreinhaltung Forschungsbericht,* 104, 02, 615, Umweltbundesamt Berlin, 1983.

60. Umwelttechnik Berlin, Märkte, Entwicklungen, Lösungen, Altlasten, Bodenreinigung hat in Berlin einen hohen Stellenwert, June 1988, 3, and 16, 1988.

61. **Vetter, H.,** Umwelt und Nahrungsqualität, W. Heyne Publishers, Munich, 1980.

62. **Vinogradov, A. P.,** The geochemistry of rare and dispersed chemical elements in soils, New York, 1959, 209.

63. **Rankama, K. and Sakama, Th. G.,** *Geochemistry,* The University of Chicago Press, 1950.

64. **Scharpenseel, H.-W. et al.,** 1981.

65. Senator for City Development and Environmental Protection, Berlin, 1988.

66. **Kloke, A.,** Richtwerke 80 — Orientierungsdaten für tolerierbare Gesamtgehalte siniger Elemente in Kulturböden. Mitt. VDLUFA, H. 1-3 and 9-11, 1980.

67. **Ulrich, B.,** Okologische Gruppierung von Böden nach ihrem chemischen Bodenzustand, *Z. Pflanzenern., Bodenkunde,* 144, 289, 1981.

68. Leidraad Bodemsanering. Ministerie van Volkshuisvesting, Ruimtelijke Ordening en Milieubeheer (VROM), Directoraat-Generaal voor de Milieuhygien, Staatsuitgevery, S'Gravenhage (in Dutch), 1983.

69. **Lux, W., Hintze, H., and Piening, H.,** Heavy metals in the soils of Hamburg, in *Contaminated Soil 88, Conference Hamburg,* Wolf, K., van den Brink, W. J., and Colon, F. J., Eds., Kluwer Academic Publishers, Dordrecht, 1988, 265.

Chapter 9

EFFECTS OF NATURAL IONIZING RADIATION

Eiliv Steinnes

TABLE OF CONTENTS

I. INTRODUCTION

Radiation in the environment from natural sources is the major source of radiation exposure to man. The assessment of the radiation doses from natural sources in humans is therefore an important task. In this chapter, the various sources contributing significantly to this radiation exposure, their relative importance to the effective absorbed dose, and their geographical variability are briefly discussed.

The natural radiation sources are classified into external and internal sources. The external sources consist of one extraterrestrial component in the form of cosmic rays and one of terrestrial origin, i.e., radioactive nuclides present in the earth's crust, in building materials, and in the air. The internal sources comprise the naturally occurring radionuclides which are taken into the human body. In areas of normal background, the internal sources contribute approximately two thirds of the effective radiation exposure to humans.[1]

The radiation exposure is most conveniently expressed in terms of effective dose equivalent (unit: Sievert, Sv), which is obtained after modifying the absorbed radiation dose to account for differences in the relative biological effectiveness of radiations of different quality. Thanks to very extensive research efforts over the last decades, the relative contributions from different radiation sources to the exposure of the general population is quite well known in many parts of the world. While the differences in radiation exposure between populations on a natural scale appear to be moderate,[1] variations on a local scale may be much larger, mainly due to the uneven geographical distribution of primordial radionuclides in rocks and soils. The naturally occurring radioactive elements should therefore be considered important targets for further geomedical research.

The contributions from different natural sources to the annual effective dose equivalent in areas of normal background, as estimated by UNSCEAR 1982,[1] are presented in Table 1.

II. NATURAL RADIATION SOURCES

A. COSMIC RAYS

The high-energy radiation entering the earth's atmosphere from outer space is known as primary cosmic rays, most of which originate outside of the solar system. These primary galactic cosmic rays consist largely of protons (about 90%) and α-particles. When the particles enter the atmosphere, they produce secondary cosmic rays by undergoing nuclear reactions with nuclei of atoms present in the air. This also leads to a variety of radioactive reaction products, the most significant which are ^{3}H and ^{7}Be. The secondary cosmic rays include neutrons, which may eventually be captured by stable ^{14}N to produce radiactive ^{14}C, an unavoidable source of internal radiation. Still, the total effect of cosmic rays is mainly in external irradiation, where it constitutes about one half of the effective dose equivalent. The internal irradiation dose to man from ^{14}C and other radioactive reaction products of cosmic rays corresponds to less than 1% of the total exposure.

The effects of cosmic rays show a rapid increase at increasing altitude. At 3000 m, for example, the effective exposure is about four times that at sea level. This means that external irradiation from cosmic rays may be more significant for groups living at high altitudes than for the general population.

B. PRIMORDIAL RADIONUCLIDES

The main primordial radionuclides are ^{40}K (half-life, 1.28×10^9 year), ^{87}Rb (half-life, 4.7×10^{10} year), and the members of the two radioactive series headed by ^{238}U (Table 2) and ^{232}Th (Table 3), which have existed in the earth's crust throughout its history. Other primordial nuclides, such as the members of the ^{235}U decay series, contribute very little to the total dose from the natural background and are not discussed here.

TABLE 1
Per Capita Annual Effective Dose Equivalents (μ Sv) from
Natural Irradiation Sources in Areas of Normal Background,
as Estimated by UNSCEAR 1982[1]

Source	Irradiation		Total
	External	Internal	
Cosmic Rays	300		300
Cosmogenic radionuclides		15	15
Primordial radionuclides			
^{40}K	120	180	300
^{87}Rb		6	6
^{238}U series	90	954	1044
$^{238}U \rightarrow {}^{234}U$		10	
^{230}Th		7	
^{226}Ra		7	
$^{222}Rn \rightarrow {}^{214}Po$		800	
$^{210}Pb \rightarrow {}^{210}Po$		130	
^{232}Th series	140	186	326
^{232}Th		3	
$^{228}Ra \rightarrow {}^{224}Ra$		13	
$^{220}Rn \rightarrow {}^{208}Tl$		170	
Total (rounded values)	650	1340	2000

1. Potassium-40

Potassium is an essential element to man and is under close homeostatic control in the body. Although the potassium concentration in the body shows some variation with age and sex, the internal radiation dose, which is of the order of 10% of the total exposure, is not expected to vary much among individuals. The external dose from ^{40}K is likely to vary much more, depending on variations in the local geological environment, and may exceed the internal dose, e.g., for people living in granitic terrain.

2. Rubidium-87

The behavior of rubidium in man is not very well known, a fact which also affects dose estimates of ^{87}Rb. Nevertheless, the contribution, which is mainly an internal exposure, is probably well below 1% of the total dose from natural radiation.

3. Uranium-238 Series

Uranium-238 is the head of a series of 15 principal nuclides (Table 2) covering most elements in the Periodic Table between uranium and lead. For discussion of human health effects, it is convenient to divide the series into five subseries: (1) $^{238}U \rightarrow {}^{234}U$, (2) ^{230}Th, (3) ^{226}Ra, (4) $^{222}Rn \rightarrow {}^{214}Po$, and (5) $^{210}Pb \rightarrow {}^{210}Po$. Except for ^{222}Rn, the head nuclide of each subseries has a half-life longer than the biological half-life of the respective elements, and the probability of human intake depends to a considerable extent on the physiochemical properties of the elements in question. After uptake, a radioactive equilibrium between the head nuclide and the daughter nuclides of the subseries in question will, in most cases, be attained within a relatively short time, increasing the radiation dose relative to that caused by the head nuclide separately.

While the Uranium-238 series yields a moderate contribution to the mean external irradiation dose (Table 1), it plays a predominant role as far as internal exposure is concerned. The relative significance of each of the above subseries and the predominant route of intake of the head nuclides, however, varies considerably. For the ^{238}U and ^{232}Th subseries, dietary intake provides most of the exposure, and the same is normally the case for ^{226}Ra, although

TABLE 2
Uranium-238 Decay Series

Nuclide[a]	Half-life[b]	Principal decay mode
^{238}U	4.47×10^9 years	α
^{234}Th	24.1 d	β
^{234m}Pa	1.17 min	β
^{234}U	2.45×10^5 years	α
^{230}Th	8.0×10^4 years	α
^{226}Ra	1600 years	α
^{222}Rn	3.82 d	α
^{218}Po	3.05 min	α
^{214}Pb	26.8 min	β
^{214}Bi	19.7 min	β
^{214}Po	1.64×10^{-4} s	α
^{210}Pb	22.3 years	β
^{210}Bi	5.01 d	β
^{210}Po	138.4 d	α
^{206}Pb	Stable	—

[a] Radionuclides produced in less than 1% of the transformations are not included.
[b] Data from Lederer, C. M. and Shirley, V. S., *Table of Isotopes*, 7th ed., John Wiley & Sons, New York, 1978.

in areas where the drinking water supplies are drawn from ground waters, this may be an equally important route of intake.

These three subseries are, however, quite insignificant in this respect compared to ^{222}Rn and its daughter products, which constitute a major part of the effective radiation dose from internal irradiation. Radon-222, being in gaseous form under normal conditions, is mainly supplied to the human body by inhalation. Most of this exposure is from indoor air. In equatorial regions where domestic conditions are different, the indoor level of ^{222}Rn and its daughters is likely to be considerably lower than in the Northern temperate regions.

Also, for the ^{210}Pb subseries, inhalation is a significant source of human exposure, but in this case consumption of food is normally the most important route of entry. Populations living on reindeer meat in the subarctic regions of the Northern Hemisphere are particularly vulnerable in this respect[2] because in the winter the reindeer graze on lichens which accumulate airborne ^{210}Pb and ^{210}Po quite efficiently.

4. Thorium-232 Series

Thorium-232 is the head of a series of 12 principal nuclides (Table 3), which may, in a manner similar to the ^{238}U series, be divided into three subseries: (1) ^{232}Th itself, (2) ^{228}Ra → ^{224}Ra, and (3) ^{220}Rn → ^{208}Pb. The annual effective dose due to external irradiation from the ^{232}Th series is somewhat higher than that from the ^{238}U series in areas of normal background, while the dose due to internal irradiation is only about one fifth of that of the uranium series. As far as the ^{232}Th and ^{228}Ra subseries are concerned, the considerations are very similar to those presented above for the corresponding links of the ^{238}U series. The ^{220}Rn series, however, contributes an internal radiation dose less than 20% of that caused by the ^{222}Rn subseries under normal background conditions, although thorium is about four times as abundant as uranium in the earth's crust (Table 4). One reason for this is the much shorter half-life of ^{220}Rn (55 s), compared with that of ^{222}Rn (3.82 d), making diffusion of ^{220}Rn into the air less probable.

TABLE 3
Thorium-232 Decay Series

	Nuclide[a]	Half-life[b]	Principal decay mode
	^{232}Th	1.41×10^{10} years	α
	^{228}Ra	5.76 years	β
	^{228}Ac	6.13 h	β
	^{228}Th	1.913 years	α
	^{224}Ra	3.66 d	α
	^{220}Rn	55 s	α
	^{216}Po	0.15 s	α
	^{212}Pb	10.64 h	β
A.	^{212}Bi	60.6 min	β (64%)
	^{212}Po	3.04×10^{-7}s	α
	^{208}Pb	Stable	
B.	^{212}Bi	60.6 min	α (36%)
	^{208}Tl	3.05 min	β
	^{208}Pb	Stable	

[a] Note the two alternative decay branches of bismuth-212.
[b] Data from Lederer, C. M. and Shirley, V. S., *Table of Isotopes*, 7th ed., John Wiley & Sons, New York, 1978.

TABLE 4
**Average Composition of Some Major Rock Types
with Respect to the Main Elements Contributing to
Natural Terrestrial Radioactivity**

Rock type	U (ppm)	Th (ppm)	K (%)	Rb (ppm)
Igneous rocks				
Basalt	0.43	1.6	0.83	37
Granite	4.4	23	3.34	150
Sedimentary rocks				
Shale	3.7	12	2.45	160
Limestone	2.2	1.7	0.31	(52)
Sandstone	0.45	3.8	1.50	46
Crustal average	2.4	12	2.10	90

Note: Data from Bowen, H. J. M., *Environmental Chemistry of the Elements*, Academic Press, London, 1979.

III. OCCURRENCE OF RADIOGENIC ELEMENTS IN THE ENVIRONMENT

A. ROCKS

The ultimate sources of the basic primordial radionuclides are the earth's crust and its underlying mantle. The energy from radioactive decay plays a key role in providing the necessary heat to bring about the melting of rocks connected with magnetic differentiation processes. As a body of molten magma cools, silicates with gradually changing mineralogical and chemical composition are formed. In an early stage, the silicates tend to be rich in iron and magnesium, basalt being an example. Later on, calcium and aluminium become more prominent, and the last major silicates to crystallize are those which contain most of the potassium and rubidium, such as granites.

Neither thorium nor uranium are compatible with the crystal structure of major silicate minerals. Moreover, both elements are normally present in too small concentrations to have a strong tendency to form specific thorium or uranium minerals. They therefore tend to concentrate in the residual magma to form miscellaneous accessory minerals, along with other "incompatible" trace elements such as zirconium and rare-earth elements. Some uranium also occurs in an "interstitial" state, i.e., along grain boundaries and in defects in crystal lattices.

The above means that all four basic radiogenic elements (K, Rb, U, and Th) are likely to be present at much higher levels, e.g., in granitic rocks rather than in basaltic ones. The average composition of some major rock types with respect to these elements is given in Table 4.

The most important oxidation states of uranium under natural conditions are $+4$ and $+6$. In minerals, $+4$ is the normal state. Some uranium minerals are, however, relatively soluble, and U^{+4} is readily oxidized to the $+6$ state, in which uranium is quite mobile. This means that uranium is present in higher concentrations than most other heavy elements, e.g., in sea water. Some sedimentary rock types formed in sea water may sometimes show very high uranium values, such as shales formed in anoxic basins, where U tends to be concentrated in organic matter, and phosphorites, where the ability of U to form phosphate complexes plays an important role.

Thorium, on the other hand, occurs only in the $+4$ state under natural conditions, and all Th minerals are relatively insoluble.

B. SOILS

The radioactivity of soil is not necessarily the same as that of the rock from which it was derived. It may be diminished by the leaching action of moving water and augmented by sorption and precipitation of radionuclides from incoming water or from the atmosphere. Moreover, the soil material may have been transported a considerable distance away from the source rock area or it may be a mixture of material from different rock types. For example, the soil content of radium (^{226}Ra) may not at all correspond to the uranium content of the same soil. Over areas enriched in U, appreciable amounts of ^{226}Ra can go into solution and migrate over a considerable distance before being fixed on soil material. This differential movement of radium is important not only for the geographical pattern of ^{226}Ra uptake in crops, but, even more importantly, for the flux of ^{222}Rn from soil to air.

C. WATER

The hydrosphere is important to the assessment of background radiation in that it provides a mechanism for the movement and redistribution of radionuclides in the environment. Furthermore, radionuclides contained in sources of drinking water may constitute significant radiation exposure.

The higher mobility of uranium compared with thorium under natural conditions may be illustrated by their abundances in sea water, which are 3.2 and 0.001 $\mu g\, l^{-1}$, respectively.[3] More significant in a geomedical context is the fact that radium is sometimes present in ground waters at very high concentrations.[4] The levels of the radon daughter isotopes are expected to relate to those of radium. Population groups depending on ground water for their drinking water supply may therefore, in many cases, experience considerably higher radiation doses than the general population.

D. AIR

A major part of the activity from radionuclides in the atmosphere is due to ^{222}Rn and its daughter products. The main removal process for ^{222}Rn is radioactive decay, while the daughter products tend to associate with the ambient aerosol, and it is the aerosol which controls the further behavior of the radionuclides. The members of the uranium and thorium series prior to radon are normally present in air in very low concentrations and only in the form of soil or dust particles.

IV. AREAS OF HIGH NATURAL RADIOACTIVITY

As evident from the preceding text, the effective radiation dose due to primordial radionuclides may show considerable geographical variability, depending not only on geological factors, but also on differences in living conditions, in particular with respect to the housing situation. Considerable effort has been made in many countries to map the background radiation, and occurrences of anomalously high levels have been identified in a large number of cases. Sometimes these natural radioactive anomalies extend over considerable areas. The most well-known cases are the monazite-bearing areas in Kerala on the southwest coast of India, where thorium is the main radioactive element, and similar areas in Brazil.[5] Additional examples of areas with high natural radioactivity are discussed in the literature.[4-6] Epidemiological studies have been conducted in the high background areas in both India and Brazil, but the output appears to have been of somewhat limited use, e.g., because of small exposed populations and inadequate medical records.[5]

In Scandinavia, high concentrations of radon in indoor air, mainly penetrating into houses from the underlying rock, have been considered to constitute a significant health problem in recent years. It has been estimated that radon may be a contributing factor in 10 to 30% of lung cancer cases in Norway.[7]

Natural radioactivity is likely to become an important parameter in future geomedical research. Particularly in countries with advanced medical registers, epidemiological studies correlating parameters associated with natural radiation levels to mortality or morbidity data on a geographical basis may provide significant new insight into this field.

REFERENCES

1. UNSCEAR 1982, Ionizing Radiation: Sources and Biological Effects, United Nations Scientific Committee on the Effects of Atomic Radiation, 1982 Report to the General Assembly, United Nations, New York, 1982.
2. **Lederer, C. M. and Shirley, V. S.,** *Table of Isotopes,* 7th ed., John Wiley & Sons, New York, 1978.
3. **Bowen, H. J. M.,** *Environmental Chemistry of the Elements,* Academic Press, London, 1979.
4. National Council on Radiation Protection and Measurements, *Natural Background Radiation in the United States,* NRCP Rep. No. 45, Washington, D.C., 1975.
5. **Cullen, T. L. and Franca, E. P., Eds.,** *International Symposium on areas of High Natural Radioactivity,* Academia Brasileira de Ciencias, Rio de Janeiro, 1977.
6. **Bowie, S. H. U. and Plant, J. A.,** Natural radioactivity in the environment, in *Applied Environmental Geochemistry,* Thornton, I., Ed., Academic Press, London, 1983.
7. **Sanner, T., Dybing, E., and Stranden, E.,** Risk of lung cancer by indoor exposure for radon, Report 1988:3, National Institute for Radiation Hygiene, Oslo, 1988 (in Norwegian, English summary).

Chapter 10

EFFECTS OF RADIOACTIVE RADIATION CAUSED BY MAN

P. Becker-Heidmann and H. W. Scharpenseel

TABLE OF CONTENTS

I. THE EFFECTS OF RADIOACTIVE CONTAMINATION ON MAN

The radiochemical and radiobiological effects of nuclear radiation of man-produced nuclides are principally not different from those of natural radioactivity. Measurement of activities is, in most cases, no problem, contrary to the prediction of the effects on living beings, which act first on the microscopic scale. The concept of "dose", as a measure of the energy absorbed in tissue by radiation, should overcome this gap. The official factors for the conversion of deposition to dose are 1.4×10^{-12} Svh^{-1}/Bqm^{-2} for ^{137}Cs and 3.8×10^{-12} Svh^{-1}/Bqm^{-2} for ^{134}Cs.[1] These factors are valid for a 1-m average exposition height, e.g., for adults, and neglect any depth distribution of nuclides or their translocation during the 50 years within the soil. The simulation model ECOSYS, which considers the biogeochemical cycling of the radionuclides, predicted a dose of 1 to 2 mSv for adults and 2 to 3 mSv for preschool children for ^{137}Cs and ^{134}Cs, caused by the Chernobyl fallout in the region of Munich.[2] Transferring these results to the contamination level of Hamburg results in 0.3 to 0.7 mSv for adults and 0.7 to 1 mSv for preschool children.[3] In comparison, the population of the European part of the U.S.S.R. will be externally exposed to a dose of 50 mSv and internally to 28 mSv on average during the next 50 years. The external dose of the population within the 30-km zone was estimated to be 118 mSv.[4] The problem is that there is no exact prediction possible on how many or which persons will suffer from what diseases as a result of this average dose. This is due not only to the statistical character of nuclear radiation itself, but also to the limited database, especially in the case of little doses. In spite of all efforts, e.g., defining and refining maximum permissible levels of doses and hazard assessment, there is still the character of general uncertainty while using nuclear power. From a geomedical viewpoint, the rising and dispersing contamination of the bio-geosphere is disquieting.

II. BIOGEOCHEMICAL CYCLING OF RADIOACTIVE POLLUTANTS

A. SOURCES

As sources of radioactive radiation caused by man, one can assess a wide range of nuclides which differ in half-life, specific activity, kind and energy of emitted radiation, and physico-chemical and physiological behavior. They can be coarsely grouped into naturally occurring isotopes being released by mining, fission products, and activation products. The following discussion identifies the main occasions on which these nuclides are used and introduced into the biogeochemical cycles.

1. Energy Production

The introduction of radioisotopes into the biogeochemical cycles by man via energy production starts with mining, not only of uranium and thorium, but also with the minor contribution of coal. One kg of uranium ore contains about 1.1 kBq ^{235}U and other nuclides, such as 21.6 kBq ^{230}Th, 21.8 kBq ^{226}Ra, and 1.0 kBq ^{231}Pa.[5]

Coal contains about 37 Bq/kg of ^{238}U, 18.5 Bq/kg of ^{226}Ra, 26 Bq/kg of ^{210}Pb, 30 Bq/kg of ^{210}Po, 18.5 Bq/kg of ^{232}Th, 7.5 Bq/kg of ^{228}Th, and 130 Bq/kg of ^{40}K.[6] Brown coal shows similar concentrations.

These long-living isotopes are also present in the dust of mines and, together with radioactive decay products, e.g., the noble gas ^{222}Rn, are inhaled by mine workers as well as distributed into the atmosphere. Therefore, uranium miners suffer more frequently from cancer of the lungs.[7]

The fabrication of reactor fuel leads to a large volume of waste. According to an estimate of the European Community, 37 to 370 TBq of radon gas and 5×10^6 m^3 of fluid containing 370 GBq from isotopes of uranium, thorium, and their decay products were generated in 1970 in its member states.[8] As uranium ore contains only 0.2% uranium, the tailings comprise the major part of the ore. At the end of 1986, nuclear power plants with a total productivity of about 285.7 GWe were installed worldwide.[9] For every GWe, 1.5 to 2×10^8 kg of uranium ore is required each year.[10]

By the fission of ^{235}U, ^{233}U, or ^{239}Pu, two fragments and two or three neutrons are generated. About 40 different pairs of nuclides are known, with atomic numbers between 30 and 66. Most of these are radioactive and decay through several intermediate steps until a stable nuclide is reached. The most probable pairs of fission products have mass numbers around 95 and 140. They cannot be retained completely in the fuel rods because diffusion through defective or corroded cladding material into the cooling medium is possible, especially for the radioactive noble gases (e.g., ^{85}Kr, ^{87}Kr, ^{88}Kr, ^{133}Xe, ^{135}Xe, and ^{137}Xe). The noble gases have to be extracted, because they disturb the chain reaction. One part of the neutrons emitted by the fission reacts with the nuclides of the cladding material and the cooling medium, which results in a number of radioactive activation products, such as ^{3}H, ^{14}C, ^{16}N, ^{19}O, ^{18}F, ^{41}Ar, ^{51}Cr, ^{54}Mn, ^{59}Fe, ^{57}Co, ^{58}Co, ^{60}Co, ^{65}Zn, and ^{65}Ni. Noble gases, fission products bound to aerosols, ^{131}J, ^{14}C, and ^{3}H, are released into the air; and fission and activation products as well as ^{3}H are released into the sewage, as they are not retained totally by filters. In 1984 in the Federal Republic of Germany, 21 nuclear power plants produced 10,542 MWe and thereby released 1.6×10^{14} Bq of radioactive noble gases with a half-life of >8 d and 2.2×10^9 Bq of ^{131}J.[11]

In 1984, the nuclear research centers of Karlsruhe and Jülich, including the reprocessing unit at Karlsruhe, together released 1.2×10^{15} Bq of noble gases, 5.5×10^9 Bq of aerosols (mostly alpha-emitters), up to 6.3×10^8 Bq of ^{131}J, up to 1.1×10^8 Bq of ^{129}J, 5.7×10^{11} Bq of ^{3}H, 4.0×10^{11} Bq of ^{14}C, and 4.0×10^7 Bq of ^{90}Sr.[12] In the same year, the German nuclear research centers released 5.2×10^9 Bq of fission and activation products and 1.1×10^{14} Bq of ^{3}H into the sewage water. The nuclear power plants released 2.6×10^{10} Bq of fission and activation products and 1.1×10^{14} Bq of ^{3}H into the sewage water. About 5.5×10^9 Bq, mostly of enriched uranium, were released by the fuel rod-producing industry.[13]

Reprocessing facilities release radioactivity as high as 10 to 10,000 times the amount released by all the nuclear power plants together, which are served by them.[14]

Besides the permanent release of radioactive nuclides during normal operation, nuclear power plants produce a large amount of radioactive waste. Introduction of these products into the biogeochemical cycles has to be prevented, but an absolutely safe method of storing it has not been found yet or is still disputed. Table 1 lists the usual contents of used fuel rods. Assuming a reprocessing capacity of 1400 t of fuel per year, the Physikalisch-Technische Bundesanstalt (PTB) in Brunswick, W. Germany, calculated the amount of radioactive waste produced in 1 year by nuclear power plants to be 53,000 MWe and that produced by research, medicine, and industry to be 2.1×10^{19} Bq/year, of which 99.9% is reprocessing waste and 80% high-activity waste. The total waste has a volume of 3.53×10^4 m^3 and would need a volume of 6.62×10^6 m^3 of rocksalt if it were deposited in a salt mine.[15] The activity of the high-activity waste will decay, following a log-log curve, from about 10^{17} Bq/t of fuel to about 10^{14} Bq/t of fuel within some 100 years and to 10^{12} Bq/t of fuel within 10^6 years.[16] The deposition of radioactive waste in geological formations of various kinds, like salt or granite, has been discussed, but nearly all nations also have been dumping parts of their waste into the ocean. For example, Great Britain, Belgium, Switzerland, and The Netherlands submerged their low-activity wastes in a small sea region near the Spanish coast at a depth of 4000 m until the "London Dumping Convention" in 1983.[17] Radioisotopes released there after the corrosion of the barrels, e.g., ^{137}Cs, ^{90}Sr, and ^{60}Co, were identified in seawater (about 1.5 mBq/l each) and plankton (about 0.4 Bq/kg).[18]

TABLE 1
Contents of Used Fuel Rods of Light-Water Reactors[a]

Nuclide	Contents (% of U fuel)	$T_{1/2}$ (a)	Activity (Bq/kg U)	Radiation
^{238}U	95.00	4.5×10^9	3.7×10^{11}	α
^{236}U	0.38	2.4×10^7	2.8×10^{11}	α
^{235}U	≈ 0.71	7.1×10^8	1.8×10^{10}	α
^{239}Pu	0.49	2.4×10^4	3.6×10^{14}	α
^{240}Pu	0.22	6.6×10^3	5.8×10^{14}	α
^{241}Pu	0.10	13	1.3×10^{17}	α, β^-
^{242}Pu	0.04	3.8×10^5	1.8×10^{12}	α
Other actinides	≈ 0.06			
Fission products	≈ 3.00			

[a] $\approx 3\%$ ^{235}U after burn down of 30 MWd/kg, 1 year after removal from the reactor core.[58,59]

Data from Weish and Gruber[58] and Herrmann.[59]

Nuclear power plants are constructed for a working period of 40 years. After closing down, the radioactive building materials have to be finally deposited. The main part of the activity then originates from activation products (^{58}Co, ^{60}Co, ^{63}Ni, ^{51}Cr, ^{54}Mn, ^{59}Fe, etc.). After 25 years, this activity has lowered to about 10%, and after 40 years to about 5%, of its initial value.[19]

Another critical issue is the transport of fuel elements and, even more dangerous, of radioactive waste using roads and railways.

Meanwhile, in addition to these "normal" releases, several accidents happened in nuclear power plants which led to severe contaminations of the environment. Some examples are mentioned below.

On October 9, 1957, a fire of fuel and graphite in reactor no. 1 at the Sellafield/Windscale plant liberated, among other nuclides, $\approx 7.4 \times 10^{14}$ Bq ^{131}J, $\approx 4.4 \times 10^{14}$ Bq ^{132}Te, 2.2 $\times 10^{13}$ Bq ^{137}Cs, 3×10^{12} Bq ^{89}Sr, and 7.4×10^{10} Bq ^{90}Sr. The radioactive cloud was distributed over all of Great Britain and western Europe.[20]

At the end of 1957 or beginning of 1958, a nuclear waste deposit plant exploded near Swerdlowsk (U.S.S.R.).[21-23] An area of about 1600 km², including 14 lakes, the industrial city of Kasli, and 30 towns had been severely contaminated.[24] An extraordinarily large amount of radioactivity was released, mostly ^{90}Sr (about 40% of the total activity) and ^{137}Cs (about 10% of the ^{137}Cs inventory). Soil samples of the contaminated area contained 6.7 $\times$ 10^7 to 1.3×10^8 Bq/m² ^{90}Sr.[25]

On June 18, 1978, about 145 tons of slightly radioactive steam with 9.3×10^{12} Bq were released from a leak at the Brunsbüttel nuclear power plant in the Federal Republic of Germany.[26]

A core melting occurred at the Three Mile Island II nuclear power plant in Harrisburg (U.S.) on March 28, 1979. About 110 m³ of water, which is more than one third of the contents of the primary cooling cycle, flew out into the swamp reservoir. Via the filters, ^{131}J, radioactive krypton, and xenon were released to the atmosphere, and radioactive water reached the Susquehanna River. A gas bubble which had built up inside the containment was partly released into the atmosphere in order to avoid an oxyhydrogen gas explosion, which would have destroyed the containment. By this action, about 2% of the noble gas inventory and 10^{-7} of the ^{131}J inventory reached the atmosphere.[27]

The reactor accident at Chernobyl (U.S.S.R.) on April 26, 1986, led to the largest contamination of the environment of all known accidents in nuclear power plants until that

TABLE 2
Inventory of the Reactor Core and Released Parts of the Radionuclides During the Accident of Chernobyl[a]

Nuclide	Half-life (d)	Inventory (Bq)	Released part (%)
^{85}Kr	3.93×10^3	3.3×10^{16}	Up to 100
^{133}Xe	5.27	1.7×10^{18}	Up to 100
^{131}J	8.05	1.3×10^{18}	20
^{132}Te	3.25	3.2×10^{17}	15
^{134}Cs	750	1.9×10^{17}	10
^{137}Cs	1.1×10^4	2.9×10^{17}	13
^{99}Mo	2.8	4.8×10^{18}	2.3
^{95}Zr	65.5	4.4×10^{18}	3.2
^{103}Ru	39.5	4.1×10^{18}	2.9
^{106}Ru	368	2.0×10^{18}	2.9
^{140}Ba	12.8	2.9×10^{18}	5.6
^{141}Ce	32.5	4.4×10^{18}	2.3
^{144}Ce	284	3.2×10^{18}	2.8
^{89}Sr	53	2.0×10^{18}	4
^{90}Sr	1.02×10^4	2.0×10^{17}	4
^{239}Np	2.35	1.4×10^{17}	3
^{238}Pu	3.15×10^4	1.0×10^{15}	3
^{239}Pu	8.9×10^6	8.5×10^{14}	3
^{240}Pu	2.4×10^6	1.2×10^{15}	3
^{241}Pu	4.8×10^3	1.7×10^{17}	3
^{242}Cm	164	2.6×10^{16}	3

[a]   Activities are related to May 6, 1986.

Data from Summary Report on the Post Accident Review Meeting on the Chernobyl Accident, International Atomic Energy Agency, Vienna, 1986.

date. A nuclear explosion followed by graphite fire and core melting caused the release of radionuclides with about 3.8×10^{18} Bq, 50% of which were noble gases, within 10 d. About 3.5% of the fuel was thrown out of the reactor building, 0.3 to 0.5% was distributed within the area of the plant, 1.5 to 2.0% within a distance of 20 km, and 1.0 to 1.5% outside 20 km.[28] Table 2 summarizes the inventory and the released parts of the different radionuclides. The specific features of the release, particularly its relatively long duration and the altitude reached by the radioactive plume, about 1000 to 1500 m, favored a widespread distribution of activity, mainly across Europe. A contributing factor was the variation of meteorological conditions and wind regimes during the period of release. Activity transported by the multiple plumes from Chernobyl was measured not only in northern and southern Europe, but also in Israel, Canada, Japan, and the U.S.[29] Besides noble gases and aerosols, highly radioactive, microscopic-sized fragments of the fuel rods, so-called hot particles, were part of the fallout not only in the U.S.S.R., but also in other European countries.[30]

It also should be mentioned that proliferation of reactor plutonium cannot be prevented. The number of states which thus can develop their own atom bombs has been rising in spite of all safety regulations. These countries have been testing their bombs partly in the atmosphere, thereby contaminating further the environment.

2. Medical Applications

Besides X-rays, the radiation of radioisotopes is being used for diagnostic purposes. Examples are ^{131}J or ^{99}Tc as tracers for investigations on the metabolism of the thyroid gland, ^{55}Fe for the detection of blood loss in suspected cases of cancer of the intestines, and ^{201}Tl for localization of a heart attack. The nuclides are injected and γ-radiation determined after

TABLE 3
Most Commonly Used Radiation Sources in
Technical Applications

Nuclide	Half-life	Radiation	Radiation energy
^{147}Pm	2.7 a	β	0.23 MeV
^{85}Kr	10.7 a	β	0.67 MeV
^{90}Sr	28 a	β	0.54 MeV
^{137}Cs	29.8 a	β, γ	0.66 MeV
^{60}Co	5.3 a	β, γ	1.17/1.33 MeV
^{241}Am/Be	458 a	n	

some time. The irradiation is sometimes high, but normally restricted to the patients, and, with the exception of the waste, there is no contamination of the environment.

Radiation therapy uses safely enclosed radionuclides, e.g., ^{60}Co or ^{137}Cs for external or, if implanted, internal irradiation of cancer. This kind of radiation source is used widely. In some cases, such sources got into the hands of people ignorant of the danger. The destruction of the cladding then led to a contamination of the environment and to severe injuries. A recent accident happened in Goiania (Brazil) on September 12, 1987, with a 100-g ^{137}Cs source of high activity. About 40 g of the material reached the soil in different places of the city before the initial damages were recognized. Four persons died, 40 required intensive medical care, another 200 were injured, and some 100 tons of soil were contaminated.[31]

3. Agricultural Applications

For agricultural applications, mostly enclosed radionuclides are used, e.g., neutron probes for the determination of soil moisture in irrigation optimization. Increasingly, food is irradiated for conservation purposes. Insects are being sterilized by irradiation and sterile males released into the environment in order to control pests. High-yield or resistant varieties are being bred by irradiation-induced mutations. With the mining of potassium for fertilizer production and with fertilizing, the release of natural ^{40}K into the environment is increased.

In agricultural research, radioisotopes such as ^{3}H, ^{14}C, ^{32}P, ^{35}S, ^{54}Mn, ^{55}Fe, and ^{99}Mo are used as tracers for finding the pathways and reactions of percolating water, fertilizers, organic matter input, etc. Neutron probes and ^{137}Cs sources as gamma probes are used in the measurement of soil moisture and bulk density, and are now commonly used tools. The release of nuclides to the environment in tracer experiments is restricted to small areas only and is subject to safety regulations.

4. Technical Applications

Radiation of encapsuled radioisotopes is used for material testing, the detection of leaks, measurements of the filling level of closed containers, and the determination of quantities and thicknesses of products on conveyor belts. It also helps sterilize medicinal bulk articles, e.g., one-way syringes and bandages, and is influencing polymerization processes in plastics production. Smoke and fire detectors contain radioactive sources. Commonly used nuclides are listed in Table 3. Radionuclides are also used as electrical energy sources by decay heat, e.g., in satellites.

Nonencapsulated radioactive nuclides are included in watches, luminous paints, etc. Used as tracers, they allow the determination of process flow, mixing and turnover rates, and leakages within containers or systems of pipelines.

5. Scientific Applications

Some of the analytical applications have already been mentioned above. Sources of

TABLE 4
Total Radioactivity Released Into the Atmosphere by Nuclear Weapon Tests Through 1973

Nuclide	$T_{1/2}$ (years)	Activity (Bq)
^3H	12	1.1×10^{20}
^{85}Kr	11	4.8×10^{16}
^{90}Sr	28	7.8×10^{17}
^{106}Ru	1.0	1.6×10^{19}
^{129}J	1.6×10^7	4.4×10^{11}
^{135}Cs	2.0×10^6	1.7×10^{13}
^{137}Cs	30	1.2×10^{18}
^{239}Pu	2.4×10^4	
^{240}Pu	6.6×10^3	1.5×10^7

Data from Haury, G. and Schikarski, W., *Global Chemical Cycles and Their Alterations by Man*, Stumm, W., Ed., Dahlem Kanferenzen, Berlin, 1977, 165.

radioactivity to be accounted for here are accelerators for the study of elementary particle physics and research reactors for the production of intense neutron fluxes. The latter are used for trace-element analysis by neutron activation and for the production of radioactive isotopes for various purposes. Radioecological research uses tracer methods within limited areas, in addition to tracing the contamination of the environment caused by other polluters.

6. Military Applications

The sources of contamination of the environment with radionuclides of military origin are diverse and include, for example, nuclear bomb production, nuclear bomb testing, and nuclear reactors in military ships, submarines, and satellites. There is no doubt that the military, especially of classic atomic bomb-producing countries, has the most detailed and comprehensive experience in the field of radiation. However, because most of the reports are classified, little is known about what kinds and quantities of radioisotopes these head-quarter activities have introduced into the biogeochemical cycles. But it is well known, that, besides the tremendous contamination of the whole biogeosphere by the atmospheric bomb tests of the 1950s and early 1960s, numerous accidents have happened since the beginning of the Manhattan project. For example, as a result of the collision of two airplanes above Palomares (Spain) on January 17, 1966, four atom bombs fell to the ground. Two of the were destroyed and released plutonium, contaminating about 2.3 km².[32]

Exact data about the contamination of military plants are not available. However, it must be considerable as, for example, eliminating the environmental damage caused by U.S. plant production of nuclear weapons will cost between $91 and 200 billion and will last between 20 and 60 years, according to an estimate by U.S. Senator John Glenn.[33]

B. THE COMPARTMENTS
1. Atmosphere

The inventory of atmospheric contamination caused by man relies heavily on estimations of the input by nuclear weapon testing, nuclear power plant operation, and severe nuclear accidents like Chernobyl. Table 4 shows the amount of long-living radionuclides released into the atmosphere by nuclear weapon tests through 1973. These values were derived from limited data on the explosion strength of the fission and fusion devices and hence are only valid within one order of magnitude.[34] The table does not show the amount of ^{14}C, which has a half-life of 5730 years. The atmospheric concentration of this nuclide nearly doubled through 1963 and now is about 120% of the initial value.[35] The mixing time in the atmosphere is about 2 years, which led to nearly uniform concentrations in the Northern as well as in the Southern Hemisphere. However, the level of the Southern Hemisphere was considerably lower for a long time because the exchange between the hemispheres is slower.

The yearly input of radionuclides produced by fission and activation may be calculated from the values reported in Section II.A.1. Derived from the data reviewed by Haury and Schikarski, with 285.7 GWe installed worldwide, nuclear power plants are producing 3.3 $\times$ 10^{14} Bq/year ^{14}C, i.e., 33% of the natural production, which is about 1 $\times$ 10^{15} Bq/year.[36] The spatial distribution of aerosol-bound nuclides, contrary to the noble gases with half-lives >2 years like ^{85}Kr, is not homogeneous due to the low emission level.

The release of fission products caused by the Chernobyl accident is listed in Table 2. Due to a chimney-like effect, the aerosols rose 1000 to 1500 m, from where they were transported over long distances, but they were not distributed homogeneously throughout the atmosphere like the nuclear bomb products. Therefore, most of the radioactivity has already reached the ground.

2. Hydrosphere

The sea and the lakes receive most of their artificial radioactivity by precipitation. The range of nuclides is similar to the atmospheric one; typical nuclides are ^{95}Zr, ^{95}Nb, ^{106}Ru, ^{137}Cs, ^{141}Ce, and ^{144}Ce.

The activity of the important nuclides decreased after the test ban treaty and in 1983, for example, was about 1 mBq/l in the Rhine River. Some rivers transport both precipitation and sewage waters from nuclear plants, thus, the spectrum is slightly different and includes ^{58}Co, ^{60}Co, and ^{134}Cs. The ^{3}H content of most German rivers was about 10 Bq/l in 1985. Suspended matter is always enriched, as are the sediments. In the Rhine, Mosel, and Weser rivers, the ^{137}Cs activity of the sediments was about 30 Bq/kg or less during the last 10 years.[37] The mixing process in rivers is fast, which results in relatively homogeneous concentration profiles, whereas in lakes and the sea, the radioactive substances subside slowly, which leads to an enrichment in the sediments.

Fission products in seawater were found in different constitutions, e.g., ^{89}Sr and ^{137}Cs soluted; ^{95}Zr, ^{95}Nb, ^{144}Ce, and ^{147}Pm precipitated; and ^{90}Sr soluted as well as precipitated. As there is a slight exchange between surface and deep sea water, a measurable radioactive contamination by ^{90}Sr could be detected, even below a depth of 1000 m. The Atlantic Ocean contains the highest and the Pacific Ocean the lowest ^{90}Sr contamination. Changes with time are different in different oceans, e.g., in 1961, the ^{90}Sr contamination of seawater was higher in the Northern than in the Southern Hemisphere. Shallow seas have higher ^{90}Sr activities than open oceans of the same latitude.[38]

The surface water of the North Sea at the position of the light-ships "P8" resp. "Deutsche Bucht" showed a maximum concentration of about 50 mBq/l ^{137}Cs and about 30 mBq/l ^{90}Sr, due to the bomb fallout, and several higher peaks between 1970 and 1984, with values of up to 191 mBq/l ^{137}Cs (in 1979) and about 40 mBq/l ^{90}Sr (in 1983), which originate from the nuclear fuel reprocessing plants Sellafield/Windscale, Dounreay, and LaHague. In August 1980, the radioactivity was between 10 and 416 mBq/l for ^{137}Cs + ^{134}Cs, between 9 and 52 mBq/l for ^{90}Sr, between 0.69 and 2.45 Bq/l for ^{3}H, and between 4 and 239 μBq/l for ^{239}Pu + ^{240}Pu. In the surface water of the Baltic Sea at the Schlei estuary, ^{137}Cs activity decreased from 80 mBq/l in 1964 to about 30 mBq/l in 1968 and rose again to about 40 mBq/l since 1978.[39]

Ground water normally is depleted in radioisotopes, compared to the precipitation, as the precipitation is filtered by passing through the soil. This is of special importance as ground water serves as the major part of the drinking water supply. While ^{137}Cs is strongly bound and retained in the soil, this does not hold true for ^{90}Sr. The average radioactivity of the drinking water of Berlin from 1969 to 1984 was between 3.7 and 14 mBq/l ^{90}Sr, while

TABLE 5
Maximum Values and Ratio of Maximum to Average of Total Deposition of ^{137}Cs,
^{134}Cs, and ^{131}J Following the Chernobyl Accident

Country	Maximum deposition (kBq/m²)		Ratio of maximum to average deposition	
	Total cesium	^{131}J	Total cesium	^{131}J
Sweden	190	950	23	22
Norway	>100	Not given	>9	—
Italy	100	≈500	≈15	≈16
Federal Republic of Germany	65	160	11	10
Austria	≈60	700	≈2.6	5.8
Switzerland	41	180	5.1	4.9
Finland	≈30	190	≈3.3	3.7
Greece	28	60	5.3	2.6
Ireland	22	16	4.4	2.3
U.K.	20	40	14	8.0
Netherlands	≈9	26	≈3.3	1.2
France	7.6	Not given	4.0	—
Luxembourg	7.3	≈40	1.8	≈2.1
Denmark	4.6	4.2	2.7	2.5
Belgium	3.0	10	2.3	2.6
Turkey	0.9	8.0	11	9.1
Japan	0.41	3.8	3.2	3.2
Iceland	0.10	Small	—	—
Canada	0.065	0.24	1.5	2.4
Spain	0.041	0.09	10	9
Portugal	0.012	0.013	4	2.6
U.S.	Small	>1.9	—	>13

Data from *The Radiological Impact of the Chernobyl Accident in OECD Countries,* Nuclear Energy Agency, Organization for Economic Cooperation and Development, Paris, 1987, 23.

^{137}Cs was between <0.1 and <2.6 mBq/l.[40] The same behavior was also noted for isotopes of Ru and Rh.[41]

The amount of radioactive nuclides released into the hydrosphere during a nuclear accident lead to local peak values, e.g., on May 3, 1986, the maximum concentration of ^{131}J in the area around Chernobyl (Kiev reservoir) was 1.1 kBq/l.[42]

3. Pedosphere

The main part of artificial radioactivity within the pedosphere is caused by sedimentation of the nuclides, which have been released into the atmosphere by the nuclear bomb tests. In Germany, the contamination of the soil was about 800 to 3400 Bq/m² for ^{137}Cs and about 300 to 1700 Bq/m² for ^{90}Sr in 1967 and is decaying very slowly.[43]

^{14}C enters the soil via the vegetation by photosynthesis and litter fall. Within a soil profile, one can see its infiltration over time. One part of the fresh ^{14}C-containing organic substances penetrates quickly through the soil and may finally reach the ground water table, whereas the bigger part remains at different depths within the topsoil.[44]

The deposition of ^{137}Cs and ^{134}Cs as well as ^{131}J by the Chernobyl fallout in the Organization for Economic Cooperation and Development (OECD) countries is listed in Table 5. The total deposition of ^{137}Cs differs very much not only between different countries and regions, but also between adjacent soils. In the region of Hamburg, the values range from 1300 to 6330 Bq/m².[45] One reason is the nonhomogeneous distribution of the precipitation which contained the activity. The radioactivity of the precipitation itself might be variable, and the ratios of dry to wet depositions were different. The different vegetation cover of the soil is another factor; woods are better filters for aerosols. The highest activities were found in the fresh litter fall and organic layers of the forest soils. Moisture stage and texture had the greatest influence on the amount of deposition in the agricultural soils. Clayey soils have a higher runoff than sandy soils. Peat soils behaved like a sponge. Thus, ^{134}Cs, which was an indicator of Chernobyl fallout because of its comparatively short half-life, was found down to 20 cm, whereas in other soils the maximum depth was 15 cm. The slope of the surface also influences runoff. In a spruce stand, only 30% of the total ^{137}Cs deposition of the Chernobyl fallout reached the soil at once; the other 70% was brought to the ground during the following year by rain and litter fall.[46]

The behavior of long-living isotopes, i.e., translocation and fixation, within the soil is important. Both processes are dependent on the chemical bonds of the nuclide, certain characteristics of the soil, and precipitation.

As an example of different chemical bonds, the infiltration of the Chernobyl ^{137}Cs was three to six times faster than the atom bomb test-derived ^{137}Cs deposited about 20 years ago.[47,48] Low pH, high sand and low clay content, and high contents of competing elements result in a high mobility of the radionuclides and vice versa. In agricultural soils, their type of management is also very important. If the soil is freshly plowed and dry, it has large, deep cracks into which the rainwater can penetrate. These relations are often neglected when calculating nuclide migration velocities in soil.[49] Earlier investigations show velocities of the magnitude of some millimeters to several centimeters per year.[50] Main modifiers of the system are pH, E_h, texture, and organic matter content.

4. Biosphere

The biosphere is characterized by the phenomenon of the food chain, which often results in a stepwise enrichment of incorporated radionuclides. This is especially true for creatures living in the hydrosphere. Microorganisms like plankton partly filter the radionuclides out of the water. Small animals eat plankton and are themselves eaten by larger fish, etc. The enrichment factors range from some hundreds to several thousands. Fish in Red Lake in Minnesota (U.S.) were found to have an average radiocesium concentration 2760 times larger than the water.[51] Enrichment depends, for example, on the concentration of competing, not radioactive, trace elements. For example, the ^{90}Sr activity of fish increases with a decreasing Ca concentration of the water in which they live. A similar relationship was found for ^{137}Cs and K.[52] In general, the isotope concentrations are higher in freshwater than in saltwater fish. In the Kolksee, a lake in Schleswig-Holstein, West Germany, the highest contamination level in pike (*Esox lucius*) was found in 1965, being about 400 Bq/kg.[53]

For the terrestrial part of the biosphere, the situation is more complex. Plants take up radionuclides from precipitation via the stomata, e.g., ^{95}Zr, ^{103}Ru, ^{106}Ru, ^{140}Ba, ^{140}La, ^{144}Ce, and ^{144}Pr, as well as from the soil via their roots. The uptake is different for the different nuclides and depends on physiological as well as environmental conditions. The isotopes then are distributed to different parts of the plant. ^{91}Y, ^{95}Zr, ^{106}Ru, ^{127}Te, ^{129}Te, ^{132}Te, ^{141}Ce, ^{144}Ce, and ^{239}Pu are mostly accumulated in roots, whereas ^{90}Sr and ^{137}Cs are mainly concentrated in the above-ground parts.[54]

C. POOL TRANSFER, PERSISTENCE, AND SINKS

The main long-lasting problem with contamination by radioisotopes is their mobility between the different compartments of the biogeochemical cycle, which leads to a constant danger to living beings. This includes the possible enrichment related to pool transfer, as stated above. On the other hand, the persistence of some isotopes (e.g., in soil) results in a long-lasting external irradiation of animals and man. And last but not least, there is the hope that there are natural sinks, e.g., deeper layers of soil, or that man will find a safe artificial sink for man-made radioactivity, e.g., in salt mines. Related investigations on

TABLE 6
Soil-Plant Transfer Factors for Cs and Sr[a]

	Cs	Sr
Leaf vegetable	0.075—0.9	0.08—7.8
Potatoes	0.023—0.16	0.015—0.38
Root vegetable	0.0025—0.15	0.55—21
Grass	0.0011—14	0.018—9.8
Clover	0.004—33	0.22—7.4
Vegetation[b]	0.05	0.4

[a] Ratio of Bq/kg of fresh plant material to Bq/kg of soil dry matter.

[b] Data from Allgemeine Berechnungsgrundlage für die Strahlenexposition bei radioaktiven Ableitungen mit der Abluft oder in Oberflächengewässer (Richtlinie zu § 45 StrlSchV), RdSchr. d. BMI, RS II 2-515603/2, Bundesministerium des Inneren, Bonn, 1979.

Data from Schmidt, M., Teufel, D., and Höpfner, V., *Die Folgenvon Tschernobyl,* IFEU, Heidelburg, 1986, 29.

radioecology use the terms ''mean residence time'', defined as the ratio of content to input resp. output in steady-state systems, and ''transfer factor'', the ratio of the radionuclide activity concentration of plants to that of the soil.

For example, the mean residence time of ^{90}Sr in the stratosphere, released by nuclear weapon tests, was calculated to be about 16 months. Through the year 2000, the total amount of ^{90}Sr sedimented to the ground is estimated to be 1.39 GBq/km^2.[55]

The transfer factors which have been found in different investigations are reviewed for different plants in Table 6. The transfer factors of Cs depend more on soil conditions than on plant type. High values were reported in sandy soils with low pH and low exchange capacity, and low values in soils with high pH, high clay content, and high content of exchangeable potassium.[56] The concept of transfer factors is also used for the plant-animal pool transfer. In living beings, metabolism causes a biological half-life of the incorporated nuclides. Transfer factors as well as biological half-life depend on the feeding habits and may therefore be manipulated easier and to a greater extent than the transfer factors of the soil-plant pathway. For example, a local contamination of the soil may be handled by the exchange of surface soil, whereas the lower, permanent pollution of the pedosphere cannot.

III. PERSPECTIVES

Together with the permanent introduction of man-made radioactive nuclides into the global biogeochemical cycle by nuclear weapon production and tests, as well as by the growing nuclear energy industry, the problems deriving from nuclear accidents and misuse of nuclear power are increasing. These problems do not affect only radiobiology, radioecology, or science. Through 1986, 135,000 persons were evacuated from Pripjat and other places around Chernobyl in order to keep their doses below the maximum permissible level.[57] Most of these people will never be able to return to their home.

REFERENCES

1. **Bundesministerium des Inneren,** Allgemeine Berechnungsgrundlage für die Strahlenexposition bei radioaktiven Ableitungen mit der Abluft oder in Oberflächengewässer (Richtlinie zu § 45 StrlSchV), RdSchr. d. BMI, RS II 2-515603/2, Bonn, August 15, 1979.
2. **Gesellschaft für Strahlen- und Umweltforschung, Umweltradioaktivität und Strahlenexposition in Südbayern durch den Tschernobyl-Unfall,** GSF report 16/86, Neuherberg, Federal Republic of Germany, 1986.
3. **Becker-Heidmann, P., Bauske, B., and Scharpenseel, H. W.,** Studies on the migration of ^{137}Cs from the reactor accident of Chernobyl in soils in the region of Hamburg, in *Proc. 19th Annu. Meet. European Society of Nuclear Methods in Agriculture (ESNA),* OEFZS Report 4489, Austrian Research Center, Seibersdorf, Vienna, 1987.
4. **Bundesminister für Umwelt, Naturschutz und Reaktorsicherheit, Ed.,** *Auswirkungen des Reaktorunfalls in Tschernobyl auf die Bundesrepublik Deutschland,* Gustav Fischer, Stuttgart, 1987.
5. **Weish, P. and Gruber, E.,** *Radioaktivität und Umwelt,* Gustav Fischer, Stuttgart, 1986, 65.
6. **Bonka, H. and Horn, H.-G.,** Anwendung der Berechnungsgrundlagen auf die Emission radioaktiver Stoffe aus Kohlekraftwerken, in *Radioökologie und Strahlenschutz,* Aurand, K., Gans, I., and Rühle, H., Eds., Erich Schmidt Verlag, Berlin, 1982, 203.
7. BEIR-Report, The effects on Populations of Exposure to Low Levels of Ionizing Radiation, report of the Advisory Committee on the Biological Effects of Ionizing Radiation, NAS, Washington, D.C., Nov., 1972.
8. **Benes, J.,** *Radioaktive Kontamination der Biosphäre,* VEB Gustav Fischer, Jena, German Democratic Republic, 1981, 75.
9. **Roth, L. and Weller, U.,** *Radioaktivität,* Ecomed, Landsberg, Federal Republic of Germany, 1987, 92.
10. **Weish, P. and Gruber, E.,** *Radioaktivität und Umwelt,* Gustav Fischer, Stuttgart, 1986, 65.
11. **Bundesminister für Umwelt, Naturschutz und Reaktorsicherheit,** *Umweltradioaktivität und Strahlenbelastung, Jahresbericht,* Bonn, 1984, 267.
12. **Bundesminister für Umwelt, Naturschutz und Reaktorsicherheit,** *Umweltradioaktivität und Strahlenbelastung, Jahresbericht,* Bonn, 1984, 268.
13. **Bundesminister für Umwelt, Naturschutz und Reaktorsicherheit,** *Umweltradioaktivität und Strahlenbelastung, Jahresbericht,* Bonn, 1984, 270.
14. **Vogt, K.,** Dispersion of Airborne Radioactivity Released from Nuclear Installations and Population Exposure in the Local and Regional Environment, ZST report, Kernforschungsanlage Jülich, Federal Republic of Germany, 1973, 179.
15. **Herrmann, A. G.,** *Radioaktive Abfälle,* Springer-Verlag, Berlin, 1983, 52.
16. **Haug, H. O.,** Zerfallsrechnungen verschiedener mittelaktiver und actinidenhaltiger Abfälle des LWR-Brennstoffkreislaufes, Teil I: Modellmässig abgeleitete Basisdaten, Aktivität und Wärmeleistung KfK-3221 Kernforschungszentrum Karlsruhe GmbH, Federal Republic of Germany, 1981, 11.
17. **World Resources Institute and International Institute for Environment and Development,** Eds., *World Resources 1986-88,* Basic Books, New York, 1988, chap. IV-2.8.
18. **Feldt, W., Kanisch, G., and Lauer, R.,** The radioactive contamination of the NEA dumping sites, in *Proc. Int. Symp. Impacts of Radionuclide Releases into the Marine Environment,* International Atomic Agency, Vienna, 1981, 465.
19. **Herrmann, A. G.,** *Radioaktive Abfälle,* Springer-Verlag, Berlin, 1983, 33.
20. **Roth, L. and Weller, U.,** *Radioaktivität,* Ecomed, Landsberg, Federal Republic of Germany, 1987, 29.
21. **Medvedev, Z.,** Two decades of dissidence, *New Sci.,* 72, 264, 1976.
22. **Medvedev, Z.,** Facts behind the Soviet nuclear disaster, *New Sci.,* 74, 761, 1977.
23. **Medvedev, Z.,** *Nuclear Disasters in the Urals,* W. W. Norton, New York, 1979.
24. **Bertell, R.,** *No Immediate Danger,* Women's Press, London, 1985, part 3, chap. 1.
25. **Kriwolutzkij,** cited in Roth, L. and Weller, U., *Radioaktivität,* Ecomed, Landsberg, Federal Republic of Germany, 1987, 32.
26. **Roth, L. and Weller, U.,** *Radioaktivität,* Ecomed, Landsberg, Federal Republic of Germany, 1987, 8.
27. **Roth, L. and Weller, U.,** *Radioaktivität,* Ecomed, Landsberg, Federal Republic of Germany, 1987, 34.
28. **International Atomic Energy Agency,** *Summary Report on the Post Accident Review Meeting on the Chernobyl Accident,* Safety Ser. No. 75, report by the International Nuclear Safety Advisory Group, Vienna, 1986.
29. Nuclear Energy Agency of the OECD, The Radiological Impact of the Chernobyl Accident in OECD Countries, Organization for Economic Cooperation and Development, Paris, 1987, 19.
30. **Van der Veen, J., Van der Wijk, A., and Mook, W. G.,** Core fragments in Chernobyl fallout, *Nature,* 323, 399, 1986.
31. **International Atomic Energy Agency,** *The Radiological Accident in Goiania,* IAEA, Vienna, 1988.
32. **Roth, L. and Weller, U.,** *Radioaktivität,* Ecomed, Landsberg, Federal Republic of Germany, 1987, 8.
33. (Reuter) *Frankfurter Rundschau,* January 6, 1989.

34. **Haury, G. and Schikarski, W.,** Radioactive inputs into the environment; comparison of natural and man-made inventories, in *Global Chemical Cycles and Their Alterations by Man,* Stumm, W., Ed., Dahlem Konferenzen, Berlin, 1977, 165.

35. **Tans, P.,** A compilation of bomb 14C data for use in global carbon model calculations, in *Carbon Cycle Modelling,* SCOPE 16, Bolin, B., Ed., Wiley, Chichester, 1981, 131.

36. **Haury, G. and Schikarski, W.,** Radioactive inputs into the environment; comparison of natural and man-made inventories, in *Global Chemical Cycles and Their Alterations by Man,* Stumm, W., Ed., Dahlem Konferenzen, Berlin, 1977, 165.

37. **Mundschenk, H.,** Oberflächenwasser, Schwebstoff und Sediment der Binnengewässer, in *30 Jahre Überwachung der Umweltradioaktivität in der Bundesrepublik Deutschland,* Bundesminister des Innern, Ed., Bonn, 1986, 26.

38. **Benes, J.,** *Radioaktive Kontamination der Biosphäre,* VEB Gustav Fischer, Jena (GDR), 1981, 139.

39. **Eicke, H.-F.,** Meerwasser und Meeressediment, in *30 Jahre Überwachung der Umweltradioaktivität in der Bundesrepublik Deutschland,* Bundesminister des Innern, Ed., Bonn, 1986, 37.

40. **Rühle, H. and Gans, I.,** Trinkwasser, in 30 Jahre Überwachung der Umweltradioaktivität in der Bundesrepublik Deutschland, Bundesminister des Innern, Ed., Bonn, 1986, 64.

41. **Aurand, K., Kerpen, W., Matthess, G., Wolter, R., and Zakosek, H.,** Gefährdung von Grundwasservorkommen durch radioaktive Kontamination, in *Radioaktive Stoffe und Trinkwasserversorgung bei nuklearen Katastrophen,* report of the working group ''Trinkwasser-Kontamination'' on behalf of Bundesminister des Innern, Bonn, 1971, annex 10.

42. **U.S.S.R. State Committee on the Utilization of Atomic Energy,** *The Accident at the Chernobyl Nuclear Power Plant and its Consequences,* information compiled for the IAEA Experts Meeting, Vienna, August 25 to 29, 1986.

43. **Bundesminister für Umwelt, Naturschutz und Reaktorsicherheit, Ed.,** *Auswirkungen des Reaktorunfalls in Tschernobyl auf die Bundesrepublik Deutschland,* Gustav Fischer, Stuttgart, 1987, 171.

44. **Becker-Heidmann, P. and Scharpenseel, H. W.,** Thin layer δ^{13}C and D^{14}C monitoring of ''Lessive'' soil profiles, *Radiocarbon,* 28, 383, 1986.

45. **Becker-Heidmann, P., Bauske, B., and Scharpenseel, H. W.,** Studies on the migration of ^{137}Cs from the reactor accident of Chernobyl in soils in the region of Hamburg, in *Proc. 19th Annu. Meet. European Society of Nuclear Methods in Agriculture (ESNA),* OEFZS Report 4489, Austrian Research Center, Seibersdorf, Vienna, 1987.

46. **Bunzl, K., Schimmack, W., Kreutzer, K., and Schierl, R.,** Interception and retention of Chernobyl derived ^{134}Cs, ^{137}Cs and ^{106}Ru in a spruce stand, *Sci. Total Environ.,* 78, 77, 1989.

47. **Bunzl, K., Schimmack, W., Kreutzer, K., and Schierl, R.,** Interception and retention of Chernobyl derived ^{134}Cs, ^{137}Cs and ^{106}Ru in a spruce stand, *Sci. Total Environ.,* 78, 77, 1989.

48. **Livens, F. R. and Baxter, M. S.,** Chemical associations of artificial radionuclides in Cumbrian soils, *J. Environ. Radioact.,* 7, 75, 1988.

49. **Becker-Heidmann, P., Bauske, B., and Scharpenseel, H. W.,** Studies on the migration of ^{137}Cs from the reactor accident of Chernobyl in soils in the region of Hamburg, in *Proc. 19th Annu. Meet. European Society of Nuclear Methods in Agriculture, (ESNA),* OEFZS Report 4489, Austrian Research Center, Seibersdorf, Vienna, 1987.

50. **Miller, J. R. and Reitemeier, R. F.,** The leaching of radiostrontium and radiocesium through soils, *Soil. Sci. Am. Proc.,* 27, 141, 1963.

51. **Gustafson, P. F.,** Comments on radionuclides in aquatic ecosystems, in *Radioecological Concentration Processes,* Åberg, B. and Hungate, F. P., Eds., Pergamon Press, Oxford, 1966, 853.

52. **Feldt, W.,** Die Entwicklung der radioaktiven Kontamination in See- und Sü wasserfischen bis zum Jahre 1965, report II/66 to the Bundesministerium für Wissenschaftliche Forschung, Bad Godesberg, Federal Republic of Germany, 1966.

53. **Feldt, W.,** Fische und Produkte des Meeres und der Binnengewässer, in 30 Jahre Überwachung der Umweltradioaktivität in der Bundesrepublik Deutschland, Bundesminister des Innern, Ed., Bonn, 1986, 79.

54. **Benes, J.,** *Radioaktive Kontamination der Biosphäre,* VEB Gustav Fischer, Jena (GDR), 1981, 164.

55. **Benes, J.,** *Radioaktive Kontamination der Biosphäre,* VEB Gustav Fischer, Jena (GDR), 1981, 116.

56. **Kerpen, W.,** Bioavailability of the radionuclides cesium-127, cobalt-60, manganese-54 and strontium-85 in various soils as a function of their soil properties. Methods applied and first results, in Proc. Sci. Semin. Appl. Distribution Coefficients to Radiological Assessment Models, Louvain-la-Neuve, Belgium, October 7 to 11, 1985.

57. **U.S.S.R. State Committee on the Utilization of Atomic Energy,** The Accident at the Chernobyl Nuclear Power Plant and its Consequences, information compiled for the IAEA Experts Meeting, Vienna, August 25 to 29, 1986.

58. **Weish, P. and Gruber, E.,** *Radioaktivität und Umwelt,* Gustav Fischer, Stuttgart, 1986, 73.

59. **Herrmann, A. G.,** *Radioaktive Abfälle,* Springer-Verlag, Berlin, 1983, 39.

60. **International Atomic Energy Agency,** Summary Report on the Post Accident Review Meeting on the Chernobyl Accident, Safety Ser. No. 75, report by the International Nuclear Safety Advisory Group, Vienna, 1986.

61. **Haury, G. and Schikarski, W.,** Radioactive inputs into the environment: comparison of natural and man-made inventories, in *Global Chemical Cycles and Their Alterations by Man,* Stumm, W., Ed., Dahlem Konferenzen, Berlin, 1977, 165.

62. **Nuclear Energy Agency of the OECD,** The radiological impact of the Chernobyl Accident in OECD Countries, Organization for Economic Cooperation and Development, Paris, 1987, 23.

63. **Schmidt, M., Teufel, D., and Höpfner, U.,** *Die Folgen von Tschernobyl,* 3rd ed., IFEU, Heidelberg, 1986, 29.

64. **Bundesministerium des Inneren,** Allgemeine Berechnungsgrundlage für die Strahlenexposition bei radioaktiven Ableitungen mit der Abluft oder in Oberflächengewässer (Richtlinie zu § 45 StrlSchV), RdSchr. d. BMI, RS II 2-515603/2, Bonn, August 15, 1979.

Chapter 11

PHYSIOLOGICAL ASPECTS OF GEOMEDICINE

Kaare Rodahl

TABLE OF CONTENTS

I. INTRODUCTION

Man has not changed materially during the last 100,000 years. He is still, as was the case when he was a cave dweller, made to live not in space, but on land, preferably close to sea level, and certainly not in water. In fact, our respiratory muscles are not strong enough to maintain normal breathing at a depth greater than about 1 m below sea level when breathing through a snorkel connected to the mouth. This is because the pressure difference between the air pressure inside the lungs and the additional pressure of the water at the surface of the thorax becomes so large that the inspiratory muscles no longer have the strength to overcome the external pressure, and natural breathing becomes impossible (see Reference 1). At the other end of the pressure scale, he can barely crawl up to the top of the earth, to the summit of Mount Everest, without the use of supplementary oxygen as long as the conditions are ideal, i.e., during periods of high barometric pressure.[17] In fact, the limit of permanent human habitation is 4,500 m. It is thus clear that our natural existence on earth has definite environmental limitations.

Less than 20 million years ago, our forerunners, the hominids, eventually abandoned their life in the trees and started to forage and hunt on the ground. By that time, they had initiated a bipedal adaptation to terrestrial life.[18] Their pelvis permitted an upright posture with a bipedal gait, which freed their arms and hands for carrying their necessities with them as they walked. By the time of the Neanderthal man, appearing more than 100,000 years ago, the unique postural abilities, manual dexterity, and range and character of movement typical of modern man had been developed. This includes the ideal proportions of the human hand, with its opposable thumb, making it ideally suited for manual work, and the use of tools. Similarly, the anatomy of the human brain reflects the successful adaptation both for manual and intellectual skills. The superiority of the human brain is clearly evidenced by the proportion of the total energy production devoted to the central nervous system. While adults of most vertebrate species devote less than 8% of their basal metabolism to the central nervous system, our central nervous system consumes 20% of our basal metabolism.[8] The superior human brain capacity not only makes us ideally suited for intellectual work, but also highly suited for skilled manual and physical performance. Consequently, an exceptionally large area of our motor cortex is devoted to the motor control of our hands and fingers. It is a remarkable fact that, from all indications, this is not a recent evolutionary feature. Most likely, a human being living more than 50,000 years ago had the same potential for physical and intellectual performance, including constructing and operating a computer, as anyone of us living today.

It thus appears that ancient as well as modern man is basically made to employ his body physically in his struggle for survival. Indeed, the human body is ideally suited for physical work. We are still born with a mass of muscles equivalent to more than half our body weight for the purpose of carrying our body around and doing muscular work. These muscles are capable of transferring chemical energy in the food we eat into mechanical work, with a mechanical efficiency comparable to that of a modern automobile engine, i.e., about 20%.

As is evident from Figure 1, the factors affecting our performance include the supply of nutrients to provide fuel, proteins, vitamins, minerals, and trace elements. Since some of these elements are lacking in certain areas of the earth, this may become an environmental problem. Iodine deficiency and the development of endemic goiter is one example. Another essential factor for our aerobic metabolic processes is the oxygen content of the air we breathe. It has taken some 2,000 million years to develop an atmosphere in which one of every five molecules is oxygen at sea level. As we ascend higher into the atmosphere, the oxygen content gradually diminishes, another environmental problem. So is the environmental temperature we are exposed to, as well as humidity, wind velocity, conditions of the surface of the ground on which we move, the pressure with which we are surrounded, electromagnetic fields, and gravity.

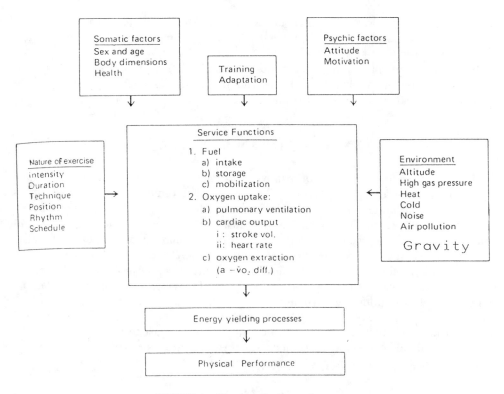

FIGURE 1.   Factors affecting performance.

II. GRAVITY

The fact that man was made to stand on his feet and to lead an upright existence in the force of gravity is also evident from what happens to him when he is put to bed and subjected to uninterrupted horizontal bed rest for 3 to 6 weeks. This causes a demineralization of his bones, due to the fact that he is now excreting twice as much calcium through his kidneys as he is taking in from his food because his long bones no longer are subject to the pressure of his body weight. And if he suddenly tries to stand up, he will faint because the mechanism regulating his blood pressure, as he constantly changes his body position while he is up and about, has not been in use while maintaining his uninterrupted horizontal position in bed.[16]

III. LOCOMOTION

If it can be assumed that the remaining primitive tribes, such as the Bushmen in Africa, the Aborigines in Australia, and the Eskimos in the Arctic, reflect the behavior of our ancestors, a great deal can be learned about our functional biology from watching them and the way they engage their endowed body potential in their struggle for survival in their particular environment. One feature which they all seem to have in common is that they move about at a fairly moderate pace when they hunt or gather food. They may stand or sit patiently for hours waiting for the game to appear, but seldom run, at any rate not for very long distances. From a strictly physiological point of view, it is evident that man was not made for marathon running, but to move about at moderate speed for prolonged periods in varied terrain, not on asphalt roads. Our most economical rates of energy expenditures are intensities below about 70% of our maximal work capacity. This will allow us to utilize our aerobic metabolic processes, using stored energy in the form of fat, of which we have an abundant amount stored in our body under the skin. This is a major advantage since our body store of fat may be at least 50 times larger than the amount of stored carbohydrates (for references see Reference 1).

It is equally true that our locomotor apparatus is ideally suited for use in varied terrain involving the varied use of muscles and joints, not for repetitive stereotype motions on paved roads which, in addition, may be harmful to the joints because of the impact against the hard surface.

IV. CIRCADIAN RHYTHMS

Another physiological feature of man which is related to his environment is that many of his vital functions affecting his performance are geared according to his circadian rhythms. These rhythms are in turn cued to the shift between light and darkness as a consequence of the rotation of the earth. These circadian rhythms, which, generally speaking, show a dip during the night, rising gradually during the day to reach a maximum in the early afternoon, include functions such as body temperature, heart rate, blood pressure, hormonal functions, etc. They are closely related to performance, generally being lowest around 4 a.m. and rising during the course of the day (for references, see Reference 14). It is thus evident that man was made to work during daylight hours and to sleep at night, not to work around the clock as modern shift workers do. The same applies, to some extent, to intercontinental travel and to changing local time on board ships travelling east to west or west to east, causing transmeridian dissynchronization. All this suggests, one may say, that, ideally speaking, man was made to remain in one place, not to roam about so much.

V. FOOD

Man, like all beasts, is a product of what he eats. His gastrointestinal tract is sufficiently flexible to allow him to live on a vegetable diet, a meat diet, or a combination of both.

The earliest humans consumed a considerable amount of meat, as is evident from the findings of large accumulations of animal remains where they lived and from the tools they used, mainly designed for the processing of game. Furthermore, their living sites were selectively located in areas where there was likely to be an abundance of large grazing animals. Widespread use of aquatic food appears to be a recent phenomenon, since shells and fish bones are infrequently found in archaeological material older than 20,000 years. From all indications, big game hunting attained increased significance with the appearance of truly modern human beings. However, during the period shortly before the introduction of agriculture and animal husbandry, there was a shift away from big game hunting toward a broader spectrum of food sources. This is apparent from the remains of fish, shellfish, and small game, as well as tools suitable for processing plant foods, such as grindstones and mortars. With the introduction of agriculture some 10,000 years ago, the proportion of meat in the diet declined drastically, while vegetable foods eventually accounted for some 90% of the diet. However, since the industrial revolution, the animal protein content of Western diets has once more increased (for references, see Reference 5).

On the basis of the available evidence, Eaton and Konner[5] have estimated the probable daily nutrition of paleolithic human beings. They conclude that the paleolithic people generally ate much more protein and far less fat than we do. Their diet contained more essential fatty acids and much higher ratios of polyunsaturated to saturated fats than our diet, although their cholesterol intake, by all indications, appeared to be high. Evidently, their calcium intake far exceeded even the highest estimates of a minimal daily requirement. Meat and entrails from game animals provided them with high iron intakes. Their intake of dietary fiber was much higher than ours, while their sodium intake was remarkably low. By all indications, their vitamin intake greatly exceeded ours.

On the whole, one is left with the impression that the ancient diet of our paleolithic ancestors was superior to ours in terms of promoting health. This impression is further strengthened by the observations that coronary heart disease, hypertension, and diabetes are relatively unknown among the few surviving hunter-gatherer populations whose way of life and eating habits most closely resemble those of preagricultural human beings.[5] Even among the more isolated Eskimos living in Alaska 35 years ago, diabetes, hypertension, and coronary heart disease were extremely rare.[11]

Because of its caloric density, fat or triglycerides is the most efficient form of stored fuel for the human body.[4] Fat is not only high in energy content, it is also stable, yet readily mobilized. The amount of energy held per unit weight of any molecule depends on its content of oxidizable carbon and hydrogen. Oxygen in a molecule of stored energy merely adds dead weight since oxygen atoms can be obtained from the air as needed. Fat, therefore, approaches maximal storage efficiency since it contains as much as 90% carbon and hydrogen and has an energy density of 9.3 kcal/g. It is much superior in storage efficiency to carbohydrates, which contain only 49% carbon and hydrogen and have an energy density of only 4.1 kcal/g. The difference is even greater when one considers that fat is deposited in droplets, while carbohydrates are deposited together with an appreciable amount of water — in mammals, 2.7 g water per gram dried glycogen.[19] In other words, hydration dilutes the energy density of stored glycogen to about 1 kcal/g. Thus, carbohydrates are a rather inferior material as an energy store in terms of portability. Consequently, in migrating fishes like the salmon and the eel, and in migratory birds, fat constitutes the main source of energy. Weis-Fough[19] points out that, owing to the impaired weight economy, carbohydrates cannot sustain a bird in flight for more than a few hours, and the recorded endurance of 1 to 3 d observed in some typical migrants therefore depends almost exclusively on the utilization of fat mobilized from stored triglycerides.

Thus, it is understandable that ancient man, through evolution, had developed the ability to store large amounts of fat, largely as subcutaneous fat depots, to tide him over periods of food shortage. The fact is, however, that modern man has retained these abundant subcutaneous fat stores even though he no longer needs them with all the supermarkets in his neighborhood.

VI. TOO MUCH AND TOO LITTLE OF ESSENTIAL NUTRIENTS

Studies of primitive tribes such as Eskimos a few decades ago[12] have shown that, to a remarkable extent, they have learned to eat what is good for them without any knowledge of nutrition. Intuitively, they chose to eat the parts of their game animals, such as the entrails, which subsequent chemical analysis has shown are especially rich in vitamins. Their ancient methods of preserving their food, as in seal pokes, have been found to preserve the vitamin content of the stored food remarkably well.[9] They have also learned to stay away from food items which may be toxic. An example of this is the liver of polar bear, which is "taboo" to the Eskimo, causing a condition called ORUTIARTUQ in the Eskimo language. It is characterized by acute gastrointestinal upset, followed about a week later by the peeling off of the surface layer of the skin, the epidermis. We now know that this condition is caused by the polar bear liver's huge content of vitamin A, causing a condition known as hypervitaminosis A when eaten in sufficient quantities.[10] The reason for this accumulation of vitamin A in the polar bear liver is that the bear eats a great deal of seal liver, which is very rich in vitamin A. The seal, in turn, feeds on vitamin A-rich fish liver, and the fish accumulate their vitamin A from other vitamin A and carotin sources in the sea. Without any knowledge of these facts, the Eskimos have always made sure that the livers of polar bears are disposed of in such a way that they may not be eaten by anyone, not even by their dogs.[10]

Another remarkable feature of primitive tribes is their ability to obtain essential elements of nutrition which may be lacking in their own local environment, such as iodine, the lack of which causes endemic goiter. An example of this was the tribe of inland Eskimos at Anaktuvuk Pass in Alaska, which we had the opportunity to study from 1950 to 1957. Originally, they were groups of nomadic Eskimos living around the headwaters of the Colville and Utukok rivers in the Endicott mountain area in the Brooks Range in Northern Alaska.[13] Here they lived as nomadic caribou hunters during the winter, living in skin tents and travelling by dog teams. In the spring, when the ice broke up on the rivers, they loaded their families and dogs into their skin boats and drifted down to the coast, where they spent the summer among their relatives and friends, hunting seal and living on marine animals and plants. In the fall, they returned to the hills.

It so happened that the food and water sources in this mountain area were extremely poor in iodine content. But this was remedied by their summer stay on the coast, living on a diet extremely rich in iodine. This prevented them from developing endemic goiter. By the time we met these inland Eskimos, they no longer took the trouble to travel to the coast every summer, but, rather elected to remain in the Endicott mountains around the area Anaktuvuk Pass the year round. When we first visited them in 1950, they were not using salt, and certainly not iodized salt, in their diet.

The geological features of the region would indicate that the natural sources were low in iodine. During the summer, drinking water was collected from the mountain stream running through the tent camp. During winter, drinking water was obtained from ice collected from a small lake or from snow in the vicinity of the camp. Samples of the drinking water contained 0.1 to 0.6 μg iodine per liter.[15]

The diet of this inland Eskimo tribe was also suggestive of iodine deficiency, their main diet being caribou, especially the meat of caribou. At times during the fall, they subsisted on mountain sheep, and occasionally they ate snowshoe hare and ptarmigan for a change. No fish were consumed during the winter by any of these people. In the summer, however, some freshwater fish, such as grayling, lake trout, and whitefish, were consumed. On the basis of an analysis of the iodine content of their most important food items, it was estimated that less than 50 μg iodine were obtained per person per day, as against a daily requirement of about 150 μg/d.

In 1956, 16 out of 27 persons about 17 years of age had definite thyroid enlargement, and thyroidectomies had to be performed on three of the women.

Four of these Eskimos were then given 600 μg potassium iodide daily for 3 months. Five other Eskimos who received no potassium iodide supplement were used as controls. The thyroid function was measured by isotope technique before and after the treatment. The results showed a statistically significant improvement in the group which was treated with iodine.[15]

VII. TEMPERATURE

Unlike furred animals, man is born nude, unprotected against climatic extremes. In this sense, man may be considered a tropical animal inasmuch as he requires an ambient temperature of about 28 to 30°C in order to maintain thermal balance at rest in the nude. In other words, without clothing and shelter, man could only survive in a fairly limited zone along the Equator. No wonder, therefore, that this was where his original cradle eventually was to be found. Subsequently, as he developed ways to protect himself against more hostile environments, he began to venture into colder regions. Consequently, as pointed out by Burton and Edholm,[3] the center of population has now shifted to a region nearly 5°C colder than where civilization arose.

The protected human being now may well tolerate variations in environmental temperature between $-50°C$ and $+100°C$. But a person can tolerate a variation of only about $4°C$ in deep body temperature without impairment of optimal physical and mental performance. The maximal limits which the living cell can tolerate range from about $-1°C$ at one end of the scale, when the ice crystals formed during freezing break the cell apart, to about $45°C$ at the other end of the scale, when thermal heat coagulation of vital proteins in the cell occurs. They can only tolerate an internal temperature exceeding $41°C$ for short periods of time. In fact, many animals, including humans, live their entire lives only a few degrees removed from their fatal thermal limit (for references, see Reference 1).

As pointed out by Hardy,[6] the hot end of the scale is more problematic than the cold end, for people can protect themselves more easily against overcooling than overheating. Consequently, the controlling mechanism for temperature regulation is particularly geared to protect the body tissues against overheating.

The physiological and biochemical processes underlying our performance are temperature dependent. Impairment of psychomotor function has been observed in subjects exposed to cold. Manual dexterity is impaired in the cold, largely due to reduced nervous conduction velocity (for references, see Reference 1).

Prolonged exposure to cold may cause local cold injury in the exposed parts of the body, such as the face, hands, and feet, either due to the freezing of tissue and formation of ice crystals or by vasoconstriction causing deprivation of the blood circulation to the exposed parts, leading to ischemic cold injury. Local cold injury of the eyes, i.e., transitory epithelial damage of the cornea with the formation of corneal edema and blurred vision, has been observed in cross-country skiers and in speed skaters competing at very low temperatures.[7] Prolonged general exposure to cold may lead to hypothermia, characterized by body temperatures below $35°C$.

While some degree of local acclimatization to cold may occur as the result of regular exposure to cold, characterized by our increased local skin temperature, it is still an open question whether or not a general acclimatization to cold may occur in man. At any rate, any such general acclimatization to cold may be of limited practical value. In our studies of Alaskan Eskimos, we were unable to find any evidence of general acclimatization to cold, and concluded that the reason why the primitive Eskimos were able to get along better in the Arctic regions than we do was because they had become used to the problems of the North and had adapted their habits accordingly.[11]

Generally speaking, it appears that the limit for efficient outdoor activities under exposed conditions is in the vicinity of $-20°C$. Below $-40°C$, life becomes more difficult, even for the Eskimo. At this point, even animal activity is markedly reduced, and with increasing cold, it ceases almost completely.

Most, if not all, of the increased energy cost of physical activity in the cold is due to the hobbling effect of cold weather clothing and the extra effort required to move through snow-covered terrain. Even when using skis to travel over level ground, carrying a load or pulling a sled may at times require almost as much energy as walking on foot (unpublished results).

Studies which we made on Eskimos living in tents or primitive native houses in Alaska revealed that they maintained a semitropical microclimate inside their clothing during the day and underneath their cover of furs during the night.[11]

While, as already mentioned, the possibility of a general acclimatization to cold is questionable, there is no doubt that man can become acclimatized to heat. After a few days' exposure to a hot environment, one is able to tolerate the heat much better than when first exposed. The improvement in heat tolerance is associated with increased sweat production, a lowered body temperature, and a reduced heart rate. The effect of heat acclimatization persists for several weeks following heat exposure, although some impairment in heat tolerance may be detected a few days after cessation of exposure (for references, see Reference 1).

VIII. BAROMETRIC PRESSURE

While it is quite evident that our human existence started more or less at sea level close to the equator, we not only have spread across the earth, both to the north and to the south, but also into the higher altitudes. This we have done, to some extent, at the expense of impaired physical performance. The impairment is already evident at an altitude of about 1200 m in the case of heavy exercise engaging large muscle groups for periods longer than a couple of minutes. A corresponding altitude is experienced by passengers in an ordinary airplane cabin. This, we know, occasionally may cause minor complications, especially circulatory, during prolonged flights. It was only 100 years ago that it was recognized that the detrimental effects of high altitude were due to the diminished partial pressure of oxygen at reduced barometric pressure.[2] The barometric pressure at a given altitude depends on the weight of the air column over the point in question. The atmosphere is compressed under its weight. Its pressure and density are therefore highest at the surface of the earth and decrease exponentially with altitude. The air temperature is, on the whole, lower, the higher the altitude. The air also becomes increasingly dry with increasing altitude. Therefore, water loss via the respiratory tract is higher at high altitudes than at sea level. This may cause dehydration in people who are physically active at high altitudes. Solar radiation is more intense at high altitudes, and ultraviolet radiation may give rise to sunburn and snow blindness. Finally, the force of gravity is reduced with the distance from the earth's center. This may improve performance in athletic events involving jumping and throwing.

Prolonged exposure to reduced oxygen pressure in inspired air is gradually compensated by an increase in pulmonary ventilation, an increased hemoglobin concentration in the blood, and tissue changes such as increased capillarization, increased myoglobin content, and modified enzyme activity. All these adaptive changes are reversible, but it may take several weeks before they return to sea-level values in the case of sea-level dwellers who have stayed at high altitudes for a month or more (for references, see Reference 1).

IX. SOCIAL IMPLICATIONS

This brief review of the physiological consequences of the wide range of environmental conditions which we may be exposed to may have served to underscore our inherent abilities to adjust to them. It remains to be seen if we are equally well equipped to cope with the social implications, and if we use or exploit the geo-resources available to our best advantage.

The ability of the human brain to invent, create, and construct is astonishing. It remains to be seen whether or not man's brain has developed its capability of ethical conduct and a responsible application of its endowed potential.

Originally, our forerunners lived in groups of limited numbers, in tribal flocks based on family units — a structure ideally suited for the care of the human young, who, in contrast to most other animals, remain helpless for so long. In the basic family unit, as we have seen it in some of the surviving native Eskimo tribes, daily chores and duties are shared and specifically divided among family members. This applies to providing game for the daily meals, carrying water or ice, nursing the fire, and caring for the young. Above all, they maintain discipline, obey the rules, and respect their leaders and those in authority. They have to honor the old, for they need their experience and advice. And they look after each other and see to it that no one is left alone to go too far astray. Their cultural heritage is passed down from generation to generation by word of mouth. They live on food which is the natural product of the land and the sea around them, and which is restored by natural processes as long as they are not overexploited. They kindle their fire by wood which is regrown. They dress in fur from the animals they kill to eat, and which will remain as a source of food and clothing as long as the species remain.

Today, the family unit is in danger of developing into a fragile cradle for future generations, which have to face a world far more complicated than that of our forefathers. Both parents may have to work in order to meet living expenses, and this may have to be done at the expense of coming generations and the children's sense of security. The greater independence of women, largely caused by more women having an education as good as or better than that of the men, has lead to more families breaking up, which, apart from the emotional upheavals involved and the negative effect on the children, increases the need for housing since more people live apart. Because of greater attention being devoted to those who are less endowed at the expense of those who have the ability to become our leaders and supporters, it may no longer appear worthwhile for the latter to do their utmost for the good of us all, as long as it is not rewarded. In our anxiousness, we have used our miraculous technical aids to exploit our natural resources, exterminating fish and mammals in the sea around us, which traditionally has been the basis for much of our coastal habitation. We no longer limit our fuel utilization to replenishable wood, but, rather, tap our limited resources of oil, which provides us with a unique fuel for such means of transportation as the airplane, which is going to be difficult to replace once the oil resources are depleted.

At the time our ancestors roamed in small bands, any destructive consequence of their activity was quite limited. But today, because of social developments and technical innovations, basically the same human brain is capable of turning man into a self-destructive monster. Since there seems to be some undesirable elements in the nature of most of us, all it takes for such tendencies to develop is for society to encourage it. This we have seen in the past, and we see it in different parts of the world today. The question is whether or not society as a whole will do something about it in time. This applies not only to the use of deadly weapons in global conflicts, but also to genetic experimentation and the possibility of producing certain types of individuals more or less at will.

In our urge to exploit our natural resources, we have destroyed rain forests and turned fertile land into deserts, and we have started to destroy part of the protective ozone layers of the atmosphere. All this we have done with our eyes open, fully aware of what we are doing.

Yet, one is left with the impression that man has a remarkable ability to adapt, to acclimatize, to live with a wide range of environmental stresses, and to survive in spite of all odds against him. Even getting to the moon is no longer impossible, although it may not be a place for man to spend his life. The fact remains, however, that the evolutionary process continues, and more recent mammalian history has seen a wave of extinction, particularly severe for large mammals, including the hominids. It has been estimated that of the 2 billion species which have appeared on earth during the last 700 million years, only 0.1% have survived, including *Homo sapiens*. This extinction, evidently, is merely due to natural selection and occurred long before we had the threat of the atomic bomb. Thus, there is no guarantee that the human species will survive forever. At any rate, with the slow but steady cooling of our planet which is taking place, it will, in all probability, eventually be no better to live here than on the moon, even though it may take a billion years or two.

REFERENCES

1. **Astrand, P.-O. and Rodahl, K.,** *Textbook of Work Physiology,* 3rd ed., McGraw-Hill, New York, 1986.
2. **Bert, P.,** *La Pression Barométrique,* Masson et Cie, Paris, 1878.
3. **Burton, A. C. and Edholm, O. G.,** *Man in a Cold Environment,* Edward Arnold, London, 1955.
4. **Dole, V. P.,** Fat as an Energy Source, in Rodahl, K. and Issekutz, B., Jr. Eds., *Fat as a Tissue,* McGraw-Hill, New York, 1964.
5. **Eaton, S. B. and Konner, M.,** Paleolithic nutrition, *N. Engl. J. Med.,* 312(5), 283, 1985.

6. **Hardy, J. D.,** Central and peripheral factors in physiological temperature regulation, in *Les Concepts de Claude Bernard sur le Milieu Intérieur,* Masson et Cie, Paris, 1967.
7. **Kolstad, A. and Opsahl, R.,** Cold injury to corneal epithelium, a cause of blurred vision in cross-country skiers, *Acta Ophthalmol.,* 48, 789, 1979.
8. **Mink, J. W., Blumenschine, R. J., and Adams, D. B.,** Ratio of central nervous system to body metabolism in vertebrates: its constancy and functional bases, *Am. J. Physiol.,* 241, R203, 1981.
9. **Rodahl, K.,** *Vitamin Sources in Arctic Regions,* Norsk Polarinstitutts Skrifter, Nr. 91, Oslo, 1949.
10. **Rodahl, K.,** *The Toxic Effect of Polar Bear Liver,* Norsk Polarinstitutts Skrifter, Nr. 92, Oslo, 1949.
11. **Rodahl, K.,** *North, the Nature and Drama of the Polar World,* Harper & Row, New York, 1953.
12. **Rodahl, K.** *The Last of the Few,* Harper & Row, New York, 1963.
13. **Rodahl, K.,** *Akiviak,* W. W. Norton, New York, 1979.
14. **Rodahl, K.,** The Physiology of Work, Taylor & Francis, London, 1989.
15. **Rodahl, K. and Bang, G.,** Endemic goiter in Alaska, *Arch. Environ.,* Health, 4, 11, 1962.
16. **Rodahl, K., Birkhead, N. C., Blizzard, J. J., Issekutz, B., Jr., and Pruett, E. D. R.,** Fysiologiske Forandringer under Langvarig Sengeleie, *Nord. Med.,* 75, 182, 1966.
17. **Sutton, J. R., Jones, N. L., Griffith, L., and Pugh, C. E.,** Exercise at altitude, *Ann. Rev. Physiol.,* 45, 427, 1983.
18. **Valentine, J. W.,** The evolution of multicellular plants and animals, *Sci. Am.,* 239(3), 67, 1978.
19. **Weis-Fogh, T.,** Metabolism and weight economy in migrating animals, particularly birds and insects, in Blix, G., Ed., *Nutrition and Physical Activity,* Almquist & Wiskell, Uppsala, 1967, 84.

Chapter 12

ZOONOSES AND GEOMEDICAL FACTORS

Per Slagsvold

TABLE OF CONTENTS

I. INTRODUCTION

Zoonosis literally means animal disease. By wider usage, it has come to mean diseases and infections which are naturally transmitted between vertebrate animals and man.[1] Zoonoses comprise the most significant group of communicable diseases and this emphasizes their prime importance in public health. A wide variety of infections are classified as zoonoses. They are caused by helminths and arthropods and by such microorganisms as protozoa, fungi, bacteria, rickettsia, and viruses. The relationship between diseases in humans and animals due to an imbalance in nutrients on the one hand and geomedical factors such as elements in the soil and climate on the other is discussed later in this book.[2] Some zoonoses are not much influenced by geomedical factors, for instance, rabies and trichinosis. They are spread over the most varied geographic and climatic conditions on the earth. These diseases can be found in a polar fox in the arctic zone as well as in a jackal or a rat in the tropics. Man and animals can be attacked. Influenza is another zoonosis that has a wide geographic distribution and can affect different species, among them man and domestic animals, especially pigs. Mouth and foot diseases also have a wide, often pandemic distribution. They infect through contact between animals and their excreta and secreta as well as through dead materials. The infection may sometimes be transported by the wind, for instance, from the European continent to Norway.[3] Most zoonoses are more influenced by geomedical factors other than the wind. The agents causing them may have a development cycle with stages living in hosts requiring special conditions. It may be an aquatic or amphibian snail or a land snail living on lime rocks. It could also be tse-tse flies, with their breeding places on swampy, humid ground in the forests and fields along the shores of the great lakes or on the banks of the rivers of Africa. Sticking mosquitoes (Culicidae) breed in slow-moving or stagnant water all over the world, while the entire development of the small Simuliidae, from egg through larva and chrysalis, is bound to flowing water.[4,5] Below will be given a few examples to illustrate the relationship between geomedical factors and zoonoses.

II. SPECIAL VECTOR GROUP, EXAMPLES

A. TREMATODES
Trematodes, such as the cattle fluke and the lancet fluke, are dependent on snails for their development. The same is true of *bilharzia* in man. They all set off their eggs with the stool of their hosts to pastures or to water.

1. Cattle Fluke *(Fasciola hepatica)*
The cattle fluke is distributed universally. The frequency of its local distribution depends on the local living conditions of the host, the living amphibian snail (*Lymnae trunculata*). In this host, the fasciola goes through the development to an infective larva. The living conditions for the snails is dependent on the humidity of the ground. The number of snails is great in areas with high annual precipitation, but may also be high in areas with low precipitation if the area is flooded for some time of the year. Clay areas are especially exposed.[6] Ruminants grassing such pastures will have their livers full of liver flukes, and man, by chewing grass, can ingest the infective larva, the metacercaria, and get the disease.[4]

2. Lancet Fluke
The lancet fluke, *Dicrocoelium lanceolatum*, lives in the temperate zone of the Northern Hemisphere. It requires two intermediate hosts for its development, a land snail and an ant. The snail requires a rocky soil rich in lime.[5] The infected ants crawl up on the grass straw and are eaten by ruminants, which become infected. Man can also be affected, by chewing straw.

3. *Bilharzia* or *Schistosomiasis*

Bilharzia or *schistosomiasis* is widely distributed in the tropics and does much harm to man and animals. Eggs from the feces of infected persons or animals are brought into shallow water where the larva, mirazidium, invades a living aquatic snail of the genera *Biomphalaria, Bulinus,* or *Onchomelania.*[4] The infective metacercaria, which invades man and domestic animals through the skin or digestive tract during wading, bathing, or drinking, is produced by a sexual reproduction. The disease affects 200 million people annually and uses dogs, cattle, and sheep as reservoir hosts. The worms reside in the blood vessels and affect the bowel, bladder, and intestine.[1] It is a disease of rural development, where artificial lakes and channels are constructed and become the living places of the water snails. The disease would be less of a problem if the society were not so dependent on the irrigation channels for washing, ablution, and recreation.

B. CESTODES

Cestodes or tape worms are dependent on one or more intermediate hosts in their development and they may require special geophysical conditions. *Diphyllobotrium latum* is the largest of the tapeworms found in man (maximum length, 20 m). It lives in the temperate and subarctic zones of the Northern Hemisphere. As the worm grows, it gives off the old segments full of eggs (the proglottids) to the feces of the host. If the proglottids enter fresh water, the larva (korazidium) becomes free. It looks for a small crustacian (cyclops), where it develops into the next stage, the procercoid. If the cyclops is eaten by a freshwater fish, a later stage, the plerocercoid, develops. When the fish is eaten raw by a dog or a cat or man, a new tapeworm can start its reproductive life. This tapeworm takes much energy as well as special nutrients. Some percent of infected humans will suffer from a deficiency of vitamin B12. Weakness and dizziness may appear as well as stomach pain and diarrhea. The main objective here is to prevent feces from entering the water.

C. NEMATODES

Nematodes cause many zoonoses. Filariasis is a common name for most of the well-known and closely related groups. Around 300 million people annually have this disease. Primates, dogs, cats, and other mammals are reservoir hosts.[1] The mated female resides in the lymphatic vessels and glands and generates elephantiasis, with usually painful swellings. The worm may also cause river blindness. The disease is transmitted by different arthropods, mainly mosquitoes belonging to the genera *Anopheles, Aedes*, and *Mansonia*. The prevention and control of the disease includes eradication and control of the vectors. In man and selected animals, insect repellents are used as well as mosquito nets for man.

Dracunculiasis, or infection with guinea or medina worm — Around 80 million people annually are affected by this disease.[4] The infection occurs mainly in an arid climate, such as the Middle East, central Asia, India, and tropical South America. Humans are the main host, but reservoir hosts are carnivores, monkeys, cattle, and horses. The full-grown female is 1 to 2 m long, while the male is only 4 cm. Infection starts by drinking unclean water, which may contain crustacean cyclopes infested with an infective larva of the medina worm. After uptake in the final host, the worms grow and reach sexual maturity. After mating, the male dies. The female moves in the subcutis of the limbs and breaks through the skin on the foot or leg. The female empties a great amount of larvae, which may enter the water at the washing place. Some may reach new cyclopes and start a new cycle. This is a painful disease, and it can only be controlled by keeping infected individuals away from water sources and by boiling or filtering all drinking water.

D. PROTOZOA

Protozoa produce some very dangerous diseases in domestic animals and man. Many of them occur only in the tropics and the subtropics. Some are pathogenic to man and animals and are included among the zoonoses. Others, such as many types of malaria, have man as the main host.

Babesia bovis has, as its name implies, its main host in cattle, where it causes high fever, increased heart rate, fatigue, and discolored mucous membranes and urine. Older, mismanaged animals have the least resistance against the disease. *Babesia bovis* also occurs in the temperate zone. In Norway, the disease is seen along the southwest coast, but is never found in the hinterland.[7] In Germany, the disease also is seen at some distance from the sea, partly in swampy forests and bush vegetation. In both countries in the fall, the occurrence of the disease is clearly related to the living conditions of the tick *Ixodes ricinus*, the main transmitter of babesia bovis. The larva, the nymph, and the imago suck blood, which is necessary for their further development and function. Animals raised in a babesia area get the disease with milder symptoms than do cattle raised in a disease-free area. Where possible, bush and other breeding and hiding places for the ticks should be eliminated. The tick even transmits the disease to man, and is seen in Mexico, the U.S., Ireland, Scotland, France, Yugoslavia, and the Soviet Union.

Trypanomosiasis is any disease caused by a trypanosome. Many of these diseases are zoonoses. They are transmitted by different arthropods. According to Greenham,[1] trypanosomiasis affected 10 million people, and has dogs, pigs, cattle, game animals, and many rodents as reservoir hosts. Chagas disease in South America is produced by a trypanosome whose target organ is the heart in man, while the different trypanosoma of the brucei group in Africa have as their main target organ the central nervous system of man. In Africa, the main transmitters of the diseases are different tse-tse flies (*Glossina*), horseflies, (Tabanidae), stomoxys, and some mosquitoes may to some extent be transmitters. In Africa, some large breeding places of the tse-tse fly have been cleared of bush and dusted with pesticides. But the fight has been hampered by the emergence of insecticide-resistant flies. At the same time, ecological problems have resulted from the use of insecticides. It is important to remove bushes and trees near the living places of man to keep the tse-tse flies away. In tropical America, trypanomiasis is mainly transmitted by assasin bugs (*Triotoma infestans*). The elimination of the Chagas is a matter of personal and public hygiene and housing. The economic and social improvement of common man is perhaps most important for a solution of the Chagas problem.

E. VIRUS DISEASES

Many virus diseases are zoonoses. They can be transmitted by direct contact between animal and man or via body liquids, excrement, feedstuffs, and emballage. The disease may also be transmitted by flies, mosquitoes, ticks, or rodents. There are usually many modes of transmission, and viruses do not go through any special development in the vector, as was shown for different worms. The geomedical interest in viruses is therefore small, compared to that in worms. But the possible transport of mouth and foot disease by wind was already mentioned.

F. BACTERIA

Bacteria-produced diseases possibly represent the greatest group of all zoonoses. Much the same may be said about methods of infection, as was described above for viruses. This can infect through direct contact from animal to man and vice versa, and indirectly through passive contact or through a lot of vectors. But the infective agent is usually not changed. Below is mentioned a case where a climatic change seemed to weaken the resistance of the animal or man, with the subsequent increased virulence of the bacteria. This happened with *Rhusiopathia suis*, the bacteria causing Red fever (swine erysipelas or diamond disease). The bacteria is common in soil, mussels, crustaceans, fish, birds, and mammals. The change from a temperate dry climate during spring to a hot, humid one during summer breaks down the resistance of the pig so much that the earlier resting bacteria can enter the organism from the pharynx or from the intestine. The disease breaks out and is transmitted to other pigs in the same bin or stall. The virulence is steadily increased by passing to new weak animals, and can then infect pigs kept under better conditions. Herdsmen, veterinarians, and workers at abattoirs and fishing shops are infected through wounds on their hands.

III. CONCLUSION

Generally, it can be said that geomedical aspects are not the most important factors in zoonoses. Still, they are quite important for the outbreak and transmission of some diseases. Lastly, it should be pointed out that deficiencies or imbalance in elements can break down the general resistance. It is, however, more difficult to associate an infection with a deficiency of a special element.

REFERENCES

1. **Greenham, L. W.,** The zoonoses, in *Microbes, Man and Animals,* Linton, A. H., Ed., John Wiley & Sons, New York, 1982.
2. **Slagsvold, P.,** Geomedical approaches in education of veterinarians, in *Handbook of Geomedicine,* Läg, J., Ed., CRC Press, Boca Raton, FL, 1989, chap. 14.
3. **Slagsvold, L.,** *Lectures in Special Pathology and Therapy,* Norges Veterinaerhøgskole, Oslo, 1940.
4. **Krauss, H. and Weber, A.,** *Zoonosen,* Deutscher Arzte Verlag, Köln, 1986.
5. **Hutyra, F., Marek, J., and Manninger, R.,** Spezielle Pathologie und Therapie der Haustiere, Gustav Fischer, Jena, 1941.
6. **Økland, F.,** Utbredelse og hyppighet av den store leverikte, *Norsk Vet. Tidsskr.,* 55, 317, 1934.
7. **Slagsvold, L.,** Husdyrsykdommer som er knyttet til beitene, *Norsk Landbruk,* 44, 252, 1954.

Chapter 13

GEOMEDICAL APPROACHES IN THE EDUCATION OF PHYSICIANS IN NORWAY

Olav Hilmar Iversen

TABLE OF CONTENTS

I. INTRODUCTION

The well-known old doctrine of tellurism, or the miasma theory, took for granted that diseases that were endemic in certain areas were caused by noxious emanations from the soil. After the breakthrough of microbiology, tellurism was put aside because of the popularity of the infection theory. In many ways tellurism, with its miasma theory based on the lack of knowledge at that time, was still the conceptual forerunner of modern geomedicine.

Modern geomedicine deals with knowledge about the influence of geofactors on the geographic distribution of diseases. Geo-factors are all influences from the external environment, i.e., earth, water, air, climatic conditions, and irradiation.

Many other factors, however, may influence the panorama of health and disease and give endemic prevalence of diseases. Among these are the man-made changes of the environment. A recent example is the Chernobyl catastrophe which contaminated air, water, and soil across large districts, and for some years has remained as a geo-factor that in theory may become a disease-provoking influence. Concentrations of heavy industry may have similar effects in some areas. Since external factors may influence all biological life in an area, thorough knowledge of the transport of energy and substances from plants to animals and to man is necessary. Accordingly, there is no sharp borderline between geomedicine proper, and other geo-factors that determine endemic occurrence of diseases or influence the frequency of diseases and the state of human and animal health in the area.

II. EDUCATION

Norway has four medical schools. The oldest and largest one is located in Oslo, the others in Bergen, Trondheim, and Tromsø. Each university has its own curriculum, but the content of the teaching at the medical faculties is largely the same, even if the methods of teaching vary somewhat.

There is no systematic teaching of geomedicine at the Norwegian medical schools. There exists, however, a textbook in preventive medicine which is probably used at all the four medical faculties, *viz., Preventive Medicine*, Vol. 1 and 2, by Haakon Natvig, Tor Bjerkedal, Arne T. Høstmark, and Odd D. Vellar. The latest edition was printed in 1982. In Volume 2 there is a chapter on geomedicine, written by Dr. Eystein Glattre at the Cancer Registry. This chapter covers 11 pages, and starts with a definition. Then follow some ecological remarks, some examples of geo-factors' influence on health and disease, a section on how to produce maps as geomedical information, and how to test hypotheses based on epidemiological evidence. In Trondheim there are plans for a chair in epidemiology at the Institute of Community Medicine, and geomedicine is included in the planned teaching and research program.

However, even if geomedicine is not taught as a separate entity under a separate heading, and even if there are presently no chairs devoted to this subject, geomedical problems are naturally discussed in many other contexts. In general pathology geomedicine is touched upon when the causes of diseases are mentioned, especially deficiency diseases, oncology, cardiology, and nephrology.

Some geomedical factors that are discussed in the education of physicians in Norway are, e.g., lack of iodine in the soil in certain areas, with endemic goiter as a result, and perhaps too much iodine in the food from fish consumption by the coastal population (especially in the northern districts of Norway) which may influence the incidence of thyroid cancer. We have the problem of fluoride and tooth decay, variations in trace substances in soil and water of, e.g., selenium, and more or less intense background radiation from radon. There are also great differences in the intensity of UV radiation in various districts of Norway.

The production of maps has during recent years become popular for characterizing geomedical factors and their consequences in Norway. The Norwegian Cancer Registry, in cooperation with Director Björn Bölviken at the Geological Survey of Norway, Trondheim, has published a map on cancer frequency in the various communities of Norway, and in 1988 a Scandinavian map of cancer incidence was published. Similar maps are worked out for heart diseases.

Geomedical aspects of infectious diseases have always been important, especially due to the influence of geo-factors on the spreading of contagious agents. Geomedicine also provides information to people who wish to travel to other areas where geomedical factors and endemic diseases are characteristic or prevalent. This concerns immunoprophylaxis, chemoprophylaxis, and hygiene. There are plans to produce a textbook on microbiology in which deficiency of nutrients, pollution as geomedical factors, influence of common environmental factors on zoonoses, and specific geomedical problems in different parts of the world are planned as separate chapters. Such a textbook will considerably increase the possibility for more emphasis on geomedicine at the medical schools.

In internal medicine, factors like calcium in soil and water, aluminum, iodine, etc. are touched upon as influencing health and disease. In radiation biology and radiation hygiene, emphasis is put on UV radiation and skin cancer, and on long-term low-dose radiation from radon. Various areas of Norway have quite a heavy radiation from radon due to specific conditions in the bedrock on which houses are built. Great interest has also been paid to the possible after-effects of the fallouts of the Chernobyl catastrophe.

In dermatology, many geomedical factors of great importance are taught to the students. Norway has a heavy oil-drilling industry, which may cause skin cancer and eczema. UV radiation is of importance, especially in connection with photosensitation, which is part of botanical dermatology, and hence is also geomedicine. Dermatology also considers the aquatic factors, with toxic and allergic effects of corals, plants, fish, and other organic material in lakes and salt water.

In oncology, The Cancer Registry has, as mentioned above, produced maps of cancer frequency, and these are discussed in relation to geomedical factors such as selenium, calcium in the drinking water, solar radiation, possible carcinogenic substances in vegetables, meat and fish, iodine, etc.

III. SUMMARY

In summary, students are exposed to geomedical problems in their graduate education, but there is in Norway no institute or chair in geomedicine at the medical schools. However, a textbook on preventive medicine and a planned textbook in microbiology specifically mention geomedicine as being of great importance for the panorama of diseases in Norway, in Europe, and globally. Thus, Norwegian graduate physicians are aware of geomedical problems and viewpoints.

ACKNOWLEDGMENT

The author is grateful for information given by Professors Tor Bjerkedal, Kjell Bøvre, and Per Thune at the University of Oslo; Dr. Jostein Halgunset at The University of Trondheim; Dr. Knut P. Nordal at the National Hospital; and Dr. Jon B. Reitan at the Institute of Radiation Hygiene.

Chapter 14

GEOMEDICAL APPROACHES IN EDUCATION OF VETERINARIANS

Per Slagsvold

TABLE OF CONTENTS

I. INTRODUCTION

Blakiston defines geomedicine as the study of the effects of climate and other environmental conditions on health and disease.[1] Låg defines it as the science dealing with the influence of ordinary environmental factors on geographic distribution of pathological and nutritional problems of human and animal health.[2] This paper deals with an approach to geomedical disease in a more narrow sense, mineral deficiencies, and, to some extent, mineral intoxications or imbalance.

A connection between disease and soil, climate and vegetation has been known for centuries, even if the relation was not clearly understood. Jens Bjelke, a civil servant who lived nearly 400 years ago, mentions that farmers in southwest Norway thought osteomalacia in cattle was due to grazing of the plant, bog asphodel, which made the skeleton weak and brittle.[3] He named the plant Gramen ossifragum (now Narthecium ossifragum). Nearly 200 years later Pontoppidan reported from the same area that farmers could prevent and cure the disease by giving their animals crushed bones.[4] This experience was known by J. H. L. Vogt, who made a geological survey in the Egersund-Sogndal area in 1888 and found the rock had a very low content of phosphates.[5] Nearly ten years earlier Dircks had found low figures for Ca and phosphates in hay from Sørlandet.[6] He suggested that there might be a possible connection between osteomalaci and deficiency of Ca and phosphates.

The teaching of veterinary medicine started in the last part of the 18th century. For many years geomedical aspects have been taught in veterinary education in different subjects, such as animal nutrition, pathology, and internal medicine. But in the last part of the 19th and the beginning of this century, following the great discoveries of bacteriology and virology, most veterinary education was concerned with infection. But some teachers understood that all diseases in animals could not be caused by bacteria and vira or by attacks from animal parasites, and thus all diseases could not be combatted by vaccines, serum, or drugs.[7,8]

In addition to the animal diseases in Norway that veterinarians early recognized as geomedical diseases, goiter was reported by A. Løken in 1912,[9] and osteomalacia in cattle was reported by Per Tuff in 1921.[7,10] Some geomedical approaches to diseases will be described below.

The following chapter is a summary of some general observations and reflections (nearly 60 years old) on the dependence of ruminants on their environment, from a paper by L. Slagsvold in 1933.[8]

II. RUMINANTS AND THEIR ENVIRONMENT

Ruminants have in most areas been mainly or totally dependent on local feedstuffs. Their supply of nutrients decreases with low content in the soil. Drought during the growth season can further lower the content of important minerals, such as phosphorus in the grass. A late harvesting time for hay reduces the content of many nutrients, and wet weather during the harvesting season further decreases the content of digestible organic and inorganic matter.

In years with favorable conditions, deficiency diseases are limited to the most exposed areas, while the frequency and evil of the diseases increase under bad conditions. Earlier, most Norwegian herds of cattle, goats and sheep lived on great areas in the mountains during summer. The animals could select among herbs, grasses, and shrubs from different soils. This could compensate for shortages of nutrients in the winter. But indoor feeding lasted in many valleys from October to June. Deficiency diseases would appear frequently in late winter and spring and did not disappear until the animals came out to pasture. Farmers would dry bundles of leafy branches from trees and bushes to increase their amount of feed, and in the late winter and spring they brought in twigs and buds. People in many mountain valleys gave their animals lichen to economize with feed stuffs. At the coast they used fish waste and kelp. The minerals and vitamins in many of these feeds may have been more important than the energy and protein. Introduction of modern agriculture and plant cropping has given more, and in many areas better, hay and grass. This, in connection with introducing roots and concentrates, has led to higher production and the disappearance of many diseases, e.g., osteomalacia and puerperal hemoglobinuria.

On the other hand, the introduction of high yielding cattle and intensive cropping in other areas has led to metabolic imbalance and has caused some deficiency diseases of an intensity and frequency that was never seen earlier. Fertilized areas produce one-sided timothy and other grasses, but not the legumes and herbs with roots reaching the deeper layers of the soil rich in valuable nutrients which are washed out of the top layers. The simultaneous ending of summer grazing in the mountains and the collection of edible parts from trees and shrubs gave a qualitatively poorer ration. In addition, cultivation of new ground often must be done on inferior soil with a low content of minerals, e.g., swamp and sand.

III. THE GEOMEDICAL APPROACH IN VETERINARY MEDICINE

Farmers were the first to encounter the geomedical diseases. In our century they could share the problems with veterinarians and advisors at the veterinary research institutions. The first step was often to exclude other causes of disease, e.g., infections with bacteria, vira, or parasites.

The approach mainly involved clinical and pathological examinations,[7,8,10,11] but sometimes one could get real help from agriculture chemistry, e.g., Tuff.[7,10] Before the last war, our research institutions had neither the qualifications nor the capacity for quantitatively analyzing feeds, blood, or animal products for many major mineral elements, and still less for trace elements. L. Slagsvold,[8] writes about his fumbling attempts to classify and deal with many metabolic diseases. In spite of this, anamnestic information, disease history, and knowledge of local soil treatment have helped in diagnostic work as well as given more-or-less effective remedies found empirically.

IV. EXAMPLES OF GEOMEDICAL DISEASES

A. MAJOR ELEMENTS
1. Phosphates

Osteomalacia has been known in Norway for centuries,[3-8,10,12] especially in the valleys in the hinterland and also in some districts along the coast. It occurred mainly in late winter and spring and disappeared gradually on pasturing. The symptoms were stiffness, tenderness, lameness of the limbs, and swelling of the joints. Young animals were found to have rachitis.[7,8,10] Tuff found the disease in areas with a low content of phosphates in the soil. The disease occurred especially frequently in winters following summers with drought. His survey was made in districts when concentrates and artificial fertilizer were very little used. He further observed the positive results many veterinarians had by treating sick animals with sodium phosphate, and he drew the conclusion that phosphate deficiency caused osteomalacia.

This conclusion has since been confirmed and used in education in our country as in other countries in the world, e.g., Palmer in the U.S. and Theiler's group in South Africa. Today phosphate problems do not have the same veterinary importance, as the modern farmer uses more fertilizer and concentrates with phosphate than his predecessors.

2. Magnesium

''Grass staggers'' is a disease which, in the acute phase, expresses itself with convulsions.

It is relatively new in Norway and a disease that appears on the spring pasture in high-yielding cows. The symptoms appear in cows with a very low blood magnesium level when blood calcium suddenly decreases. This was first shown by Dutch workers in 1930.[13] There is great individual variation in resistance against low blood magnesium values among cows. The pasture in spring is a poor source of available magnesium. The content is low, and a high content of potassium and protein seems to inhibit the magnesium absorption.[14] This disease can now be prevented by prophylactic nutrition and veterinary treatment and is less harmful than earlier. The pioneering work in our country was done by Ender and coworkers.[15,16]

B. TRACE ELEMENTS
1. Iodine
The goiter was in Norway, as in most countries, mainly found in the hinterland far away from the coast and in areas where mountains block direct wind from the ocean and decrease the precipitation, e.g., in Gudbrandsdalen, Valdres and Hallingdal, and around the great lakes.[8,9] The goiter would give obstetric problems both in man and domestic animals, decreased fertility, and weak or stillborn young.[8,9]

Practitioners in such districts knew well about endemic goiter, symptoms, factors that interfered as well as cure and prevention. Later the addition of iodine salts in mineral licks and in concentrates has made goiter rare. Today overfeeding with iodine through supplementing adequate rations with extra minerals or kelp represents greater goiter problem for pet animals and horses than deficiency of iodine. The veterinary duty is more to warn against unresponsible advertising.

2. Cobalt
Acobaltosis may be a pure deficiency of cobalt or a combination with other deficiencies which interfere with the clinical symptoms of the diseased ruminants. To the first belongs pica, to the latter dry sickness where a deficiency of phosphate, protein and easily fermentable carbohydrates may be as important as the deficiency of cobalt.

The clinical picture — pathology, prevention, and treatment — of these diseases was described by L. Slagsvold in 1933, some years before the effect of cobalt was known.[8]

a. Pica
Pica occurred most often on the sandy soils and wet climates on our south, west, and north coasts. In the areas where clay composed the growing bed of the fodder plants, the ruminants kept free of the disease. Frequency and intensity increased with the use of fertilizer and with the height of milk yield.

The pica could appear as early as December, with licking, lost appetite, pelleted faeces, and loss of calves. Young animals were anemic, dwarfy, had hard stool which might shift to diarrhea. Untreated, the disease continued until the animals were let out, but if the pasture was located in the pica area, the animals would only slowly recover. Molasses and kelp were used for prevention and cure, as suggested by Holmboe.[17]

Grass and hay showed a lower content of alkalis, especially potassium, in the pica areas than in healthy areas, but there was no difference in content of other minerals. (The analyses did not include trace elements, authors comment.)

Because the pica areas were lying near the coast and fertilized with kelp, only reluctantly could pica be accepted as an alkali deficiency.[8] But animals were cured with ash of grass from healthy areas high in alkali, and still faster results were achieved with ash from deciduous trees. In a few days the appetite, the rumination, peristaltic movement, and the stool was normal.

Slagsvold writes: "This alkalinizing of the feed quite soon changes the fermentation in the reticulorumen as well as the secretion in the intestines. Unfortunately no comparison has been made of microflora and microfauna in the reticulorumen before and after the treatment.''

The healing agent in molasses and ash was still not known. In 1933 the Dutch scientist Sjolemma described a pica that was caused by a low copper content in the herbage.[18] He could prevent and even cure the disease by giving copper salts. Copper deficiency in plants had been found in Norway. Ender discovered a low copper content in areas endemic for pica, and thought this was the cause of the disease.[19] But the effect of copper salts was not constant or complete. The effect was better when he tried cobalt. This cure had been worked out in Australia in 1933 to 1935 by Filmer and Underwood.[20,21,22] Founded on diagnostic treatment trials and chemical analysis, Ender and co-workers made maps showing the areas deficient of cobalt in soil and herbage as a help for veterinarians.[23,24,25] There were high cobalt values in the mountain pasture. Home remedies such as molasses had a fair content of cobalt, and the ash from branches of young leaf trees had a still higher content. Today cobalt is included in the recipe of mineral mixtures, salt licks, and concentrates for ruminants, and pica is a rare disease in the former pica areas.

b. Dry Sickness

Dry sickness has been known for a very long time in the eastern valleys of the Norwegian hinterland and the neighboring areas of Sweden.[8] The disease appeared in recently calved cows in the last part of the winter. The best milkers were frequently attacked, and lost appetite. If they had received any concentrates or roots, these were first rejected. The temperature was low, gastrointestinal activity reduced, and the stool dry, pelleted, and dark.

Young animals became lean, with flat belly and long and bristly hair. Sheep on the same farms ate each other's wool and had bare patches. Especially exposed were animals kept in an old-fashioned way — in the summer on pasture in the forests and in winter eating late-harvested hay from swampy or sandy ground. The disease was most frequent when bad harvesting conditions had washed out soluble organic and inorganic substances. The disease could be prevented by supplying good quality hay and well-composed concentrates. Other means were roots and mash or concentrates. Molasses was shown to cure this disease. The curative effect of sodium phosphate shown by Svanberg[26] of Sweden was to some extent confirmed.[8]

Ender found low content of cobalt in the hay from farms with dry sickness and he had good curative effect by giving sick animals cobalt tablets, and a small extra effect when combined with copper.[24] Introducing new organisms through fresh rumen mass from healthy cows was another cure (Flatla[27]).

Dry sickness is an example of a geomedical disease with a complex etiology. Organic and inorganic nutrients in the feeds interfere with the vital biological life of bacteria, fungi, and protozoa in the reticulorumen.

3. Copper

Copper, like cobalt, is an element necessary for fermentation in the rumen, but it is also necessary for many enzymatic processes inside the animal body, e.g., hematopoiesis, structure of hair and wool, production of pigment in black wool, growth process, bone structure, myelin isolation in nerve tissue, reproduction, etc. There exist two kinds of copper deficiency in animals: primary copper deficiency (copper deficiency where the forage has a low content of copper[18,19]), and secondary copper deficiency (apparently the ration contains normal amounts of copper, but show disease symptoms which can be cured by copper salts[28,29-31]). In these cases other factors decrease the absorbtion of copper, e.g., molybden, sulfur, zinc, iron, and lead.[32] Grønstøl states that the relation between copper and molybden in the ration should not be less than 2 to avoid deficiency, and not be above 20 to avoid intoxication.[32] Acute intoxication is mainly accidental, as when the herd eats grass contaminated by copper containing wash against fungi in the garden, or they eat copper-stained seed produced for planting, or use a pasture too early after fertilizing with a copper-containing fertilizer.

Chronic copper intoxication is presently the most frequent intoxication in sheep.[33] They must be protected against mineral mixtures and concentrates intended for other domestic animals to avoid intoxication in the flock. Frøslie[31] found sheep in the inland district of Norway heavily loaded with copper, and was of the opinion that the copper poisoning in sheep is related to this loading,[30] but he found no relationship between content of copper in grass and sheep. The copper poisoning in sheep possibly was caused by the low level of molybden in grass in inland districts. As copper toxity in sheep is a clinical problem, it may be a help to balance the copper in mineral licks and mineral mixture with molybdenum.[33]

4. Selenium

A myosites has occurred in lambs and calves (sometimes also in foals) in nearly the same areas as already described for goiter. Sick calves were found, with sore and swollen muscles, stiff movement, and increased respiration rate and heart frequency.[11]

Severely diseased animals had difficulty with standing and were lying down, still the appetite could be quite good. In the carcasses of dead calves were found degeneration of body and heart muscles. In the same district, as early as 1890, veterinarians reported cases where apparently normal calves slaughtered at the abbatoirs had white-to-yellow degenerated muscles[34] (or lamb stiffness[35]). Meat inspectors called the findings "fish meat," while it clinically is called "Rheumatisme," now white muscle disease (WMD). Lambs were found with nearly the same symptoms and postmortem picture. The disease was most frequent in areas where the mother animals were kept inside the barn over very long parts of the year and home-produced fodder dominated the ration.

The disease appeared on dry farms with sandy soil as well as on farms with inadequately drained swampy soil. The frequency was highest in lambs and calves born in springs following very dry or cold summers with a wet harvesting season. The authors thought the disease was due to metabolic disturbances caused by shortage of necessary nutrients.[11] They could not cure the disease, but could reduce the outbreaks by giving a mineral mixture to mother sheep and cows in the last part of the pregnancy. They were near a solution, writes Grønstøl.[32] But finding the solution should take time. The cause was a deficiency of selenium, an element first shown to be essential in 1957 by Schwarz and Foltz.[36] The preventive effect against white muscle disease in lambs and calves is reported in Maynard and Loosli.[37] In 1961 a combination product of selenium and vitamin E, was licensed by Tocosel and in veterinary use for cure and prevention.

If the disease had not progressed too far, most animals could now be rescued by an injection and oral treatment.[32] Westermarck of Finland used selenium in experiments with white muscle disease in heifers in 1962.[38] The same year Wasa Mills in Finland introduced commercially a mineral mixture with added selenium, and in 1969 the Fins added selenium to commercial concentrates. Most Nordic scientists were shaking their heads and thought inadequate mixing and/or separation of batches could lead to inhomogen concentrates and intoxications. But Westermarck writes: "So far not a single case of selenium poisoning has occurred in Finland in spite of the common selenium feeding and treatment of various animal species.[39] As students in 1940, we heard from the U.S. about alarming intoxications; acute blind staggers with irregular breathing, heart rate and rhythm with sudden death; a chronic "alkali disease" with loss of epidermal tissue such as hair and hooves; apathy; and liver injury.[40] But this was only for our knowledge and not intended for practical use. But conditions change; in 1987 an author writes in a compendium for students about intoxications of animals in Norway. These occur as a result of erratic dosing with selenium or of letting out animals too soon into pastures fertilized with selenium. Accidents may occur when selenium is supplied independently from more sources, because the toxic dose is only ten times the therapeutic dosage.[32] Use of this harmful poison in fertilizers, as a supplement to the diet of animals or as medicine, requires attention and care.

5. Fluoride

Fluoride is an element distinctly beneficial with respect to tooth decay in man, and many municipalities add fluorine to their water supply. But for animal feeding, the harmful effect of fluorine is one of practical importance.[37] In Queensland, the Rift Valley of East Africa, and other volcanic areas such as Iceland, the earth is rich in fluorine. In the water in drilled wells in Kenya an average content of 17 ppm of fluorine was found.[41] In connection with high levels of fluorine, clinical dental fluorosis was found in 37.6% in an examination of 4,562 people (age 7 to 18 years). In the Machacos district 2.7% of all cattle had both tooth and bone lesions with marked lamenes and poor condition. In Norway the fluorine content of the bedrock is low, and the ground water has between 0.06 and 1.3 ppm.[42] In spite of this low fluorine, the country has had a practical fluorine intoxication problem in at least two generations. Industrial aluminum plants are polluting their environment with fluoride-containing smoke dust or soot. Many trees are killed, while most grass still can survive. But the leaves are covered with fluoride-containing dust. Sheep and cattle eating the contaminated grass get chronic fluorine deficiency. Veterinary investigations started at the veterinary clinic in 1930.[43] Slagsvold found anemic animals with irritated eye mucosa, leanness and later emaciation, rigid spine and sore, stiff limbs often with fractures. After death the bones in the head could easily be cut with a knife, and exostoses were found along the mandibles.

When the Norwegian aluminum industry expanded after World War II, Slagsvold tried to prevent this pollution from the new factories by getting the biochemist F. Ender to look after the level of fluorine release with the smoke. But the director of the States Aluminium Plants had more political strength than the director of veterinary services. The pollution continued for many years, but is now considerably reduced, a fact of importance not only for animals, but also for workers and people living in the local areas, and for vegetation.

V. PREVENTIVE MEDICINE IN ANIMALS

The old generation told us that the cow is a product of the soil where she lives. This is even more true with sheep. Although commercial farming uses fertilizer and often brings in concentrates and mineral mixtures, local feedstuffs still dominate in many areas. There are also ideological movements, such as biodynamic agriculture, where fertilizer and most concentrates are banned. As there are many places with deficiencies, excess, or imbalance in the local soil and crops, it would be unwise not to correct these imbalances. And supplements are necessary to compensate for elements lost through sold products. On the other hand, overuse of artificial fertilizer can lead to crops of lower content in essential elements, and fodder of lower quality for animals and humans.

''Prevention is better than cure,'' according to the old saying. To know enough of probable deficiencies, veterinarians must study maps of elements in soil and feedstuffs. These are distributed by geological, agricultural and veterinary institutions, e.g., the maps of copper and cobalt in Norway by Ender[19,23,25] and for the separate veterinary districts by Havre et al. for Setesdal.[28]

But in many countries the soil and environment can vary so much within a district that it still may be difficult for new veterinarians. Here a local agricultural officer with knowledge of plant pathology could be a great help.

A lot of elements essential for man and animals are also essential for plants; deficiency in plants may give quite clear symptoms.[44] It will be a great help for the veterinarians themselves to know the symptoms of plant diseases due to imbalances in elements that are essential for animals in their district.

Finally, let us stress the necessity to combat ideologically decided feeding, e.g., biodynamic agriculture, where necessary elements may be excluded. Another dangerous idea is that kelp and products from the sea are safe additives, even when the concentrates already contain generous amounts of iodine.

VI. SUMMARY AND CONCLUSION

Geomedical approaches and understanding are important for the veterinary practitioner in his daily life. Some examples are described.

Our pioneer-scientists in this research were trained in pathological and clinical work, and made a good approach both in defining the diseases, their symptoms and pathology, and in finding remedies for prevention and treatment. Although their biochemical background was small, in many cases they could teach their students and give the practioners advice for their work. However, this was only an approach[8] to meet the immediate problems. Much work and cooperation with other sciences has been necessary, e.g., geology, soil science, plant science, chemistry, and biochemistry. Still new approaches steps are necessary especially where more elements are combined.

REFERENCES

1. **Blakiston,** *Gould Medical Dictionary,* 4th ed., McGraw Hill, New York, 1979.
2. **Låg, J.,** Plant-available mineral elements in Nordic soils, with special reference to geomedical problems, *Mineral Elements,* Vol. 80, Helsinki, 1981, 53.
3. **Bjelke, J.,** Gramen Ossifragum 1580-1659; quoted from **Låg, J.,** Hundreårsjubileum for en geomedisinsk oppdagelse, sammenheng mellom fosformangel og benskjørhet, *Jord Og Myr,* 12, 174, 1988.
4. **Pontoppidan, E.,** Det förste forsóg på Norges naturlige historie, Forste deel, København 1752; quoted from **Låg, J.,** *Jord Og Myr,* 12, 174, 1988.
5. **Vogt, J. H. L.,** Norske ertsforekomster. Titanjernforekomsterne i Noritfeltet ved Ekersund-Soggendal; quoted from **Låg, J.,** *Jord Og Myr,* 12, 174, 1988.
6. **Dircks, W.,** Nogle undersøgelser af hø; quoted from **Låg, J.,** *Jord Og Myr,* 12, 174, 1988.
7. **Tuff, P.,** *Osteomalacie Hos Storfe,* Vol. 2, Nordiska Veterinärmøtet, Stockholm, 1921, 403.
8. **Slagsvold, L.,** *Stedbundne Mangelsjukdommer Fremkalt Ved Stoffmangler Eller Abnorm Fórsammensetning,* Vol. 4, Nordiska Veterinärmøtet i Helsingfors, 1933, 189.
9. **Løken, A.,** Struma, *Norsk Vet. T.,* 24, 178, 1912.
10. **Tuff, P.,** Osteomalacia and its occurrence in cattle in Norway, *Proc. World Dairy Congr.,* Vol. 2, Washington, D.C., 1494, 1923.
11. **Slagsvold, L. and Lund Larsen, H.,** Myosit hos lam, kalver og ungfe, *Norsk Vet. T.,* 55, 529, 1934.
12. **Låg, J.,** Some geomedical problems in connection with regional research; **in Låg, J.,** *Geomedical Research in Relation to Geochemical Registrations,* Universitetsforlaget, Oslo, 1984, 10.
13. **Sjolemma, B.,** On the nature and therapy of grass staggers, *Vet. Rec.,* 10, 425, 1930.
14. **Payne, J. M.,** Magnesium; in **Payne, J. M.,** *Metabolic Disease in Farm Animals,* William Heinemann Medical Books Limited, London, 1977.
15. **Ender, F., Halse, K., and Slagsvold, P.,** Undersøkelser vedrørende kramper og lammelser hos kyr. Hypomagnesemi fremkalt under kontrollerte betingelser, *Norsk Vet. T.,* 60, 1, 48, 1948.
16. **Breirem, K., Ender, F., Halse, K., and Slagsvold, L.,** Experiments on hypomagnesemia and ketosis in dairy cows, *Acta Agric. Suec.,* 3, 827, 1949.
17. **Holmboe, F. V.,** Erfaringer væsentlig fra Ytre Ryfylke over saakaldt beinskjørhet på kvæg, *Norsk Vet. T.,* 41, 159, 1929.
18. **Sjolemma, B.,** Kupfermangel als Ursache von Krankheiten bei Pflanzen und Tieren, *Bioch. Zeitschr.,* 267, 151, 1933.
19. **Ender, F.,** Undersøkelser over slikkesykens etiologi, *Norsk Vet. T.,* 54, 3, 78, 137, 1942.
20. **Filmer, J. F.,** Enzootic Marasmus in cattle and sheep; quoted from *Nutr. Abstr. and Rev.,* 3, 1200, 1933-34.
21. **Filmer, J. F. and Underwood, E. J.,** Enzootic Marasmus. Treatment with Limonite Fractions; quoted from *Nutr. Abstr. and Rev.,* 4, 696, 1934-35.

22. **Underwood, E. J. and Filmer, J. F.,** Enzootic Marasmus. The determination of the biologically potent element (Cobalt) in Limonite; quoted from *Nutr. Abstr. and Rev.,* 5, 842, 1935-36.

23. **Ender, F.,** Sporelementenes etiologiske og terapeutiske betydning ved spesielle mangelsykdommer hos storfe og sau i vårt land. Koboltmangel som sykdomsårsak, *Norsk Vet. T.,* 56, 173, 1944.

24. **Ender, F.,** Koboltmangelen som sykdomsårsak hos storfe og sau belyst ved terapeutiske forsøk, *Norsk Vet. T.,* 58, 118, 1946.

25. **Ender, F. and Tananger, I. W.,** Fortsatte undersøkelser over årsaksforholdene ved mangelsykdommer hos storfe og sau. Koboltmangel som sykdomsårsak belyst ved kjemiske undersøkelser av fóret, *Norsk Vet. T.,* 58, 313, 1946.

26. **Svanberg, O.,** Om afosforotiske bristsjukdomar bland boskapen, *K. Landbr. Akad. Handl. og Tidsskr* (Reprint), 1932.

27. **Flatla, J.,** Forelesninger i intern medisin, *Stoffskiftesykdommer,* Norges Veterinaerhøgskole, Oslo, 1950.

28. **Havre, G. N., Dynna, O., and Ender, F.,** The occurrence of conditional and simple copper deficiency in cattle and sheep, *Acta Vet. Scand.,* 1, 250, 1960,

29. **Frøslie, A. and Norheim, G.,** Copper, Molybdenum, Zink and Sulphur in Norwegian forages and their possible role in chronic copper poisoning in sheep, *Acta Agric. Scand.,* 33, 97, 1983.

30. **Frøslie, A.,** Geomedical aspects of animal production in Norway; in **Låg, J.,** *Geomedial Research in Relation to Geochemical Registrations,* Universitetsforlaget, Oslo, 1984.

31. **Frøslie, A.,** Kobberstatus hos sau i Norge, *Norsk Vet. T.,* 89, 711, 1977.

32. **Grønstøl, H.,** *Mikromineral-Mangelsjukdommar Og Forgiftningar,* Norges veterinærhøgskole, Oslo, 1987.

33. **Norheim, G.,** The relation between copper accumulation in sheep and the concentration of copper and molybdenum in roughages harvested in two different parts of Norway, in *Mineral Elements,* Vol. 80, Helsinki, 1981, 429.

34. **Flesvik,** Rheumatisme hos kalve, Beretn. om Vet. vesen og Kjøtt-Kontr., 1912; quoted by **Slagsvold, L. and Lund Larsen, H.,** Myosit hos lam, kalver og ungfe, *Norsk Vet. T.,* 55, 529, 1934.

35. **Taraldrud,** Innberetning om lammestivsyke i Norge, 1896; quoted by **Slagsvold, L. and Lund Larsen, H.,** *Norsk Vet. T.,* 55, 529, 1934.

36. **Schwarz and Foltz,** Selenium as an integral part of Factor 3 against necrotic liver degeneration, *J. Am. Chem. Soc.,* 79, 3292, 1957.

37. **Maynard, L. A. and Loosli, J. K.,** *Animal Nutrition,* McGraw-Hill, New York, 1962, 182.

38. **Westermarck, H.,** Selenium in the prevention and therapy in white muscle disease in calves and heifers, *Nord. Vet. Med.,* 16, 264, 1964.

39. **Westermarck, H., Rauny, P., and Lappalainen, L.,** On the effect of selenium added to the feed for ruminants and chicks as well as toxicological aspects on selenium injections to ruminants, *Mineral elements, 80, Proceedings,* Helsinki, 1981, 573.

40. **Slagsvold, L.,** *Lecture in Special Pathology and Therapy,* Norges veterinærhøgskole, Oslo, 1940.

41. **Said, A. N., Slagsvold, P., Bergh, H., and Laksesvela, B.,** High fluorine water to wether sheep maintained in pens, *Nord. Vet. Med.,* 29, 172, 1977.

42. **Soveri, J. and Soveri, R.,** On fluoride and chromium concentrations in ground water in Finland, *Mineral Elements 80.* Helsinki, 1981, 511.

43. **Slagsvold, L.,** Fluorforgiftning, *Norsk Vet. T.,* 46, 2, 61, 1934.

44. **Hansson, A.,** *Bristsjukdomar hos Kulturväxter och Exempel på Fóljsjukdomar Hos Husdjur,* Tidskriftsaktiebolaget Vext-näringsnytt, Nordisk Rotogravyr, Stockholm, 1959.

Chapter 15
SPECIFIC GEOMEDICAL PROBLEMS IN DIFFERENT PARTS OF THE WORLD

Chapter 15A

SOME GEOMEDICAL PHENOMENA AND RELATIONSHIPS IN THE WESTERN HEMISPHERE: THE AMERICAS

Francis D. Hole

TABLE OF CONTENTS

I. INTRODUCTION

In 1974 a committee of the National Academy of Sciences, U.S., explored "the relation of selected trace elements to health and disease," and concluded that appropriate data were scarce, and that research projects be designed to focus on regions and local landscapes rather than on entire countries.[1] Three factors were recognized relating to the distribution and concentration of nutrients essential to the health of animals and people: plant genetics, soil fertility, and environmental conditions. These are included, with examples from the last two categories, in the following comprehensive equation.

II. A GEOMEDICAL HEALTH FACTOR EQUATION

An equation of the factors that influence the level of health of animals and human beings provides a framework for the study of geomedical phenomena. The following is modeled after equations of factors of soil formation.[2,3]

$$LH_{a,h} = f (Pd,Sd,Rd,Cld,Od,AT, ...) \tag{1}$$

The equation means that the level (L) of health (H) of animals (a) and human beings (h) is a function (f) of geographic domains (d) of parent (geologic) materials (Pd) and derived soil domains (Sd) under the influence of terrains known as domains of topographic relief (Rd), of climatic domains (Cld), of individuals and domains of organisms (Od), including the genetic factor and including human populations and the anthropic factor (Ohd), space-time (AT) relationships,[4] and numerous specific factors (. . .) that investigators may add to the equation as appropriate in a particular study.

A few examples will be given of each category of the equation.

A. GEOLOGIC PARENT MATERIAL DOMAINS (Pd)

In the Americas, patterns of incidence of disease are clearly (Figure 1) to vaguely (Figure 2) related to patterns of geologic materials.[5,6] It is difficult to interpret geomedical patterns when relationships are highly complex, as illustrated in Figure 2.[6,7] Three specific instances of relatively clear relationships are cited, as follows.

A study of levels of carcinogenic radon-222 (and its progeny, Po-218 and Po-214) in 1700 Appalachian homes in the eastern U.S. showed that many homes built over Paleozoic metamorphic and igneous rocks had levels in the air higher than 4 p ci/l, which is the maximum permitted by the U.S. Environmental Protection Agency. Lower levels were present in homes with basement walls made of solid concrete than in homes with cinder-block basement walls.[8]

Unacceptably high levels (above 50 ppb) of arsenic in drinking water and irrigation water used on hay fields were measured in the valley of the Madison River, which heads in the geothermal region of Yellowstone National Park in Wyoming. Arsenic is a natural emission from deep-seated magma. Evaporation of river water in hay fields accounts for some of the high concentrations of arsenic there.[9]

On Staten Island, New York City, water in lakes with clay bottoms contain 0.3 million fibers of chrysotile asbestos per liter, as compared with 15 to 86 million fibers per liter in lakes with asbestos bedrock exposed on bottoms. Ingestion of high concentrations of fibers may cause formation of tumors in human beings; inhalation of the fibers may result in cancer of the lungs.[10]

B. SOIL DOMAINS (Sd)

Many properties of soils are connected with health problems.[11] In the U.S., levels of selenium and cobalt in soils have been correlated with disease in animals and people.[12]

FIGURE 1. Areas of rocks, soils, and vegetation deficient in iodine. These are also areas (black) of endemic goiter.[5]

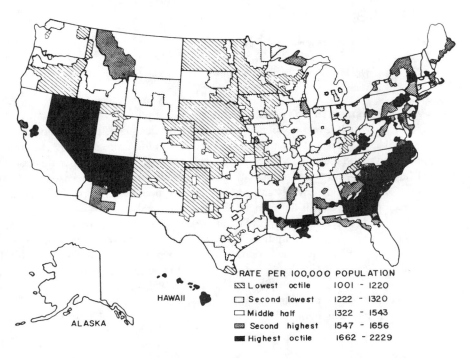

FIGURE 2. All-causes death rates for white males, age 65 to 74, 1959 to 1961, U.S. (From Sauer, H. I. and Brand, F. R., In *Environmental Geochemistry in Health and Disease*, Cannon, H. L. and Hopps, N. C., Eds., Memoir, 123, 131, 1971.)

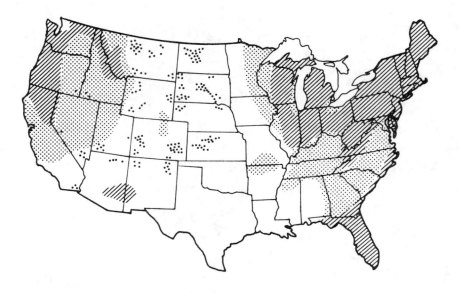

Where Se levels are too low to meet requirements of farm animals

Where Se is adequate to meet requirements of farm animals

Where Se is both adequate and inadequate in same locality

•   Where Se toxicity may be a problem

FIGURE 3.   Geography of levels of concentration of selenium in soils and plants in the contiguous U.S.[12]

1. Low Levels of Selenium in Soils (Sd, Selenium-Deficiency)

Areas of soils in which levels of selenium are deficient, adequate, a mixture of the two conditions, and toxic with respect to animal health are shown in Figure 3 for the coterminus 48 states. Deficiency will be discussed here. Selenium-deficiency in livestock in this country is usually on farms with acid soils, particularly sandy Podzols in northeastern states. Grass and protein supplements are commonly shipped to dairy and poultry enterprises of the region to offset both selenium- and cobalt-deficiencies there. Most bread, a good source of selenium for people, is produced from grain grown in Se-adequate soils in the U.S. In proper amounts, selenium may prove to be important in diets of animals and humans to prevent cancer.[13,14,15] The As/Se ratio may play a role in development of cancer in low-selenium areas in southeastern parts of the Coastal Plain. In the same area, cardiovascular disease is persistent in people. Interaction of Cd and Se is possible in connection with high blood pressure as well as with lung cancer.[16] Major nutritional syndromes in people are frequently signalled by changes in hair, skin and nails, which dermatologists are first to notice and most experienced in monitoring.[17]

2. Toxic Levels of Selenium in Soils (Sd, Selenium Toxicity)

Although selenium, in small amounts, is essential to animal and human health, it is toxic in large amounts.[18] In the San Joaquin Valley of California, drainage water from irrigated cropland was until recently directed into 12 ponds of the Kesterson Wildlife Refuge. In 1983 strange malformation and death of wildfowl were observed and traced to high concentrations of selenium in the drainage waters and adjacent wetland soils. This posed a

///	Where legumes are usually deficient in cobalt for cattle and sheep
:::	Where legumes may be marginally deficient in cobalt for cattle and sheep

FIGURE 4.   Areas (shaded) in the U.S. where soils and legumes contain low levels of cobalt.[12]

threat also to human health. The selenium came from long-weathered seleniferous pyrite in rocks of the Coast Range mountains that were uplifted in the Cretaceous period.[19] In 1986 drainage water was diverted away from the Kesterson ponds. One alternative disposal is recycling of irrigation water, with dilution by fresh water, before going on to fields.

3. Low Levels of Cobalt in Soils (Sd, Cobalt Deficiency)

Cobalt is a trace element that is required by nitrogen-fixing organisms in nodules of legumes. The element is essential to the formation of vitamin B_{12} in the rumen of cattle, sheep, and goats. Virtually no cobalt deficiency has been found in western states, where soils contained 4.2 to 37.0 ppm of the trace element.[20] In marginally cobalt-deficient areas (Figure 4) soils contain 3 to 4 ppm of the element. In severely deficient areas the level is 1 to 2 ppm. No soils with toxic levels of cobalt were reported by Kubota.

C. RELIEF OR TOPOGRAPHIC DOMAINS (Rd)

Steep lands are subject to mass wasting which can disrupt homes and systems of water supply and waste disposal, leading to health problems.[21,22] As noted above, the concentration of selenium in the San Joaquin Valley has been a result of the uplift of adjacent mountains.

D. CLIMATIC DOMAINS (Cld)

Climate has measurable impacts on human health. For example, cold temperature in a polar climate "retards bacterial multiplication and stimulates saliva flow" in the mouths of outdoor workers, who as a result have low numbers of lactobacilli in saliva and low rates of caries in teeth.[23] Climatically related disasters that disrupt environment and lead to disease and death include hurricanes, tornadoes, and floods in river valleys.

E. DOMAINS OF ORGANISMS (Od)

1. Low Levels of Selenium in Plant Materials (Od, Selenium Deficiency)

Contents of selenium in forages and grains were reported from contrasting areas in the U.S. (patterns 1, 2, and 3, Figure 3) to be 0.02 to 0.05, 0.26, and 0.8 ppm, respectively.[24] Health of plants does not seem to be affected by level of selenium in the tissues. However, selenium deficiency causes "white muscle disease" in calves and lambs.

2. Toxic Levels of Selenium in Plant Materials (Od, Selenium Toxicity)

Certain shrubs and weeds in semiarid and arid regions of the U.S. accumulate poisonous concentrations at levels of 50 ppm or more of selenium, as compared to nontoxic levels of less than 5 ppm in associated range grasses and field crops.[25] In this country, "alkali disease" and "blind staggers" are the common names for the disorder caused in cattle by this toxicity.[17] Greenhouse experiments have shown that barley, beets, tomatoes, and alfalfa are capable of accumulating selenium in amounts that are potentially harmful to animals.[25]

3. Low Levels of Cobalt in Plant Materials (Od, Cobalt Deficiency)

Determinations have been made in national and regional studies of cobalt content of legumes used for feeding cattle and sheep in the U.S.[20] In western states legumes had 0.1 to 0.2 ppm of cobalt, which is adequate for the nutrition of animals. In markedly cobalt-deficient areas (Figure 4, pattern 1) levels were less than 0.07 ppm in legume tissue. In moderately deficient areas (Figure 4, pattern 2) legume tissue contained 0.1 to 0.07 ppm of cobalt. Farm animals are more subject to nutrient deficiencies than are people, because imported foodstuffs constitute a small proportion of animal diet than of human diet.

F. SPACE-TIME RELATIONSHIPS (AT)

Spatial (A) relationships include the extent of interchange of foodstuffs between regions, to offset, as has been mentioned above, trace element deficiencies in certain landscapes. Geographic extents of fertile and infertile soils, level and steep soils, well drained and poorly drained soils influence the capacity of a landscape to support animals and people in a healthy condition. Many immigrants to the Americas during the past several centuries were attracted by the existence of vast areas of arable soils.

Temporal (T) relationships include effects of age of soils. Those people are fortunate who have access to food produced from unpolluted, young, fertile soils of alluvium, glacial drift, and volcanic materials. In contrast to these, very old soils in equatorial terrains are commonly much leached and are incapable of supporting crops, under intensive agriculture, that provide adequate nutrition for people. Soils of tropical America, in contrast with soils of tropical Africa, are extensively deficient in calcium and phosphorus.[27] Range cattle in tropical America almost always need supplemental calcium and phosphorus to prevent bone fracture. Whereas large mammals, including elephants, evolved in tropical Africa, they did not in tropical America, where soil acidity has been a prohibiting factor. The scarcity of calcium in soils of the vast Amazon basin also helps to explain the surprisingly small proportion of the basin that is subject to schistosomiasis. This major world disease, caused by *Shistosoma* spp., requires for completion of its life cycle the presence of snails. The calcium for formation of snail shells is missing in most of the basin.[28]

Another temporal relationship is that of the half-life of radioactive nuclides in various substances, as discussed below.

G. ADDITIONAL SPECIFIC ITEMS (. . .)

The dots at the end of the equation represent additional items that a researcher may wish to include in the analysis. Uses of the equation may range from the general to the specific. Some examples of specific items are as follows:[29] aluminum involvement in Alzheimer's disease; vanadium in diabetes; selenium in Keshan disease; copper in Wilson's and Menkes' diseases; mercury, lead, arsenic, fluorite, and iron in various disorders, and calcium metabolism in kidney stone formation and in osteoporosis.

III. SOME ANTHROPOGENIC PHENOMENA

A. PRODUCTION AND DISSEMINATION OF HARMFUL SUBSTANCES

1. Radioactive Substances

Some materials that constitute threats to human life are products of human activity, yet their distribution may be largely by natural processes, including movement of substances in food chains. Environmental radioactivity is well known.[30] Hydrogen-bomb tests in the Pacific region under the auspices of the U.S. Government in 1954 released ^{90}Sr, ^{137}Cs and ^{131}I into circumglobal air streams. The radioactive substances moved to the tundra of northern Canada where the caribou became contaminated, and to the U.S. where children's teeth became measurably radioactive with ^{90}Sr. ^{131}I was detected in mother's milk.[31] In 1982 scientists at Oak Ridge National Laboratory in Tennessee were developing ways to seal the surface of a body of radioactive waste buried in soil, to prevent the spread of radioactive materials by water during rain storms.[32] Monitoring of levels of concentration of radionuclides in soil continues.[33,34]

2. Certain Agricultural Chemicals

The manufacture and use of the pesticide DDT in the U.S. led to contamination of fish, birds, mammals, people, vegetation and soils.[35] As a result, Congress enacted regulations designed to insure safe use of environmental chemicals.[31] Accumulation of particles of heavy metals in soil from fall-out from smoke stacks and from vehicular traffic has made surface soils in Dayton, Ohio sources of airborne toxins for pedestrians on windy days.[36]

The pollution of well water to the point of making it unsafe for people to drink affects considerable agricultural areas in the U.S. For example, chemical fertilization of vegetable crops on irrigated sandy soils in central Wisconsin has caused downward movement of nitrates into the groundwater reservoir, raising levels of NO_3 above the U.S. drinking water standard of less than 10 mg/l[37] An insecticide, aldicarb, that was used for years to control the potato beetle in the same area, also appeared in well water. Aldicarb has been banned recently from the state of Wisconsin because it was a threat to human health. A strain of bacteria is now being used to control the beetle.[38]

A shallow, unconfined aquifer under soils bordering Lake Erie in Ohio, was severely contaminated with chemical fertilizers from commercial nurseries, brine from drilling for oil and gas, and salt that was spread on highways in winter. As a result, well water was not safe to drink.[39]

3. Soil Translocated by Anthropically Accelerated Erosion

Soil erosion that has been accelerated for a century or more by human activity has decreased food production on the land and has polluted streams and lakes in many areas of the U.S. Even military maneuvers in 1940 in the Mojave Desert in the southwestern U.S.A. have left their marks on soils in the 1980s. Where soil had been compressed by tracks of armored tanks, the soil is still too dense for vegetative growth. Where tank movement had destroyed the fragile desert pavement, the exposed soil has been subject to wind erosion ever since.[40] The incidence of soil erosion in the Americas is appalling, despite some very good work to control accelerated translocation, and to protect bodies of water and of wetlands from inwash of soil debris.

4. Some Observations on Entry of Surficial Food Contaminants Through the Mouth

Insoluble radionuclides, heavy metal particles, hydrophobic organic compounds, and soil particles can enter the human food chain without going through the roots and above-ground vascular systems of plants. These materials can splash or blow onto surfaces of plant stems, leaves, fruits, and grain as much as a meter above ground, and in part survive washing during harvesting and food preparation. "Soil particles on surfaces of grains of wheat and corn can be incorporated into food products during processing at the rate of 10 to 30% of total particles" originally deposited. Lettuce is more heavily coated than are taller plants.[41] Animals are more affected. "The ingestion of soil by grazing animals is an important biochemical pathway which by-passes barriers at the soil-plant and plant-animal interfaces. In cattle, 10 to 20% of dry matter intake may be soil." With sheep it was as much as 30%.[42]

Geophagy, the eating of soil by people, is a widespread folk practice that was engaged in by American aborigines.[43] Sacred clay tablets for human consumption were distributed widely in Central America by Mayans. Later (after 1578) a Christian church in Guatemala sponsored the manufacture and distribution of tablets for hundreds of kilometers. Blacks brought as slaves to the Americas continued their traditional practice of geophagy. Chemical analyses indicate that both macronutrients and trace elements were taken in by the practice.[44-47] Modern accelerated pollution of soil may make geophagy particularly harmful.

IV. GLOBAL BIOETHICS

Aldo Leopold (1887—1948) referred to the "biotic arrogance of *Homo americanus*" when, as a professor of Wildlife Management (1935—1948) at the University of Wisconsin, he observed the degradation of land in the Great Lakes Region by farmers. In his book, *Sand County Almanac* (1970), Leopold defined the appropriate ecological role of human beings as that of plain members and citizens of the land-based community, not conquerors and despoilers of it.[48] Potter, Emeritus Professor of Oncology at the McArdle Laboratory for Cancer Research at the University of Wisconsin-Madison, has outlined in his book *Global Bioethics* two interrelated spheres:[49] (1) that of medical bioethics and (2) that of environmental bioethics. The author stated that medical bioethic takes a short-term view and has as its goal the prolongation of the life of an individual human being, to the neglect of the health of the environment and to the neglect of population control. He concluded that the result will be, at best, a miserable survival for some people, if solely medical bioethics is practiced. Environmental bioethics he characterized as having a long-term view, with survival of the human species and health of the environment as twin goals, leading to an acceptable survival of people and supportive ecosystems.

V. CONCLUSION

The level of health of people and animals is dependent on favorable environmental conditions and proper management of resources. In the Americas the fragility of the environment and of the health of the inhabitants requires that human activities be informed, just, and benign, out of respect for and understanding of processes that affect landscapes and living organisms.

REFERENCES

1. National Academy of Sciences, U.S.A., Geochemistry and the environment, Vol. 1, The relation of selected trace elements to health and disease, Washington, D.C., 1974.
2. **Jenny, H.,** The Soil Resource: Origin and Behavior, Springer-Verlag, New York, 1980, chap. 8.
3. **Buol, S. W., Hole, F. D., and McCracken, R. J.,** Soil Genesis and Classification, Iowa State University Press, Ames, Iowa, 3rd ed., 1989, Chap. 6.
4. **Hole, Francis D. and Campbell, James, B.,** Soil Landscape Analysis, Rowman and Allanheld, New Jersey, 1985, Chap. 2.
5. World Health Organization, Ann. Epidemiological and Vital Statistics, UNESCO, 1960.
6. **Sauer, H. I. and Brand, F. R.,** Geographic patterns in the risk of dying, in *Environmental Geochemistry in Health and Disease,* Cannon, H. L. and Hopps, N. C., Eds., Geol. Soc. of Amer. Memoir 123, 1971, 131.
7. **Tourtelot, H. A.,** The health connection and its problems for geoscientists, The Clay Society, 17th Ann. Meeting, 28th Ann. Clay Minerals Conference, p. 96, 1980.
8. **Mose, D. G. and Mushrush, G. W.,** Regional levels of indoor radon in Virginia and Maryland, *Environmental Geology and Water Sciences,* 12, 197, 1988.
9. **Sonderegger, J. L. and Ohguchi, T.,** Irrigation related arsenic contamination of a thin, alluvial aquifer, Madison River Valley, Montana, U.S.A., *Environmental Geology and Water Sciences,* 11, 153, 1988.
10. **Maresca, G. P., Puffer, J. H., and Germine, M.,** Asbestos in lake and reservoir waters of Staten Island, New York: source, concentrational, mineralogy and size distribution, *Environmental Geology and Water Sciences,* 6, 201, 1984.
11. **Låg, J.,** Soil properties of special interest in connection with health problems, *Experientia,* 43(1), 63, 1987.
12. **Allaway, W. H.,** The effect of soils and fertilizers on human and animal nutrition, *Agric. Information Bull. No. 378,* Agric. Research Service and Soil Conservation Service, U.S.D.A., in cooperation with Cornell University Agric. Exper. Sta., Supt. of Documents, Washington, D.C., 1975.
13. **Shamburger, R. J.,** Selenium and cancer, in *Biochemistry of Selenium,* Plenum Press, New York, 207, 1983.
14. **Howells, G. M., Crounse, R. G., Thomas, J. M., Whitely, T. E., Finn, M., Bray, J. T., Smith, A. M., and Jones, C.,** Prolonged survival of Balb/c mice with a transplanted tumor (NOPC 467) induced by oral selenium plus immunoactivation with C. parvum, Poster at meetings, Soc. for Environmental Geochemistry and Health, Greenville, North Carolina, 1982.
15. **Clark, L. C., Graham, G., Crounse, R. G., Cremson, R., and Shy, C. M.,** A case control study of skin cancer and selenium in eastern North Carolina, Ann. Meeting, Am. Public Health Ass., Montreal, Canada, 1982.
16. **Andrews, J. W., Hames, O. G., Metts, J. C., and Davis, J. M.,** Selenium, cadmium, and glutathione peroxidase in blood of cardiovascular diseased and normal subjects from the cardiovascular belt of southeastern U.S.A., in *Trace Substances in Environmental Health,* XIV, Proceedings, D. D. Hemphill, Ed., University of Missouri, Columbia, Missouri, 1980, 38.
17. **Crounse, R. G.,** Geochemistry and human health in the 1980s, in *Applied Geochemistry in the 1980s,* Thornton, I. and Howarth, R. J., Eds., Graham and Trotman, London, 1986, 337.
18. **Anderson, M. S.,** History of selenium toxicity, in *Selenium in Agriculture,* Anderson, M. S., Ed., U.S.D.A. Agric. Handbook No. 200, U.S. Gov. Printing Office, Washington, D.C., 1961.
19. **Presser, T. and Barnes, I.,** Dissolved constituents including selenium in waters in the vicinity of Kesterson National Wildlife Refuge and West Grassland, Fresno and Merced Counties (California), Water Resources Invest. Report 85-4220, U.S.G.S., Menlo Park, CA, 1985.
20. **Kubota, J.,** Distribution of cobalt deficiency in grazing animals in relation to soils and forage plants of the United States, *Soil Science,* 106, 122, 1968.
21. **Fonner, R. F.,** Homeowner's guide to geologic hazards, Mountain State Geology, West Virginia Geological and Economic Survey, Morgantown, WV, 1987, 1.
22. **Degraff, J. V. and Agard, S. S.,** Defining geologic hazards for natural resources management using tree-ring analysis, *Environmental Geology and Water Scienes,* 6, 147, 1984.

23. **Koerner, F. C.,** Polar Biomedical Research: an Assessment, National Academy Press, Washington, D.C., 1982.

24. **Hodgson, J. F., Allaway, W. H., and Lockman, R. B.,** Regional plant chemistry as a reflection of environment, in *Environmental Geochemistry in Health and Disease,* Cannon, H. L. and Hopps, H. C., Eds., The Geological Society of America, Inc., Memoir, 123, 1971, 57.

25. **Kubota, J., Allaway, W. H., Carter, D. L., Cary, E. E., and Lazer, V. A.,** Selenium in crops in the United States in relation to selenium-responsive diseases of animals, *J. Agric. Food Chem.,* 15, 448, 1967.

26. **Wan, H. F., Mikkelsen, R. L., and Page, A. L.,** Selenium uptake by some agricultral crops from central California soils, *J. Environ. Qual.,* 17, 269, 1988.

27. **Sanchez, P. A. and Buol, S. W.,** Soils of the tropics and the world food crisis, *Science,* 188, 598, 1975.

28. **Hopps, H. C.,** Geographic pathology and the medical implications of environmental geochemistry, in *Environmental Geochemistry in Health and Disease,* Cannon, H. L. and Hopps, H. C., Eds., The Geological Society of America Memoir 123, 1971.

29. **Maier, R. M.,** personal communication, 1988.

30. **Eisenbud, M.,** *Environmental Radioactivity,* 3rd ed., Academic Press, Orlando, FL, 1987.

31. **Woodwell, G. M.,** Toxic substances and ecological cycles, in *Man and the Ecosphere,* Ehrlich, P. R., Holdren, J. P., and Holen, R. S., Eds., Readings from Scientific American, W. H. Freeman, San Francisco, 1971, 128.

32. **Huff, D. D., Farrow, N. D., and Jones, J. R.,** Hydrologic factors and ^{90}Sr transport: a case study, *Environmental Geology and Water Sciences,* 4, 53, 1982.

33. **Meriwether, J. R., Beck, J. N., Keeley, D. F., Langley, M. P., Thompson, R. H., and Young, J. C.,** Radionuclides in Louisiana soils, *J. Environ. Qual.,* 17, 562, 1988.

34. **Kiss, J. J., de Jong, E., and Martz, I. W.,** The distribution of fallout cesium-137 in southern Saskatchewan, Canada, *J. Environ. Qual.,* 17, 445, 1988.

35. **Pratt, C. J.,** Chemical fertilizers, in *Man and the Ecosphere,* Ehrlich, P. R., Holdren, J. P., and Holen, R. W., Eds., Readings from Scientific American, W. H. Freeman, San Francisco, 1971, 236.

36. **Ritter, C. J. and Rinefierd, S. M.,** Natural background and pollution levels of some heavy metals in soils from the area of Dayton, Ohio, *Environ. Geol. Water Sci.,* 5, 73, 1983.

37. **Zaporozec, A.,** Nitrate concentrations under irrigated agriculture, *Environ. Geol. Water Sci.,* 5, 35, 1983.

38. **Anderson, D.,** Natural insect controls, Quarterly of the University of Wisconsin College of Agriculture and Life Sciences, Madison, WI, 7, 1, 1988.

39. **Inglis, J. M., Martisoff, G., and Kelly, W. R.,** Pollutant transport in a shallow unconfined aquifer in Perry, Ohio, *Environ. Geol. Water Sci.,* 8, 237, 1986.

40. **Prose, D. V.,** Persisting effects of armored military maneuvers on some soils of the Mojave Desert, *Environ. Geol. Water Sci.,* 7, 163, 1985.

41. **Pinder, J. E., III, and McLeiod, K. W.,** Contaminant transport in agro-ecosystems through retention of soil particles on plant surfaces, *J. Environ. Qual.,* 17, 602, 1988.

42. **Thornton, I., Abrahams, P. W., Culbard, E., Rother, J. A. P., and Olson, B. H.,** The interaction between geochemistry and pollutant metal sources in the environment; implications for communities, in *Applied Geochemistry in the 1980s,* Thornton, I. and Howarth, R. J., Eds., Graham and Trotman, London, 1986, 270.

43. **Bourke, J. G.,** The medicine-men of the Apache, 9th Annu. Report, Bur. of Ethnology, Smithsonian Institute, Washington, D.C., 1982.

44. **Hunter, J. M.,** Geophagy in Africa and the United States, *Geograph. Rev.,* 63, 170, 1973.

45. **Hunter, J. M. and de Kleine, R.,** Geophagy in Central America, *Geograph. Rev.,* 74(2), 157, 1984.

46. **Dickens, D. and Ford, R. N.,** Geophagy (dirt eating) among Mississippi negro school children, *Am. Sociol. Rev.,* 7, 59, 1942.

47. **Vermeer, D. E. and Frate, D. A.,** Geophagy in a Mississippi county, *Assoc. Am. Geographers, Annals,* 65, 414, 1975.

48. **Leopold, A.,** *A Sand County Almanac,* Oxford Univeristy Press, London, 1970.

49. **Potter, V. R.,** *Global Bioethics: Building on the Leopold Legacy,* Michigan State University Press, E. Lansing, MI, 1988.

Chapter 15B

A GEOMEDICAL APPRAISAL OF AUSTRALIA AND NEW ZEALAND

C. R. Lawrence and M. W. Johns

TABLE OF CONTENTS

I. INTRODUCTION

This chapter summarizes the known problems of disease in livestock and humans caused by physical and chemical agents in the natural environment in Australia and New Zealand.

This appraisal has examined the literature from the viewpoint of disease as well as the characteristics of the environment. Accordingly, the survey has spanned the literature in environmental medicine,[1-5] occupational medicine, veterinary medicine, and environmental geoscience[6] in Australia and the distribution of water quality, soil type,[7] toxic native vegetation,[2] venomous animals,[3] and climate in Australia.

The Australian environment is characterized by low rainfall, high solar radiation, and extensive tracts of land that are dominated by native vegetation. However, both Australia and New Zealand have low human populations with respect to surface area: Australia has a population of 16 million and a surface area of 7.7×10^6 km^2 and New Zealand has a population of 3.3 million and a surface area of 2.17×10^5 km^2 (see Figure 1).

The chapter is divided into two sections, the first for diseases of livestock and the second for human diseases. Discussion is based on the nature of potential or established causal factors, the geographic distribution of those factors, and their associated diseases.

II. DISEASES OF LIVESTOCK

Because of their herbivorous and restricted diet, livestock are often more susceptible to the vagaries of the environment than are humans: any excess or deficiency of minerals in the plants the animals eat or the water they drink are reflections of soil composition at the particular locality where they live.

The diseases of livestock in Australia and New Zealand, including symptoms and treatment, have been discussed in detail by Hungerford.[2] The following summary leans heavily on that text.

A. MINERAL DEFICIENCY

Within many different parts of Australia and New Zealand, livestock suffer mineral deficiency diseases which reflect the soil type and, in turn, the composition of the host rock and the products of its weathering. There are significant areas where there is a deficiency in iodine, copper, phosphorus, cobalt, or selenium.

Table 1 summarizes the different types of mineral deficiency, the animal species affected, the diseases caused, and the areas of occurrence. This table is not intended to oversimplify the subject of mineral deficiency, which can involve complex chemical interactions such as a high level of molybdenum or sulfate causing a relative deficiency in copper.

B. NATURAL HAZARDOUS SUBSTANCES
1. Minerals

An excess of nitrate, fluoride, or arsenic can cause disease in livestock. This does not seem to be common as a result of food digestion, but, rather, through the drinking of mineralized water. Over large areas of Australia, particularly where rainfall is low, stock are dependent on ground water tapped by wells for drinking. The composition of this water varies considerably from one area to another and, indeed, the ground water in some areas is too salty even for stock to drink because of high concentrations of sodium chloride.

Nitrate levels are high in western Victoria and areas throughout the Northern Territory and in western New South Wales. However, there have been cattle illnesses and deaths in western New South Wales and Victoria due to nitrate-rich drinking water whose source may have been a chemical waste dumped at the ground-water intake.

Fluoride levels in stock water are generally low, but in the Northern Territory and

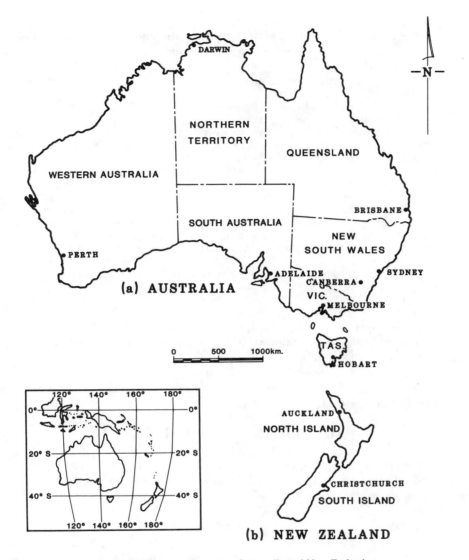

FIGURE 1. Locality maps of Australia and New Zealand.

Queensland there are tapped ground waters which contain more than 5 mg/l fluoride. These have caused fluorosis with abnormalities of bone and teeth, particularly in sheep.

Arsenic occurs in high levels in leachate from abandoned mines in some areas throughout Australia. Arsenic poisoning has been the cause of death of cattle near Bendigo and Bethanga, Victoria, where there has been a lot of gold mining in the past and where arsenical minerals such as arsenopyrite are included in the ore bodies.

2. Plant Toxins

There are hundreds of species of indigenous plants which produce toxic alkaloids in Australia. For a complete list, see Hungerford.[2]

In the high-rainfall fringe on the eastern, southeastern, and far southwestern parts of Australia, poisoning of livestock through eating toxic native vegetation is not a significant problem because of the clearing of vegetation and the establishment of pasture, however, in northern and inland Australia, where there has been little clearing and the livestock are

TABLE 1

Mineral Deficiency Diseases in Livestock in Australia and New Zealand — Their Causes and Occurrence

Element	Species	Diseases	Occurrence
Iodine (I)	Pigs, goats, sheep, cattle, horses, birds	Goiter, stillborn, weak at birth	Tasmania
Copper (Cu)	Sheep, pigs	Enzootia ataxia, hypocuprosis ("stringy wool", "staggers")	Soils on Cainozoic quartzose sands, S.E. of S. Aust. and W. Victoria
Phosphorus (P)	Cattle	Osteomalacia, osteoporosis ("stiffs", "creeps")	Riverine Plains (New South Wales), Western Australia, coastal New South Wales, on terrestrial or strongly weathered soil
Cobalt (Co)	Sheep, cattle	Enzootia marasmus, ("phalaris staggers")	S.W. & W of. W. Aust., S.E. & W. coast of S. Aust., Dandenong Ranges, Victoria
Selenium (Se)	Sheep, cattle, horses, pigs,	Muscular dystrophies, "white muscle" disease, heart conditions	Widespread in Australia and New Zealand

free ranging, there continues to be a problem. A notorious example is Kimberley horse disease from eating wedge-leaved rattlepod (*Crololaria retusa*). This disease is prevalent in the far north of Western Australia.

Also of concern is algal growth on surface-water supplies. Promoted by high levels of phosphate and nitrate in the water, some of these growths contain toxins which have been harmful and, in some cases, fatal to livestock.

III. DISEASES OF HUMANS

The environmental factors which cause disease in humans are various forms of natural radiation and adverse compositions of the food we eat, water we drink, and air we breathe. Each of the factors is discussed in turn.

A. RADIATION

Radiation is in the form of electromagnetic radiation alone when it is derived from a heat source, such as the sun and particles and electromagnetic radiation when it is derived from radioactive decay.

1. Solar Radiation

Excessive exposure to ultraviolet rays (0.2 to 0.4 μm wavelength) from solar radiation is a major public health problem in Australia and New Zealand.

Australia lies between latitudes 10 and 44°S and New Zealand, between 24 and 47°S. Both countries receive long hours of sunlight, an average of more than 2000 h/year and more than 3500 h/year in some places, such as in central Australia. The intensity of radiation varies with the angle of incidence, which depends, in turn, on the latitude, time of day, and season. The greatest intensity is at about midday during the summer months, nearer the equator.

The aboriginal and Maori populations of Australia and New Zealand, respectively, produce large amounts of a pigment called melanin in the skin which is able to absorb ultraviolet light. Unfortunately, fair-skinned people of European origin with less pigment have less protection. Yet both countries have an outdoor, recreational lifestyle and suntans are regarded as desirable. The result is that Australia and New Zealand have a high incidence of skin cancer. Indeed, the Australian population probably has the highest incidence of skin cancer in the world, with over two thirds of the population developing skin cancer during their lifetime. A public awareness campaign has been launched to encourage the use of protective clothing and chemical sunscreen applications.

The body areas affected are mostly those with maximum exposure to solar radiation — the face, arms, and shoulders. The three main types of skin cancer in Australia are basal cell carcinoma (75 to 80% of all skin cancers), squamous cell carcinoma (20%), and melanoma (2 to 3%). The incidence of malignant melanoma in Australia now is 32 per 100,000 population per year, but is even higher in Queensland and is doubling there every 15 years.

There probably will be an increase in the intensity of ultraviolet radiation reaching earth in the future. The ozone layer in the lower to middle stratosphere (15 to 35 km), which partially absorbs ultraviolet radiation, has shown a progressive decrease in concentration over a period of years, attributed to the effect of chlorofluorocarbons released into the atmosphere as a result of human activities.

2. Radioactive Minerals

Natural radioactivity is derived from cosmic radiation, which doesn't vary much over the earth's surface, and radioactive minerals, which can vary from place to place, depending on the geology. Nonetheless, the combined dosage from these sources is very small and causes no immediate damage to tissues, although it is believed to have been the major source of genetic change in organisms.

Granite rocks contribute a major share of the natural background radiation. Through radioactive decay of uranium minerals, radon gas is produced. This has been found in the basements of houses in Canada and the U.S. which overlie either granites or glacial moraine with a granitic component. This does not appear to be a problem in Australia and New Zealand, although a national survey, including monitoring, is being carried out in Australia at present.

Of greater concern in Australia is the radiation hazard associated with the mining of radioactive minerals, such as the uranium minerals in the Northern Territory and monazite, which includes thorium from beach sand deposits. Apart from direct radiation hazards increasing the risk of several malignancies such as leukemia and lymphomas, there is the risk of radioactive dust being inhaled into the lungs and causing lung cancer.

Some mineral waters, which naturally effervesce with carbon dioxide, from Lyonsville, Victoria, and Queensland contain levels of radioactivity above acceptable limits. This is in the form of radon gas. No such mineral waters are marketed currently, but 50 years ago the radon content and radioactivity of such spas was advertised as being therapeutic.

B. MINERAL DEFICIENCY

1. Iodine

A deficiency in iodine in drinking water in Tasmania and eastern Victoria, both characterized by high rainfall and high runoff, has caused endemic goiter in the past, but this has been overcome largely by dietary supplements of iodine, particularly in bread and table salt.

2. Selenium

There is evidence from soil analyses and from livestock disease that selenium deficiency occurs both in New Zealand and Australia. Thus, there is the possibility that humans could also be affected by selenium deficiency, but this has yet to be confirmed.

3. Fluoride

Most of the reticulated water supplies for the cities and towns of Australia are derived from surface runoff. The naturally occurring concentration of fluoride in these waters is usually less than 1 mg/l, which seems to be the optimal level in the prevention of dental caries in children. In some but not all areas, fluoride has been added to the water supply in recent years to provide a concentration of 1 mg/l. There has been a significant decline in the incidence of dental caries in children as a result.

C. NATURAL HAZARDOUS SUBSTANCES

Some substances which occur naturally can be a hazard to human health. These substances can be included in the diet, in drinking water, or in the air we breathe.

1. Diet

a. Nitrate

There are extensive areas of Australia[8,9] and New Zealand[14] where there are nitrate-rich ground waters.

It was pointed out by Johns and Lawrence[8] that nitrate-rich ground water containing more than 45 mg/l NO_3, and capable of causing illness or even death to infants, occurs in widely scattered areas of Australia. Such water is normally excluded from reticulated water supplies. However, private wells are a common source of water for domestic purposes in rural areas; some of these contain toxic amounts of nitrate which causes methemoglobinemia, particularly in infants being fed a milk formula which includes the nitrate-containing water.

The nitrate is converted to nitrite in the upper gastrointestinal tract by bacterial action and is absorbed into the bloodstream. The nitrite ion combines with reduced hemoglobin, modifying it to a brown pigment, methemoglobin, which is unable to transport oxygen in the blood and, hence, causing hypoxemia and cyanosis.

Nitrates and nitrites in the diet may also pose a human health hazard by the formation of carcinogenic nitrosamines in the lower gastrointestinal tract.

The nitrate-rich ground waters in Australia and New Zealand are invariably from shallow sources and in a variety of rock types, which include basalt, sand, calcrete, and limestone.

The nitrate in the ground waters of northern and central Australia is derived from N-fixing indigenous vegetation, particularly acacia, which produce high soil-nitrate levels. Some of this nitrate enters shallow aquifers and is believed to have been present in a similarly high concentration before European settlement. By contrast, the nitrate-rich ground waters of the temperate and southern part of Australia and also of New Zealand are related to man's activities, including cultivation and the disposal of sewage, dairy waste, and industrial waste.

b. Arsenic

The recommended limit for arsenic in drinking water is 0.05 mg/l. In excess of that concentration, there is the risk of hyperpigmentation of the skin, generalized weakness, and diarrhea.

In Australia, arsenical ground waters occur in the Eastern Highlands, where there are existing mine workings and where the ground waters are more acidic. For the Victorian part of the Eastern Highlands, Shugg[10] has reviewed the distribution of arsenic in ground waters. Locally, there are springs and mine leachates which are arsenic rich. A detailed study of one site in the New England district of New South Wales by Brooks and Mellveen[11] showed leachate from the abandoned Conrad Mine discharging into Borah Creek. This creates a potential hazard should such waters be consumed by humans.

c. Mercury

Mercury levels in fish, particularly those in fresh-water surface storages such as Lake Eildon and Sugarloaf Reservoir in Victoria, Australia, and Lake Taupa in New Zealand, have sometimes reached levels where the people have been advised not to eat them. These storages are in catchments where there has been mining in the past and are probably contaminated by the mercury which was used in the recovery of gold. Symptoms of mercury poisoning in humans vary widely. They include metallic taste, thirst and salivation, abdominal pain, vomiting, diarrhea, ataxia, and tremor.

2. Air Pollution

The composition of air can be adversely affected with respect to human health by naturally occurring gas or mineral particles. Usually, the mineral particles are derived from mining operations, rather than deflation.

Inhalation of rock and/or coal dust has always created a health hazard in mines and their environs. Chronic exposure to dust containing fine particles of free silica has been a particular hazard in hard-rock mining, sand blasting operations, and founderies. It produces silicosis and chronic pulmonary fibrosis which is irreversible. Anthracosis is a related pneumoconiosis in coal miners. Both were prevalent in Australia during the last century, particularly in association with tuberculosis, but have been largely overcome by better ventilation, mining practice, and the use of masks.

a. Asbestos

Another health hazard is found in the mining, handling, and processing of asbestos. The asbestos group of minerals includes blue asbestos or crocidilite, fibrous serpentine or chrysotile, and tremolite and actinolite.

Asbestos fibers are long and flexible, with individual fibers as small as 10 μm in diameter. Broken filaments are carried as dust into the lungs, causing asbestosis with pulmonary fibrosis and predisposing to bronchogenic carcinoma and pleural mesotheliomas. Diseases may take many years to develop, but can occur after only a relatively brief exposure to asbestos dust.

Areas where asbestos mining and associated asbestosis have occurred in Australia are Wittenoom (Western Australia) and Baryulgil (New South Wales). The subject of the health hazard at these sites has been the subject of government inquiries and litigation by affected individuals, and asbestos mines in Australia have been abandoned recently because of concern over the health hazard.

b. Manganese

At Groot Eylandt, off the Northern Territory coast, there is evidence of manganese intoxication in both the aboriginal and European populations at the mining site. There is a relatively high incidence of a neurological disorder resembling Parkinson's disease.

Either the inhalation or ingestion of manganese-rich dust particles can cause disease, including pneumonitis, liver disease, and extrapyreuridal tract lesions of the central nervous system.

c. Hydrogen Sulfide

In the geothermal areas of Rotorua, New Zealand, there is natural emanation of hydrogen sulfide gas into the atmosphere. This is produced by the action of steam or water at high temperatures on inorganic sulfides. Also, hydrogen sulfide is commonly present in those ground waters throughout Australia which tap deep aquifers interbedded in carbonaceous material and which include the sulfide minerals of pyrite or marcasite. Water discharging from wells tapping these aquifers can effervesce with escaping hydrogen sulfide gas.

Hydrogen sulphide is very toxic in concentrations above 10 ppm in inspired air. Even at low concentrations, it irritates the conjunctivore and lungs and can cause pulmonary edema.

d. Pollen

In both Australia and New Zealand, there are extensive grasslands, particularly of introduced species such as rye grass and cocksfoot. The pollens of such grass cause allergic (seasonal) sinusitis or hay fever and exacerbate the symptoms of bronchial asthma. This is expressed by the higher seasonal levels of hay fever and asthma during late spring and summer, when the pollen are released and dispersed, particularly in southeastern Australia.

The incidence of asthma is very high in Australia and New Zealand, compared with other countries; for example, one in four school children in Australia suffers from some degree of bronchial asthma.

Using mortality as an index, Woolcock[12] has reported that for those countries for which data are available (W. Germany, U.K., U.S., Sweden, Australia, and New Zealand), New Zealand has by far the highest mortality rate for bronchial asthma: 8 deaths per 100,000 population per year, followed by Australia with 4 deaths per 100,000 population per year.

3. Venomous Animals

There are a number of venomous vertebrates and invertebrates which are capable of injecting their toxins into humans or other animals by way of bites or stings. This is much more of a problem in Australia than in New Zealand.

Within Australia, there are 13 venomous snakes, each species with its own distribution.[13] For the southern part of the continent, snakes are most active and, hence, more of a threat in the spring and summer. By contrast, New Zealand has no venomous snakes.

There are at least two venomous spiders in Australia, the red-backed spider *Latrodectus hasselti* and *Atrax robustus*. Around the coastlines of Australia and New Zealand, there are a number of venomous coelenterates, echinoderms, and annelids. Most notorious are the blue-ring octopus *Hapalchlaena maculosa* and, for the tropical waters of northern Australia, the stonefish *Syrancja trachynis*. Antivenoms have been developed to treat the bites and stings of these venomous animals, which have caused deaths in the past in Australia. Occasional deaths still occur.

ACKNOWLEDGMENTS

The authors wish to express their appreciation for the help of Dr. John Reeve of the New Zealand Department of Health and Mr. D. C. Lawrence for the literature search, as well as the staff of a number of organizations, including the Commonwealth Radiation Laboratory, Asthma Foundation, and the Victorian Anti-Cancer Council.

REFERENCES

1. **Holland, W. M., Johannes, I., and Kostrzewski, J., Eds.,** Measurement of Levels of Health, World Health Organization, Copenhagen, 1979.
2. **Hungerford, T. G.,** *Diseases of Livestock,* McGraw-Hill, New York, 1975.
3. **McGashan, N. D., Ed.,** *Medical Geography,* Methuen & Co. Ltd., 1972.
4. **Låg, J.,** General survey of geomedicine, in *Handbook of Geomedicine,* Låg, J., Ed., CRC Press, Boca Raton, FL, 1989, chap. 1.
5. **Hut, M. S. R. and Burkitt, D. P.,** *The Geography of Non-Infectious Disease,* Oxford University Press, New York, 1986.
6. **Strahler, A. N. and Strahler, A. M.,** *Environmental Geoscience: Interaction Between Natural Systems and Man,* Hamilton, 1973.
7. **Stace, H. C. T., Hubble, G. H., Brewer, R., Northcote, K. H., Sleeman, J. R., Mulcahy, M. J., and Hallsworth, E. G.,** *A Handbook of Australian Soils,* Rellim Technical Publishers, S. Australia, 1968.

8. **Johns, M. W. and Lawrence, C. R.,** Nitrate-rich groundwater in Australia: a possible cause of methemoglobinemia in infants, *Med. J. Aust.,* 2, 925, 1973.

9. **Lawrence, C. R.,** Nitrate-rich groundwater of Australia, Australian Water Resources Council, paper, 1979, 110.

10. **Shugg, A. S.,** Distribution and cause of high arsenic concentrations in groundwater in central Victoria, Department of Water Resources, Victoria, Staff Paper, 110, 1989.

11. **Brooks, K. A. and Mellveen, G. R.,** The impact of a derelict base metal mine on the aquatic environment, in *Proc. 3rd Int. Mine Water Congress,* Melbourne, Australia, 1988, 793.

12. **Woolcock, A. J.,** Worldwide differences in asthma prevalence and mortality. Why is asthma mortality so low in the U.S.A.? *Chest,* 90(5), 405, 1986.

13. **Cogger, H. G.,** Venomous snakes of Australia and Melanesia, in *Venomous Animals and their Venoms,* Vol. 2, Bucherl, U. and Buckley, E. E., Eds., Academic Press, New York, 1971, chap. 23.

14. **Thorpe, H. R.,** personal communication.

Chapter 15C

GEOENVIRONMENT-RELATED DISEASE ENDEMICITY IN AFRICA, WITH SPECIAL REFERENCE TO TANZANIA

U. Aswathanarayana

TABLE OF CONTENTS

I. INTRODUCTION

Like most African countries, Tanzania is poor (with a low GNP per capita per year of $220), sparsely populated (24 people per km^2), and essentially rural (89% of the estimated population of 23 million live in villages), with a high rate of population growth (3.2% per year). There is very little industry — the manufacturing and mining sectors contribute 4.4 and 0.3%, respectively, to the GNP. The few industrial plants that exist are located near urban centers, and heavy metal (Cd, Hg, Pb, etc.) and other kinds of pollution due to the fertilizer, oil refining, cement, battery, plastics, metal rerolling and other industries are more or less confined to those areas. The incidence of diseases (such as silicosis, talcosis, pneumoconiosis, lung cancer, etc.) caused by the inhalation of artificial dusts arising from mining, quarrying, and blasting activities is limited. Thus, the number of persons at risk due to industrial pollutants in Tanzania should not exceed about 1% of the total population of the country.

There is a paucity of data on the relationship between the geoenvironment and disease endemicity in Africa. Consequently, it should be emphasized that the proposed models with respect to geochemical pathways, etiology, and epidemiology are speculative.

II. DISEASE ENDEMICITY IN RELATION TO THE AVAILABILITY AND SPECIATION OF SOME TRACE ELEMENTS

The endemicity of some diseases in Tanzania (e.g., goiter and fluorosis) is directly traceable to diet and water. Whereas the water drunk is almost invariably of local origin, the diet may not always be reflective of the local geoenvironment for the following reasons:

1. Part of the food (e.g., wheat or rice) may have been partly or wholly imported,
2. Extensive use of fertilizers, herbicides, pesticides, etc. tends to complicate the geochemical environment
3. Meat may not be faithfully reflective of the geochemical environment as the meat animal is an efficient buffer of trace elements that it picks up from plants
4. The processing of food can bring about changes in the relative abundance of elements (e.g., Zn/Cd ratio).

Maize and cassava, which are the staple foods in Tanzania, are mostly of local origin. The liquors brewed from maize and cassava therefore reflect the local geochemical environment. Except in the case of fluoride, diet contributes 80 to 90% of the trace elements, and water provides the rest.

The quantity of trace element that reaches man through food depends upon (1) geochemical availability (depending upon the leachability of an element and the mode of its distribution and availability in the rock or soil) and (2) bioavailablity (the fraction of the element present in the food of plant or animal origin that is available to man). Trace elements exist in nature in different chemical forms, and their speciation influences their toxicity (e.g., toxicity of Cr (VI) versus the nontoxicity of Cr (III)). Besides, trace elements may undergo biotransformations in the human body and may form metallobiocomplexes. This may involve a change in the oxidation state and, hence, in the toxicity of the trace element concerned. Two or more trace elements may act conjointly in such a manner that the beneficiality or toxicity of one element is suppressed or accentuated by the other. For instance, the presence of zinc moderates the toxicity of cadmium.

III. NUTRITIONAL STATUS AND DISEASE ENDEMICITY IN TANZANIA

The inadequacies in nutrition (in terms of quantities per capita per day) in Tanzania[9] and the recommended daily intake (RDI) levels are given below.

1. Energy: 6 to 10.5 MJ (deficiency of 30%)
2. Reference: protein: 24 to 35 g (deficiency of 10%)
3. Calcium: 310 g (very low)
4. Iron: 16 mg (adequate; RDI: 18 mg/d)
5. Vitamin A: 220 mg of retinol (very low)
6. Vitamin C: 55 mg (adequate)
7. Thiamine: 1.0 mg (low)
8. Iodine: low (estimated to be around 25 µg/d in the highland areas, versus the RDI of 150 µg/d)
9. Fluorine: high in some and adequate in most areas (RDI: 1.5 to 4 mg/d)

No information is available regarding the daily intake of essential elements like Zn, Mn, Cu, Se, Cr, Mo, etc.

According to the Tanzania Food and Nutrition Centre (TFNC), the aflatoxin content of the food items consumed in Tanzania range from 100 µg/kg for coriander seed to 800 µg/kg for cassava. Among the foods consumed in any region, cassava invariably has the highest aflatoxin content, with samples from Mtwara and Rukwa having particularly high aflatoxin contents. Incidentally, the permissible level of aflatoxin content, according to the World Health Organization, is 30 µg/kg. The possible goiterogenic and carcinogenic attributes of cassava may be related to its thiocyanate and aflatoxin contents.

Malnutrition may be the direct causative agent for diseases like anemia and avitaminosis A. Most often, undernutrition has the consequence of aggravating a variety of diseases (e.g., goiter). Thus, even in areas where a particular disease is endemic, the poor, undernourished people tend to be more affected by the disease than the affluent, well-nourished people.

IV. IODINE IN THE GEOENVIRONMENT

The concentration of iodine in rocks, soils, and water is highly variable, but generally low (of the order of ppb). The abundance levels are not known with any reasonable degree of precision because of the difficulties of determining iodine at such low levels.[4] The average concentration of iodine in igneous rocks is 40 ppb, with carbonatites having the highest concentration of 500 to 1000 ppb. Metamorphic rocks have a low iodine content of less than 10 ppb, with granulite-grade metamorphic rocks having the lowest concentration among all rocks. Iodine is closely associated with organic matter in sediments. Thus, sedimentary rocks which contain organic matter, like shales, have about 10,000 ppb (10 ppm) of iodine, whereas sandstones and quartzites which contain no or a low content of organic matter have only 50 ppb of iodine. Chilean nitrate is rich in iodine and is in fact, a commercial source of iodine. The guano-related phosphorite of Minjingu in northern Tanzania is asociated with Lake Manyara, a soda lake in the East African Rift. The phosphorite is enriched in F, Cl, and U. It may be enriched in iodine, though this has not been established.

Iodine gets concentrated several fold in soils. Hence, soils could contain a few tens of ppm of iodine, depending upon the humus content, pH and contribution from atmospheric precipitation. Iodine in the diet is directly related to iodine in the soils. In areas subject to severe erosion, the soils would have lost what little iodine they would have contained initially. Hence, the iodine content of soils (and, hence, of food grown in them) in mountainous areas like the Himalayas and the Andes tends to be extremely low. In most African countries, erosion attributable to geomorphic factors (such as slope) is being accentuated by the indiscriminate destruction of vegetation cover for firewood.

Seawater contains 0.06 mg/l of iodine; hence, aerosols arising from sea spray, sea food, and kelp products are rich in iodine. Sea salt does not, however, inherit all the iodine contained in the sea water, because of the volatility of iodine.

V. ETIOLOGY AND EPIDEMIOLOGY OF IODINE DEFICIENCY DISORDERS

Iodine deficiency disorders (IDD) arise when the intake falls below the recommended level of 150 μg per capita per day. Goiter, which is the enlargement of the thyroid gland, is the most familiar form of IDD. The thyroid gland produces thyroxine, which controls body growth and activity. Goiter may develop because of (1) a deficiency of iodine in the diet (by far the most common causative factor) and (2) consumption of goiterogenic foods (e.g., cassava), which impair the utilization of iodine by the body. Excess calcium in the drinking water could precipitate iodine, thereby making it biologically unavailable to man. Similarly, chemical or bacterial pollution or protein-energy undernutrition may aggravate goiter, though they may not be able to cause it.

A horrendous consequence of endemic iodine deficiency is the development of cretinism in children born of parents residing in goiter-endemic areas. A cretin is characterized by a large head, short limbs, deaf-mutism, mental retardation, etc. The results of fetal hypothyroidism include abortions, still births, low birth weight, congenital abnormalities such as spastic diplegia, and increased mortality and psychomotor retardation during the first 2 years of life. Kavishe[8] estimates that if the iodine intake of women in a population falls below 20 μg/d per person, up to 10% of the children born may develop irreversible cretinism. It is difficult to comprehend the enormity of the personal tragedy and socio-economic consequences of this disastrous situation.

In general, IDD tends to be more prevalent in the highland areas away from the sea coast, where the soils and waters are low in iodine. People living in the coastal areas and consuming sea food rarely suffer from IDD.

The severity of goiter is graded 0, 1a, 1b, 2, and 3. Below grade 0, the thyroid is not palpable or, if palpable, not larger than normal. Grade 3 corresponds to a very large goiter, visible from a distance (World Health Organization classification).

VI. IODINE DEFICIENCY DISORDERS IN AFRICA

According to Ekpechi,[3] 150 million people out of the total population of 400 million in Africa are at risk with respect to IDD. The bulk of the people at risk are children and women of child-bearing age. Since about half the population in African countries is below 15 years of age, the seriousness of the risk is all too evident. About 8 million people are affected by severe IDD, including 2 million cretins and 6 million mentally retarded people.[3]

In Africa, 21 countries have severe IDD, 4 have moderate IDD, and 10 have mild IDD. There is no reliable information for 15 countries. The distribution of IDD is given below.[3]

Group A — Countries or regions in the countries with severe IDD, as manifested by the prevalence of palpable goiter (grades 1 to 3) >30% or visible goiter (grades 2 and 3) >10%: Cameroons (east and west regions), Ivory Coast (five northwest towns), Guinea (northwest Founta Djallon), Algeria (northern towns), Morocco (insufficient data), Mali (Mandingo Plateau and Bandingo Cliff), Sudan (Darfur province), Zaire (northeast and northwest regions), Tanzania (Mbeya, Mufundi, Nkasi, Sumbawanga, Songea, Mpanda, Rungwe, Ngara, and Morogoro Dts.), Zambia (northwest Dt.), Kenya (highlands), Comoros (high plateau and coastal regions), Ethiopia (highlands), Lesotho, (southern districts and mountains), Senegal (eastern Dts. and Cassamance), Sierra Leone (southeast region), Libya (Fezzan Dt.), Zimbabwe (Kariba Dt.), Nigeria (parts of eastern and western regions), Central African Republic (southwest Dt.), and Tunisia (northwestern part).

Group B — Countries or regions with moderate IDD involving the prevalence of palpable goiter (grades 1 to 3) between 20 and 29% and visible goiter (grades 2 and 3) between 5 and 9%: Tanzania (Kwimba, Sangarema, and Handeni Dts.), Central African Republic (Nola region), Zambia (south and southcentral Dts.), and Swaziland.

Group C — Countries or regions with mild IDD, as manifested by the prevalence of palpable goiter (grades 1 to 3) >20% and visible goiter (grades 2 and 3) >5%: Lesotho, Malagassy Republic (south, east, and west Dts.), Ghana, (north Dts.), Burkina Faso, Tunisia (northeastern part), Niger, Chad (south Dts.), Swaziland (school children), Gabon (north Dts.), Mali (northern part).

Hetzel[5] graded the prevalance of goiter in a similar manner and indicated the ameliorative measures.

Severe IDD — Goiter prevalence of >30%, prevalence of endemic cretinism of 1 to 10%; median level of iodine in urine >25 µg/g of creatinine; requires the use of iodinated oil either orally or through injection (at a cost of $0.09 per capita per annum).

Moderate IDD — Goiter prevalence up to 30%, some hypothyroidism; mean urine levels in the range of 25 to 50 µg/g of creatinine; controllable with iodinated salt (25 to 40 mg of I per kg of salt) or iodized oil.

Mild IDD — Goiter prevalence in the range of 5 to 20% (school children); median iodine levels in urine >50 µg/g of creatinine; controllable with iodinated salt at concentration levels of 10 to 25 mg of I per kg of salt (at a cost of $0.05 per capita per annum).

It should be emphasized that the scourge of goiter is entirely preventable and at a very low expense. Where a piped water supply is available, drinking water can be iodinated at a cost of $0.05 per annum per capita.

A. IDD IN TANZANIA

Under the auspices of the National Council for the control of IDD, the Tanzania Food and Nutrition Centre (TFNC) has been actively involved in surveys for and the amelioration of IDD.[7,8] Based on an examination of 63,666 cases in 30 districts in Tanzania, they reported a goiter prevalence of 48.5% gross, 15.8% grade 1B, and 9.1% visible types in the country as a whole. The prevalence of goiter and the estimated endemic cretinism, respectively, are graded as severe in the following districts with a dietary intake of >20 µg of iodine per capita per day: Mbeya rural (88.0 and 9.0%), Mufindi (80.9 and 8.5%), Nkasi (80.9 and 8.5%), Sumbawanga rural (78.6 and 8.1%), Songea rural (75.1 and 7.4%), Mpanda (67.5 and 5.6%), Rungwe (68.1 and 7.6%), Ngara (67.7 and 4.5%), and Kasulu (64.8 and 4.1%). Iodinated oil has been given orally or by injection in these areas, with highly evident and beneficial results. The geology of these areas is widely different, but the common factor in all the areas is their high elevation, severe erosion (accentuated by deforestation), and consequent low content of iodine in the soils, waters, and food grown in the areas.

Where the dietary intake is 20 to 50 µg/d, IDD is moderate. IDD is mild where the dietary intake is 50 to 100 µg/d.

An exception to the above generalization is the Morogoro Urban District, with a goiter prevalence of 60.4% and endemic cretinism of 2.7%. Morogoro Urban is not a highland area, nor is it far from the coast. Since Morogoro is a food-exporting district, the prevalence of goiter and cretinism may be traced to the low iodine content of rocks (granulite-grade metamorphics) and, hence, of soils and food.

An increased incidence of carcinoma of the thyroid has been reported in areas of endemic goiter. This possibility exists for Tanzania and needs to be studied. Pregnancy wastage is high in areas of severe IDD (5.5 to 1.7 per 1000 of population) in Tanzania. About 10 to 15% of the infant mortality in the highland regions of Kenya is attributed to IDD.[10]

VII. ENDEMICITY OF FLUOROSIS IN NORTHERN TANZANIA

Though fluoride has been regarded as an essential element with an RDI level of 1.5 to 4.0 mg, excessive ingestion of the element will have deleterious consequences. Mottling of teeth and skeletal fluorosis (bow legs, knock-knees, stiffness of trunk, impeded movement of limbs, severe joint contraction, etc.) are attributed to the ingestion of more than 10 mg/d of fluoride over several years. The incidence of skeletal fluorosis among workers in a cryolite factory in Denmark has been traced to their ingestion of fluoride dust. On the other hand, endemic fluorosis in Tanzania is confined to areas of high fluoride waters and arises from drinking such waters over the years.

The fluoride content of the waters of the various regions in Tanzania differ markedly. The highest average contents are encountered in the northern regions (e.g., Arusha: 3.5 to 78.0 mg/l) and some central regions (e.g., Singida: 0.7 to 24.0 mg/l), whereas the contents are generally low in the coastal regions (e.g., Mtwara: 0.2 to 0.9 mg/l). It is therefore not an accident that fluorosis is endemic in northern Tanzania, but is virtually unknown in coastal Tanzania. However, if a person lived in the fluorosis-endemic areas during his/her childhood, that person will continue to have mottled teeth, even though he/she subsequently moved to, say, the coastal region!

It has been estimated that the rate of ingestion of fluoride in northern Tanzania is about 30 mg/d, contributed as follows:

3 l of water with 8 mg/l of fluoride	24 mg/d
10 g of locally grown tea with about 200 ppm (dry weight) of fluoride	2 mg/d
5 g of "magadi" with 1000 ppm of fluoride	5 mg/d
Miscellaneous (through diet)	2 mg/d
Total	**33 mg/d**

This is about 15 times more than the estimated intake of 2 mg/d of fluoride per capita in temperate countries with normal water.

Variability in the incidence of severity of fluorosis among people who draw their drinking water from the same source arises from any one or combination of the following factors. (1) inadequate intake of nutrients (such as proteins, ascorbic acid, calcium, etc.) tends to aggravate fluorosis. Thus, low-income groups are at a higher risk than well-to-do persons. (2) The nature and amounts of staples consumed influence the absorption of fluoride in the body, depending upon the content of elements such as Ca and Zn. (3) Children are more affected than adults, probably because their teeth and bones (in which F^- substitutes for OH^-) are still forming. (4) Men are more affected than women, probably because of their greater physical activity and consequent higher consumption of water. The latter factor could also explain the difference in the incidence of fluorosis between farm workers, who have to work in the hot sun, and office workers.

VIII. FLUORIDE IN THE GEOENVIRONMENT

Aswathanarayana et al.[2] proposed a geochemical model for the existence of waters and "magadi" with abnormally high fluoride contents. Two sources have been identified: (1) a steady influx of fluoride into surface and ground waters by the leaching of Quaternary volcanics of the East African Rift with fluoride contents ranging from 0.029 to 0.49% and (2) the episodic, massive influx of fluoride by the leaching of highly soluble villiaumite (NaF) from volcanic ash, exhalations, and sublimates related to Miocene to Recent volcanism (the ash of the carbonatite of Oldoinyo Lengai, which erupted in 1960, has a fluoride content of 2.7%). Fresh and brackish spring and river waters form a cluster with consistently negative values of δD and $\delta^{18}O$ that become increasingly negative with an increase in altitude, F^- content, and specific conductivity.

Soda lakes in the region, which are the habitat for hippos and flamingoes, have a generally high pH (9.2 to 10.0), very low Ca and Mg, and very high Na, K, and F contents. The waters of the soda lakes are characterized by consistently positive values and parallel enrichment of δD and $\delta^{18}O$. The relevance of soda-lake waters to endemic fluorosis arises from their "magadi" encrustations with very high fluoride contents (up to 6000 ppm) which are used by the local people for cooking lentils and as salt. It is possible that the consumption of magadi may be related to the incidence of IDD and even stomach cancer, though the etiology of this is completely unknown at the moment.

Since the East African Rift volcanics and the associated ensemble of rift lakes, magadi, and high fluoride waters are genetically related, fluorosis similar to that in Tanzania is endemic all along the Rift in eastern and southern Africa.

The natural pattern of fluoride distribution in the waters is being modified in a few places by the direct application of phosphorite of Minjingu, northern Tanzania (with about 30% P_2O_5, 40% CaO, 3% F, and 200 ppm of U), as a fertilizer-cum-liming agent, at the rate of about 500 kg/ha. Leaching of the directly applied phosphorite by rainwater tends to add fluoride and uranium to the stream waters, which already have high fluoride contents. The gross radiation dosage at Minjingu has been estimated to be about 150 μR/h (equivalent to about 25 mGray/year?). The health consequences of the mining and direct application of this phosphorite are being studied.[1]

IX. WATER-RELATED DISEASES

Though they do not arise directly from the geoenvironment, diseases caused by the unavailability of adequate amounts of clean water are by far the most serious and widespread of all diseases in Tanzania, e.g., water-borne diseases (cholera, typhoid, bacillary dysentery, infective hepatitis, etc.), water-washed (trachoma, scabies), water-based (schistosomiasis, guinea-worm), and water-related insect vectors (malaria, sleeping sickness), etc. These diseases are largely preventable if only about 30 l of clean water per capita per day is available and hygiene is maintained.

A. RECENT VOLCANIC EXHALATIONS IN NORTHERN TANZANIA AS A NATURAL ANALOG OF "ACID RAIN"

As a consequence of the industrial emissions of SO_x and NO_x (in roughly equal quantities), the pH of the precipitation in the industrialized countries tends to be highly acidic (about 4.0 and occasionally as low as 2.2). This "acid rain" and its dry equivalent make the pH of the soil and surface waters acidic, and the lower forms of life (e.g., fish) may even die. Acid rain is currently regarded as the most serious environmental problem facing the industrialized countries of North America and western Europe. The concentration of dissolved aluminium in acid waters is about 1000 times greater than in normal waters. Since the crustal abundance of Al is next only to Si, and since Al has the potential to form complexes of various kinds, the biochemical behavior of Al in an acid pH situation can have quite serious implications. Besides, acid pH can enhance the solubility of some toxic elements, reduce the solubility of some essential elements, or both.

Volcanic emissions of CO_x, SO_x, and NO_x are in no way different chemically from the industrial emissions; the only difference between them is that whereas industrial emissions are continuous, volcanic emissions are episodic. Northern Tanzania witnessed volcanic activity during the last few million years (Mt. Meru and Oldoinyo Lengai erupted in 1896 and 1960, respectively, and limited activity of Oldoinyo Lengai was reported in 1987). Thus, we seem to have in northern Tanzania a natural analog of acid rain. It remains to be seen what effect the Quaternary volcanism has on the endemicity of diseases.

Most of the N-nitroso compounds (derived from nitrates related to Quaternary volcanism?) appear to be carcinogens. A correlation has been reported between the ingestion of nitrate and the incidence of atrophic gastritis and gastric cancer. The nitrate content of well water is said to be higher in areas of high risk of carcinoma of the stomach and colon. Variation in the risk factor is attributed to the extent of urinary excretion of nitrate.[6]

X. ETIOLOGY OF GASTRIC CANCER

Hopps[6] gave a succinct account of the etiology and epidemiology of cancers in relation to the geochemical environment. Among the various kinds of cancers, those of the stomach, colon, and rectum have been found to be strongly influenced by the environment. The 30-fold variation in the rates of incidence of stomach cancer in the world attest to this. Analogous to similar situations elsewhere, the incidence of gastric cancer in Tanzania appears to be affected by four factors.

1.   Highest gastric cancer mortality in Japan, followed closely by Chile, Costa Rica, Iceland, etc. (The common factor among these countries is the Recent volcanism they witnessed — similar to that in northern Tanzania and the Rungwe province in southern Tanzania.)
2.   Peaty, ill-drained, acidic soils in Japan (similar acidic soils exist in Tanzania)
3.   The premalignant role of pernicious anemia (caused by malnutrition) and precancerous gastric lesions (caused by nitrate ingestion?)
4.   In Colombia (South America), eating of above-average amounts of corn and moras (local berries) with resultant higher prevalence of gastritis than any other combination of foods (probably due to a synergistic effect). On the other hand, consumption of above-average quantities of lettuce and fresh vegetables gives protection against the incidence of cancer.

XI. STOMACH AND LIVER CANCERS IN THE KILIMANJARO REGION OF NORTHERN TANZANIA

The Kilimanjaro region of northern Tanzania has the highest incidence of cancers in Tanzania (38.2 per 100,000 population in 1975 versus the national average of 9.4 per 100,000 population).[11] The endemicity of stomach cancer and liver cancer (characterized by the highest mortality of all cancers in Tanzania) in the region may be attributed to the nitrosoamines arising from the Quaternary volcanism in the region (Figure 1). The etiology of gastric cancer in the Tanzanian context needs to be examined in terms of the interplay between the following factors:

1.   Initiators: carcinogens and co-carcinogens (nitrosoamines, abrasive foods like inadequately ground maize, aflatoxin, and thiocyanate-bearing foods like cassava, and some types of locally brewed liquors?)
2.   Protectors: lettuce, fresh vegetables, ascorbic acid, and essential elements like Zn
3.   Accentuators: nutritional deficiencies, presence of toxic elements like Cd, etc.

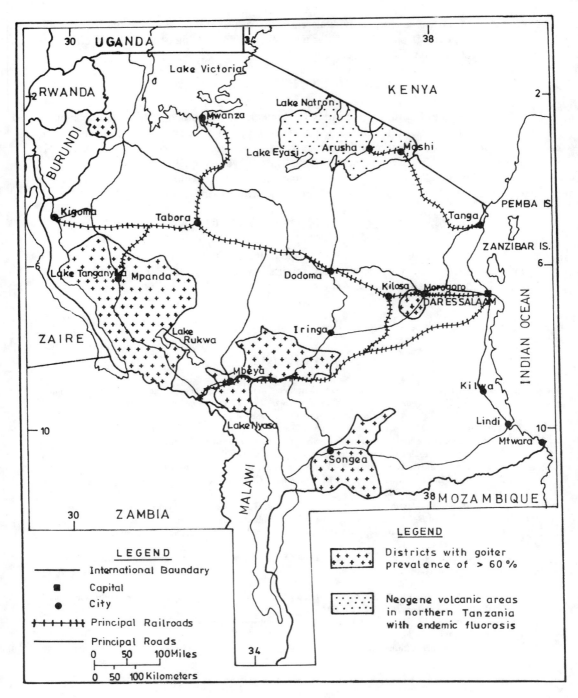

FIGURE 1. Areas of endemic goiter and fluorosis in Tanzania. (According to the Tanzania Food and Nutrition Center.)

In the case of gastric cancer, individuals carry the risk of the place where the first decade life was spent; even if one migrated during adulthood from a high-risk place of birth to a low-risk location, one may still bear a high risk.

XII. CONCLUSIONS

Both goiter and fluorosis are entirely preventable — the former through iodine supplementation and the latter through the de-fluoridation of drinking water. While it may not be possible to persuade whole populations to avoid eating goiterogenic foods like cassava, it is not an impossible task to provide iodinated salt to everybody. It would be fascinating to study the mode and consequences of the interaction between nitrosoamines and fluoride, and their bearing on the etiology of cancer in the Kilimanjaro region.

XIII. ACKNOWLEDGMENTS

The author is grateful to Dr. F. P. Kavishe of the TFNC and Dr. Luande of the Tumour Centre for kindly allowing access to their reports, and to Drs. J. T. Nanyaro, M. R. Khan, and H. B. Pratap for helpful discussions.

REFERENCES

1. **Aswathanarayana, U.,** Natural radiation environment in the Minjingu phosphorite area, northern Tanzania, in Låg, J., Ed., *Health Problems in Connection with Radiations from Radioactive Matter in Fertilizers, Soils and Rocks,* Norwegian University Press, Oslo, 1988, 79.
2. **Aswathanarayana, U., Lahermo, P., Malisa, E., and Nanyaro, J. T.,** High fluoride waters in an endemic fluorosis area in northern Tanzania, in *Environmental Geochemistry,* Thornton, I., Ed., 1986, 243.
3. **Ekpechi, O. L.,** Iodine deficiency disorders in Africa, in *The Prevention and Control of Iodine Deficiency Disorders,* Hetzel, B. S., Dunn, J. T., and Stanbury, J. B., Eds., Elsevier, 1987, 219.
4. **Fairbridge, R. W., Ed.,** Iodine, in *Encyclopaedia of Geochemistry and Environmental Sciences,* D Van Nostrand, New York, 1972, 590.
5. **Hetzel, B. S.,** An overview of the prevention and control of iodine deficiency disorders, in *Prevention and Control of Iodine Deficiency Disorders,* Hetzel, B. S., et al, Eds., Elsevier, 1987, 7.
6. **Hopps, H. C.,** Cancer, in *Geochemistry and the Environment,* Vol. 3, National Academy of Sciences, Washington, D.C., 1978, 81,
7. **Kavishe,, F. P.,** Iodine deficiency disorders (IDD) as a problem of public health significance in Tanzania, in Towards Eradication of Endemic Goiter, Cretinism and Iodine Deficiency in Tanzania, Kavische, F. P. and Mlingi, N. V., Eds., TFNC-SIDA Report, 1985, 19.
8. **Kavishe, F. P., Mlingi, N. V., and Chorlton, R., Eds.,** A National Programme for the Prevention and Control of IDD in Tanzania, TFNC-SIDA Report, 1987.
9. **Maletnlema, T. N.,** The Malnutrition Problem, unpublished report, 1975.
10. **Gitad, W.,** unpublished.
11. **Luande, ,** personal communication.

Chapter 15D

SPECIFIC GEOMEDICAL PROBLEMS IN ASIA

P. N. Takkar and S. S. Randhawa

TABLE OF CONTENTS

I. INTRODUCTION

Every year, land is being increasingly brought under intensive cultivation to meet the growing demand of the rapidly increasing Asian population for food grains, fodder, fiber, and fruits. This is causing a depletion of minerals and trace elements from the soil which, in the soil-plant-animal-human chain, is ultimately reflected in alterations of the mineral metabolism of animals and the human population. This, in turn, affects the productivity of animals and human health due to deficiency diseases. Recently, adoption of intensive agricultural practices has lead to a number of nutritional deficiencies in field and fodder crops.[46] The excess levels of some trace and heavy-metal elements, either present natively or added to soils through anthropological activity, are also causing health hazard problems in animals and human beings. Consequently, high-yielding animals are working under maximum stress conditions. They are, therefore, liable to suffer from metabolic disorders and mineral element imbalances resulting from their deficiencies or excesses in the field and fodder crops. These imbalances in the soil are ultimately reflected in corresponding imbalances in plants and animals, which are responsible for low production and reproduction among grazing livestock as well as serious deficiency and toxicity diseases in animals and human beings.

II. PHOSPHORUS

Nutritional hemoglobinuria associated with phosphorus deficiency has been found to occur in buffaloes in many parts of India, Pakistan, and Egypt.[2,27,36] Although the cases have been reported throughout the year, a large number of incidences were recorded during the spring and winter season. The disease mostly occurs in buffaloes within 3 to 4 weeks after parturition or in animals in advanced stages of pregnancy. The characteristic clinical symptoms of the disease are normal to subnormal temperature, rumen stasis, constipation, passing coffee/chocolate colored urine, and anemia. Blood biochemical analysis reveals hypophosphatemia (with an inorganic phosphorus content of 1.73 ± 0.37 mg/100 ml, compared to normal values of 6.19 ± 0.73 mg/100 of blood), which has been correlated with a low phosphorus status of the soil and fodder in these areas and is considered to be one of the main factors causing phosphorus deficiency in high-yielding buffaloes.[27] In Egypt, nutritional hemoglobinuria in buffaloes has been found to be associated with low phosphorus intake as a result of excessive feeding of berseem (*Trifolium alexandrinum*) low in phosphorus content. Also, this condition has been reported to occur in India in buffaloes fed on leguminous plants, particularly berseem.

Leguminous plants are known to accumulate large amounts of molybdenum during the summer season. Nayyar et al.[28] observed that excessive molybdenum seldom retards plant growth, but is toxic to ruminants feeding on herbages containing excessive quantities of it. Molybdenum toxicity has been observed as an endemic nutritional problem of ruminants, primarily in wet, poorly drained neutral to alkaline soils. The values of molybdenum content of berseem grown on calcareous flood plains of the six districts of Punjab, India, indicating accumulation of molybdenum in toxic concentrations in the forage, are presented in Table 1. This problem is characterized by severe diarrhea and the loss of general health of the animals; it has been thought that increased uptake of molybdenum by the animals also causes phosphorus deficiency. Dhillon et al.[10] found that the molybdenum content of blood in hemoglobinuric buffaloes was 1066 ± 538 ppm, compared to 0.1 ppm in normal bovines.

Hafex[16] reported that phosphorus deficiency in ruminants also causes ovarian dysfunctions, which, in turn, lead to depressed signs of estrus and, eventually, cessation of estrus. In Hisar, India, Verma and Gupta[49] reported a rheumatism-like syndrome associated with phosphorus deficiency in buffaloes, mostly in late gestation. Symptoms exhibited were retarded growth, varying degrees of emaciation with arched back and tucked up abdomen,

TABLE 1
Molybdenum Content of Berseem (*Trifolium alexanderinum* L.)
Grown on Calcareous Flood Plains of the Various Districts of
Punjab, India

District	No. of sites	Mo (ppm) Range	Mo (ppm) Mean	Percent toxic samples (above 10 ppm Mo)
Ludhiana	35	0.5—14.2	6.56	17.2
Ropar	14	1.2—20.7	6.11	21.4
Kapurthala	13	2.1—13.0	5.91	15.4
Hoshiarpur	7	4.3—16.0	7.10	42.9
Ferozepur	24	2.9—19.5	7.44	20.8
Amristsar	12	3.4—12.9	7.68	41.7
Mean	105	0.5—20.7	6.80	26.6

Data from Nayyar, V. K., Randhawa, N. S., and Pasricha, N. S., *J. Res. Punjab Agric. Univ. Ludhiana*, 14, 245, 1977.

reduced fertility, stiffness of joints, and a dirty brown discoloration of the skin, which had a rough and dry surface. Some of the advanced cases remained in recumbent posture and were unable to get up even with support. Hypophosphatemia correlated with low levels of phosphorus in soil and fodder samples collected from the affected areas. This was confirmed by a favorable response of animals treated for phosphorus deficiency.

III. MAGNESIUM

Magnesium deficiency in animals, i.e., lactation tetany in cattle and buffaloes characterized by clonic and tonic muscular spasms, convulsions, and death due to respiratory failure, has been recognized in India and Japan. This is usually manifested in animals who graze on swards receiving a heavy dressing of sulphate of ammonium or other nitrogenous fertilizers.[48]

IV. ZINC

Zinc deficiency in human beings has been extensively noticed in Iran and Egypt.[31] This has been attributed to result largely from the poor availability of zinc because of excessive intake of cereals in the form of bread made of wheat flour. Symptoms of zinc deficiency in the human population are marked growth retardation, dwarfism, sexual underdevelopment characterized by testicular atrophy with absence of facial, pubic, and body hair, poor appetite, mental lethargy, night blindness, rough, dry skin, and enlargement of the spleen and liver. Geophagia was the characteristic finding in subjects which had a markedly low zinc concentration in their plasma, red blood cells, and hair. Prasad et al.[30] also found that sickle cell disease in human beings in the Middle East was associated with zinc deficiency. In India, Spears et al.[42] analyzed a large number of samples of serum, red blood cells, urine, and tissues of normal surgical patients without metabolic diseases, as well as diabetic and diabetic ulcer patients, and found that zinc values were low, particularly in the diabetic and diabetic ulcer patients (Table 2).

Gauba et al.[14] found that the presence of zinc, copper, iron, and manganese in drinking water and soil correlated with dental caries in 1,516 children (7 to 17 years of age) in ten rural areas in the district of Ludhiana (Punjab), India. They found that zinc had an inverse relationship with dental caries, i.e., it proved to be a cariostatic (caries-preventive) trace element. Copper also consistently evidenced an inverse and significant relationship with

TABLE 2
Serum Zinc Levels in Different Subjects

| Condition | Serum zinc (μg/ml) | |
	Mean ± SD	Number of subjects
Normal	0.93 ± 0.12	27
Surgical	0.78 ± 0.19	33
Diabetic	0.79 ± 0.10	32
Diabetic ulcer	0.69 ± 0.16	10
Hypogonad	0.76 ± 0.13	15

Data from Spears, A. B. et al., Bull. P.G.I. Chandigarh, 8, 21, 1974.

different variables of dental caries, whereas iron and manganese in drinking water correlated positively with different variables of dental caries, i.e., they proved to be cariogenic or caries-causing trace elements. Also, there were correlations of gingival and periodontal health status with zinc, copper, iron, manganese, and fluoride in drinking water, as well as correlations of dental caries with these elements in the saliva of children.

Parakeratosis, a disease associated with bone and joint disorders in animals, generally results from their grazing on forages containing 18 to 40 ppm zinc in dry matter. Most of the forages from the Punjab and Haryana states of India contained less than 20 ppm zinc in dry matter, and the continuous feeding of these forages to growing calves and sheep affected growth and resulted in poor feed efficiency, shedding of wool, and parakeratotic skin lesions.[48] At the Central Sheep Breeding Farm in Hisar, India, Mandokhot et al.[26] found a wool shedding syndrome in Corriedale sheep. The disease was characterized by a loss of crimpness in the wool, which became thin and loose with different degrees of shedding. It was attributed to zinc deficiency and probably is aggravated by low phosphorus levels.

V. COPPER

Copper deficiency in the form of enzootic ataxia of sheep has been recorded in India and eastern Saudia Arabia.[6] It was found that lambs suffering from delayed neonatal ataxia had low concentrations of copper in the blood, liver, and kidney, consistent with a diagnosis of copper deficiency. Low concentrations of copper and higher concentrations of molybdenum and sulfates in forages appeared to contribute to the high incidence of neonatal ataxia observed in sheep. Copper deficiency syndrome in cattle resulting from a gross deficiency of Cu in some feeds and a deficiency or marginal deficiency in others has been observed at 60 farms in India. Clinical and laboratory studies confirmed "falling disease" in milch cows associated with copper deficiency. Low levels of copper in the livers of the affected animals (32.6 ppm, compared to the normal concentration of 55.7 ± 8.52 ppm in healthy cows) were recorded.[48] Leucoderma, also referred to as vitiligo and suspected to be due to copper deficiency, has been reported in buffaloes in Indonesia, India, and Pakistan.[12,40] The disease is characterized by the depigmentation of hair in patches, particularly around the brisket, neck, and abdomen, and which may spread to the flanks. In kids, copper deficiency disease-Swayback has also been recorded in India. It is indirectly due to excess molybdenum uptake from certain fodders, with symptoms very similar to a disease recorded in Australia in lambs. Molybdenum combines with copper to form copper tetrathiomolybdate, making the copper unabsorbable. In Malaysia, the incidence of this deficiency disease was as high as 80% in some areas.

VI. MANGANESE

Chronic manganese toxicity with neurological manifestations has been observed in factory workers in Egypt.[5] The disease was characterized by headache, involuntary movements; sleep, speech, and gait disturbances, and exaggerated reflexes which significantly increased with the duration of the exposure. Also, symptoms of fatigue, exhaustion, depression, hallucination, visual memory defects, or prolonged reaction time, seborrhea, and sialorrhea were observed in subjects after mild exposure to manganese in the environment. Although analysis of the levels of manganese in the blood and air failed to show a significant correlation between them, the incidence of involuntary movements, speech and gait disturbances, hallucinations, and visual memory defects was significantly higher in factory workers with the highest mean levels of manganese in their blood (2.8 μg/100 ml) and with a higher manganese concentration in the air (7.0 mg/m³). They observed a higher prevalence of central nervous system dysfunctions and psychotic manifestations among the exposed workers, compared to the control group.

VII. LEAD

A strikingly higher content of lead (3.8 to 4.5 ppm) was found in barseem and maize fodder on soils irrigated with sewage water than on soils receiving normal irrigation water (1.93 to 1.98 ppm) in India.[17] Continuous feeding of animals with such fodders may lead to serious health hazards. Lead toxicosis has been observed in human beings and animals as well.[4,21] In Punjab, India, lead toxicosis has been recorded in buffaloes and cattle due to ingestion of green fodder grown on soil contaminated with lead discharged from a factory. The consumption of contaminated fodder resulted in the development of nervous symptoms in the form of staggering gait, hyperexcitability, temporary blindness, and striking violently against the wall. The symptoms were observed intermittently for a few days, followed by death of the animals. A high rise in temperature was seen in some of the animals during a stage of excitability. Kwatra et al.[21] showed that the lead level in blood ranged from 13.8 to 35.8 ppm, compared to a normal lead concentration of less than 0.25 ppm in the blood of farm animals and 6.9 to 96.5 ppm in feces. Lead content in the liver and kidney of dead buffaloes was 96.5 and 137.5 ppm, respectively.

The highest prevalence of lead-related symptoms was recorded in lead workers. The symptoms were pain in muscles and joints, and the severity of the pain increased with an increase in the duration of exposure.[4] Prolonged exposure to lead also lead to renal dysfunction and hyperuricemia, with its subsequent manifestations as muscle and joint pains, urinary troubles, and renal stones.

VIII. FLUORINE

Fluorine is a cumulative poison both for man and animals. Its distribution in water assumes importance because the margin of safety between the deficiency and the toxicity condition, resulting from low and excessive intake, respectively, is very narrow. The incidence of fluorosis in endemic form has been observed in the Punjab, Haryana, Rajasthan, Bihar, Madhya Pradesh, and Andhra Pradesh states of India, where soils, underground water (hydrofluorosis), and forages (food-borne fluorosis) contained excessive amounts of fluorine.[1,18,19,39,41,43] Kanwar and Mehta[18] observed that 100% of the well water samples from the Hisar district of Haryana, India, and 91% of the samples from the Barnala district of Punjab, India, contained more than 1 ppm fluorine, which is considered to be the optimum limit in drinking water. Similarly, the fluorine content of well water from the Sangrur, Jind, Sirsa, and Fatehabad areas was fairly high. They also observed a significant positive correlation between the fluorine content of the water and the soil. The fluorine content of berseem and wheat crops varied from 4.75 to 52.71 ppm and 1.36 to 25.95 ppm, respectively. Somani et al[41] showed that in the Nagaur district, the fluorine content of well water varied from 1.5 to 13.3 ppm, with an average value of 5.5 ppm, whereas in the Jaipur district of Rajasthan, India, it ranged from 4.4 to 28.1 ppm, with an average value of 12.2 ppm.

In the past and presently, the problem in Asia is hydrofluorosis and food-borne fluorosis. However, in the years to follow, the problem will be augmented by industrial fluorosis. As a matter of fact, in India during the last few years, an increasing incidence of industrial fluorosis has occurred among chemical, steel, aluminium, and fertilizer factory workers.[43]

Osteomalacia and rheumatic arthritis associated with fluorine toxicosis have been observed in cattle and buffaloes. Fluorosis has been observed in animals in Andhra Pradesh, India, where the fluorine concentration in water was more than 2 ppm.[23-25,32]

In India, the optimum levels of fluoride in drinking water differ from the south to the north. While the values are 0.6 to 0.7 ppm in the south, they range from 0.7 ppm in the summer to 1 to 1.2 ppm in the winter in the north because of extreme weather conditions. However, there are a number of endemic fluoride belts traversing the various zones of India in which the fluoride level in drinking water is substantially above the optimum levels. According to Tewari,[47] 20% of the areas in India can be labeled as endemic fluoride areas, and about 5 to 8% of the Indian population actually consumes water containing 1.5 ppm or more fluoride.

Fluorine toxicity in human beings was found to cause dental mottling and skeletal fluorosis, and was associated with drinking water from wells containing toxic concentrations of fluorine. Skeletal fluorosis is characterized by osteosclerosis, ligamental calcification, and calcification of tendinous insertion and muscular attachments. New bone formation is the basic effect of excessive fluorine ingestion. The neurological manifestations are secondary to this.[19]

IX. IODINE

Iodine deficiency primarily causes goiter, characterized by enlargement of the thyroid gland and cretinism. Endemic goiter has been reported from all over Asia.[7] Iodine deficiency disorders include a spectrum of crippling conditions which affect the health and well being of man from early in fetal life through adulthood. The principle underlying cause of environmental iodine deficiency essentially is reduced iodine content of the soil. The other defects associated with iodine deficiency in human beings vary from subnormal intelligence, delayed motor milestones, mental retardation, hearing and speech defects,, strabismus, nystagmus, spasticity, and neuromuscular weakness to intrauterine death in the form of spontaneous abortion and miscarriage.

Iodine deficiency has been recorded in many areas in India, a geographical problem over a 2400-km area along the Himalayas and the Satpura and Vindyachal ranges.[34] About 9 million Indians are said to be suffering from either goiter or a subclinical thyroid function which impairs their health and mental capabilities. Clinical cases of goiter reported in goats and sheep were observed to be of the parenchymatous type.[11]

An outbreak of iodine deficiency was observed in goats in the Ludhiana district of Punjab, India.[20] The disease was characterized by the clinical symptoms of enlarged thyroid gland, alopicia, stunted growth, infertility, and still births. Schmidt[37] studied infertility of ovarian origin in Egyptian water buffalo and observed that iodine deficiency is important in depressing ovarian functions during the winter season.

X. IRON

Low levels of iron in soil and plants in the soil-plant-animal-human chain cause anemia in both human beings and animals and is widely prevalent in Asia.

XI. SELENIUM

Selenium has been established as an essential element for animals because it is specifically required for the activation of one type of glutathione peroxidase, while the other type is selenium independent. Its deficiency can lead to a number of diseases, such as "stiff limb disease" in lamb, "white muscle disease" in calves, encephalomalacia in chicks, and muscular dystrophy, which are also known as vitamin E-deficiency diseases. In certain regions of China, Keshan disease in human beings has been recorded in endemic form in rural areas and is considered to have resulted from selenium deficiency. Children less than 15 years of age and pregnant women showed clinical pathological changes in the myocardium and responded to sodium selenide treatment in dietary doses of 10 to 15 ppm selenium. A clue to this selenium-deficiency disease came from the white muscle disease prevalent in cattle in the same area.[13]

Dhillon[8] found that the selenium content in the fodders of sorghum, pearl millet, oats, cluster beans, egyptian clover, and mustard ranged from traces to 1.5 ppm, and in natural grasses/weeds, from traces to 0.54 ppm. These levels were thought in the range of adequacy or deficiency; yet not one case of selenium-deficiency disease has been recorded in animals or human beings, nor have cases been reported from other Asian countries.

Animal sickness resembling selenium toxicity was first noticed in the Far East by Marco Polo.[29] A clear-cut relationship between excess selenium in soil and plants and livestock illness was established in South Dakota.[35] Later cases of selenium toxicity have been reported from Israel,[33] some regions of the Punjab in Pakistan, and the northwest province of India.[22] Called "alkali disease" in the U.S. and "degnala" in India and Pakistan, the syndromes appear throughout the year in India and Pakistan, but more so after the rainy monsoon season. The animals develop lesions on their tail, ear tips, and limbs. Skin and hooves are the common tissues which are affected.[3] In very severe cases, the hooves fall off and the animals may ultimately succumb.

Arora et al.[3] reported a range of 0.9 to 6.7 ppm selenium in different fodders. In rice straw and rice husk, a selenium content of more than 0.5 ppm may prove toxic. Sharma et al.[38] reported 1.0 to 9.5 ppm selenium in the soil of Haryana, India.

In the state of Punjab, India, animals fed on rice straw were found to suffer from a sickness (gangrenous syndrome) resembling selenium poisoning in 1969 to 1970; the condition affected about 1400 buffaloes in different rice-growing districts of the state.[22] But some buffaloes developed the disease even when fed fodders other than the rice straw.[9] Nevertheless, an authenticated incidence of chronic selenium poisoning in domestic animals[15] and clear-cut symptoms of selenium toxicity in wheat crops as well as in buffaloes, cows, and goats in a few villages of the Hoshiarpur district were registered.[44,45] Takkar and Dhillon[45] found typical symptoms of selenium toxicity, i.e., snow-white chlorosis, in wheat, particularly in the fields which have been under rice-wheat rotation for 10 or more years (Figure 1). Also, selenium poisoning was found in animals fed on the produce and fodders from this area. The symptoms are development of cracks in the hooves, followed by their gradual detachment (Figure 2), peeling off of horns, loss of hair from the body, and necrosis of the tail. The animals do not come into heat at the right age and premature abortions take place. Even animals new to the area start developing selenium poisoning symptoms within a period of 5 to 6 months, ultimately leading to death. Also, a high content of selenium — 12.6 to 55.0 ppm in the hair, 21.2 to 32.8 ppm in the hooves, and 2.2 to 5.3 ppm in the blood — was found.

The soils of this region have formed from the alluvium deposited by the rivers of the Indus system, which have completely shrouded the old land surface. The soil are calcareous (0.25 to 4.9% $CaCO_3$), alkaline (pH 8.1 to 8.4), and have a silty loam texture. The mean total selenium content in the seleniferous soils (0 to 15 cm) was 1.34 ppm, compared to

FIGURE 1.   Selenium toxicity in wheat, Punjab, India. Wheat field, left side of water channel: severe selenium toxicity symptoms in wheat — snow-white chlorisis. The field was under rice-wheat rotation for more than 10 years. Wheat field, right side of water channel: normal wheat crop, except for selenium toxicity symptoms in small area adjoining the channel. The field was changed from rice-wheat to maize-wheat rotation only 3 years ago. Background: tube well — the source of underground water for irrigation.

0.78 ppm in normal soils. The higher amount of selenium in the former resulted from their copious irrigation with underground water, the main source in the area, particularly under rice-wheat rotation, compared to maize-wheat rotation. Also, the average selenium content in the underground water from the seleniferous area was five times greater (11.7 μg/l) than that from the normal areas (2.5 μg/l). The animals of the area drink huge amounts of this water, particularly during the summer season, which results in a large intake of selenium. Furthermore, the selenium content in 45- to 60-day-old wheat plants with symptoms of selenium toxicity or from selenium-toxic fields varied from 108 to 262 ppm, with a mean value of 191 ppm, and was 3.6-fold higher than that in the lush green plants. The fodders in the selenium-toxic areas also contained more selenium than the toxic level of 5 ppm. The average selenium content was 4.5, 8.5, 21.7, and 33.3 ppm in maize, sorghum, Egyptian clover, and oat fodders, respectively. This indicates that in the soil-water-plant-animal chain, water appears to be the root cause for causing selenium toxicity in the soils, plants, and animals of the area.

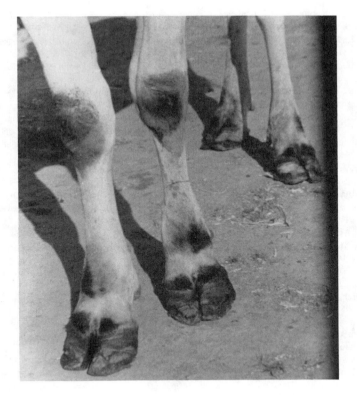

FIGURE 2. Selenium toxicity symptoms in ox, Punjab, India. Cracks and peeling-off of hooves due to selenium toxicity.

REFERENCES

1. **Anand, D., Bagga, O. P., and Mullick, V. D.,** Endemic fluorosis and dental decay — preliminary study, *Indian J. Med. Res.,* 52, 117, 1964.
2. **Awad, F. I. and Abd-El-Latif, K.,** The first record of a disease condition in Egyptian buffalo simulating postparturient haemoglobinuria, *Vet. Rec.,* 75, 298, 1963.
3. **Arora, S. P., Parvinder, K., Khirwar, S. S., Chopra, R. C., and Ludri, R. S.,** Selenium levels in fodders and its relationship with Degnala disease, *Indian J. Dairy Sci.,* 28, 249, 1975.
4. **Badawy, A. B. N., El-Kawy, A. M., Abdel-Karim, A. H., El-Sawaf, H. A., Farag, S. A., and Gad, M. M.,** Hyperuricaemia and renal calculi in lead workers, in *Proc. 5th Int. Symp. Trace Elements in Man and Animals,* Smills, C. F., Bremner, I., and Chesters, J. K., Eds., Commonwealth Agricultural Bureaux, Farnham Royal, U.K., 1985, 286.
5. **Badawy, A. B. N. and Shakour, A. A.,** Chronic manganese intoxication (neurological manifestations), in *Proc. 5th Int. Symp. Trace Elements in Man and Animals,* Mills, C. F., Bremner, I., and Chesters, J. K., Eds., Commonwealth Agricultural Bureaux, Farnham Royal, U.K., 1985, 261.
6. **Chamberlain, A. G. and Clarke, S. H.,** Copper deficiency of sheep in eastern Saudia Arabia, *J. Agric. Sci.,* 97, 213, 1981.
7. **DeMaeyer, E. M., Lowenstein, F. W., and Thilly, C. H.,** *The control of endemic goiter,* World Health Organization, Geneva, 1979.
8. **Dhillon, K. S.,** Selenium Status of Common Fodders and Natural Grasses, Masters thesis, Punjab Agricultur University, Ludhiana, India, 1972.
9. **Dhillon, K. S.,** Preliminary observations on the treatment of Deg (Dek) Nala disease in buffaloes, *Indian Vet. J.,* 50, 482, 1973.
10. **Dhillon, K. S., Singh, J., and Bajawa, R. S.,** Treatment of haemoglobinuria due to molybdenum induced phosphorus deficiency in buffaloes: a note, *Indian J. Anim. Sci.,* 42, 996, 1972.

11. **Dutt, B. and Kehar, N. D.,** Incidence of goiter in goats and sheep in India, *Br. Vet. J.,* 115, 176, 1959.

12. **Gill, H. S. and Gill, B. S.,** Vitiligo in a bull buffalo, *Indian Vet. J.,* 82, 589, 1975.

13. **Graus, L. X.,** Selenium and the prevalence of keshan disease and Kaschin-Beck disease in China, paper presented at Int. Symp. Geochemistry and Health, London, April 15 to 17, 1985.

14. **Guaba, K.,** A Study of Inter-relationship Between Trace Elements and Dental Caries in Real Life Situations, Masters thesis, Dept. of Dentistry, P. G. I., Chandigarh, India, 1983.

15. **Gupta, R. C., Kwatra, M. S., and Singh, N.,** Chronic selenium toxicity as a cause of hoof and horn deformities in buffalo, cattle and goat, *Indian Vet. J.,* 59, 738, 1982.

16. **Hafez, E. S. E.,** *Reproduction in Farm Animals,* 3rd ed., Lea & Febiger, Philadelphia, 1975.

17. **Kansal, B. D. and Singh, J.,** Influence of the municipal waste water and soil properties on the accumulation of heavy metals in plants, in Proc. Int. Conf. Heavy Metals in the Environment, Heidelberg, September, 1983.

18. **Kanwar, J. S. and Mehta, K. K.,** Toxicity of fluorine in some well waters of Haryana and Punjab, *Indian J. Agric. Sci.,* 38, 881, 1968.

19. **Krishnamachari, K. A. V. R.,** Osteoporosis in fluoride toxicity, in *Proc. 5th Int. Symp. Trace Elements in Man and Animals,* Mills, C. F., Bremner, I., and Chesters, J. K., Eds., Commonwealth Agricultural Bureaux, Farnham Royal, U.K., 1985.

20. **Kwatra, M. S.,** personal communication, 1988.

21. **Kwatra, M. S., Gill, B. S., Singh, R., and Singh, M.,** Lead toxicosis in buffaloes and cattle in Punjab, *Indian J. Anim. Sci.,* 56, 412, 1986.

22. **Kwatra, M. S. and Singh, A.,** Experimental reproduction of gangrenous syndrome in buffaloes *(Bos bubalus), Zbl. Vet. Med.,* B20, 481, 1973.

23. **Majumdar, B. N., Ray, S. N., and Sen, K. C.,** Fluorine intoxication of cattle in India, *Indian J. Vet. Sci.,* 13, 95, 1943.

24. **Majumdar, B. N. and Ray, S. N.,** Fluorine intoxication of cattle in India. II. Effect of fluorosis, on mineral metabolism, *Indian J. Vet. Sci.,* 16, 107, 1946.

25. **Majumdar, B. N. and Ray, S. N.,** Determination of fluorine in biological material and its application in fluorine intoxication studies in cattle in India, *Indian J. Med. Res.,* 35, 323, 1947.

26. **Mandokhot, V. M., Vasudevan, B., Mandal, A. B., and Yadav, I. S.,** Serum and wool mineral status of wool shedding manifestation in Corriedale sheep, in *Proc. Symp. Recent Advances in Mineral Nutrition,* HAU, Hisar, India, Mandokhot, V. M., Ed., 1987.

27. **Nagpal, M. C., Gautam, O. P., and Gulati, R. L.,** Haemoglobinuria in buffaloes, *Indian Vet. J.,* 45, 1048, 1968.

28. **Nayyar, V. K., Randhawa, N. S., and Pasricha, N. S.,** Molybdenum accumulation in forage crops. I. Distribution of molybdenum, copper, sulphur and nitrogen in Barseem (*Trifolium alexandrinum* L.) grown on calcareous flood plains, *J. Res. Punjab. Agric. Univ. Ludhiana,* 14, 245, 1977.

29. **Polo, M.,** *The Travels of Marco Polo,* Liveright, New York, 1926, 81.

30. **Prasad, A. S., Schoomaker, E. B., Ortega, J., Brewer, G. J., Oberleas, D., and Oelshlegel, F. J.,** Zinc deficiency in sickle cell disease, *Clin. Chem.,* 21, 582, 1975.

31. **Prasad, A. S. and Oberleas, D.,** *Trace Elements in Human Health,* Vol. 1 Zinc and Copper, Academic Press, New York, 1976.

32. **Ramamohan, R. N. V. and Bhaskaran, C. S.,** Endemic fluorosis. Study of distribution of fluorine in water sources of Kurnool district of Andhra Pradesh, *Indian J. Med. Red.,* 52, 180, 1964.

33. **Ravikovitch, S. and Margolin, M.,** Selenium in soils and plants, *Agric. Res. Stn. Rehovot,* 7, 41, 1957.

34. **Ray, S. N., Dutt, B., and Majumdar, B. N.,** *Final Report, Trace Elements Scheme, 1952-1959.* Animal Nutrition Division, IVRI, Izatnagar, India, 1959.

35. **Robinson, W. O.,** Determinations of selenium in wheat and soils, *J. Assoc. Off. Agric. Chem.,* 16, 423, 1933.

36. **Samad, A., Singh, B., and Qureshi, M. I.,** Some biochemical and clinical aspects of haemoglobinuria in buffaloes, *Indian Vet. J.,* 56, 230, 1975.

37. **Schmidt, K.,** Infertility of ovarian origin in Egyptian water buffalo, *Vet. Bull.,* 37, 3464, 1966.

38. **Sharma, S., Ramendra, S., and Bhattacharya, A. H.,** Perspective of selenium research in soil plant animal system in India, *Fert. News,* 26, 19, 1981.

39. **Singh, A., Jolly, S. S., Devi, P., Bansal, B. C., and Singh, S. S.,** Endemic fluorosis. An epidemiological, biochemical and clinical study in the Bhatinda district of Punjab, *Indian J. Med. Res.,* 50, 387, 1962.

40. **Sinha, B. P., Jha, G. J., and Sinha, B. K.,** Leucoderma in Indian buffaloes, *(Bubalus bubalis), Indian Vet. J.,* 53, 812, 1976.

41. **Somani, L. L., Gandhi, A. P., and Paliwal, K. V.,** Note on the toxicity of fluorine in well waters of Nagaur and Jaipur districts in Rajasthan, *Indian J. Agric. Sci.,* 42, 752, 1972.

42. **Spears, A. B., Kaufman, B. M., Mathock, M. B., Saharia, S., Wig, J. D., Sachdeva, H. S., Takkar, P. N., and Randhawa, N. S.,** *Bull. P. G. I. Chandigarh,* 8, 21, 1974.

43. **Susheela, A. K., Jha, M., Koacher, J., and Jain, S. K.,** Recent advances in fluoride toxicity — fluoride and calcified tissues, in *Proc. Symp. Recent Advances in Mineral Nutrition,* HAU, Hisar, India, Mandokhot, V. M., Ed., 1987.

44. **Takkar, P. N. and Dhillon, K. S.,** Selenium toxicity in wheat, *The Tribune,* February 15, 1984, Chandigarh, India.

45. **Takkar, P. N. and Dhillon, K. S.,** personal communication, 1988.

46. **Takkar, P. N. and Randhawa, N. S.,** Micronutrients in Indian Agriculture, *Fert. News,* 23, 3, 1978.

47. **Tewari, A.,** personal communication, 1988.

48. **Vasudevan, B.,** Mineral deficiencies syndrome among dairy cattle, sheep and goats in India, in *Proc. Symp. Recent Advances in Mineral Nutrition,* HAU, Hisar, India, Mandokhot, V. M., Ed., 1987.

49. **Verma, P. C. and Gupta, R. K. P.,** Phosphorus deficiency in buffaloes, in *Proc. Symp. Recent Advances in Mineral Nutrition,* HAU, Hisar, India, Mandokhot, V. M., Ed., 1987.

Chapter 15E

ENVIRONMENTAL GEOCHEMISTRY AND HEALTH IN WESTERN EUROPE

Brian E. Davies

TABLE OF CONTENTS

I. INTRODUCTION

The history of mankind can be perceived as a continuous and increasingly successful endeavor to become isolated from and unaffected by environmental influences. Our Palaeolithic ancestors invented clothes and discovered fire so as not to freeze to death as they roamed the tundra wastes of periglacial Europe during the Pleistocene. Now our houses are central heated and sometimes air-conditioned, our windows double or treble glazed, our doors draft proofed; all with the intention of ensuring we pass our days both dry and warm and isolated from the natural environment. Yet man cannot escape from the ecosystem of which he is part. The problem for city dwellers, and in Europe 80% or more of the population is classed as urban rather than rural, is that the environment-man link is rarely obvious. This is in contrast with rural people who have always understood that the health of farm animals is strongly influenced by dietary composition, and therefore soil properties, and have intuitively sensed that they, too, may be caught up in a similar causal relationship.

For virtually all of our evolutionary history death has most likely been the consequence of fatal injury, infectious diseases, starvation, or acute food poisoning. The discovery that infectious diseases are caused by pathogens led to improved sanitation and the provision of clean water and, later, to vaccines and antibiotics. Consequently, epidemics of diseases such as plague or cholera have passed into history in Europe and European life expectancy is now (1980 to 1985) 73.2 years, compared with only 58.9 years for the world as a whole.[1] Although much medical thought and practice is still dominated by a simple, unicausal, readily diagnosible disease model, our health problems are now multifactorial and dominated by diseases of older people: Learmonth[2] contrasts infectious ''universal'' diseases with what he terms the ''western diseases'' such as heart diseases, strokes, and cancers. These are often disorders without a single, known cause and are endemic rather than epidemic. Table 1 illustrates the disease experience of Europe, together with recent data for major world regions.

When the distributions of these diseases are mapped within any one region, variations are seen and these patterns may persist over many years.[3] It is this uneven geographical distribution that leads to the hypothesis that environmental factors may be involved in the etiologies of some diseases.

II. DIRECT LINKS BETWEEN INDIVIDUAL ELEMENTS AND DISEASE

The health of animals and man is known to be affected by the amounts and properties of the chemical elements present in food, beverages, and inhaled dust or air. Recent years have seen more elements being recognized as essential, e.g., chromium or selenium, and the role of essential elements in metabolic processes is better understood. Mertz[4] has written a comprehensive survey of those elements now considered essential to man.

There is a growing body of evidence that environmental geochemistry can affect health.[5-7] But in Europe, individuals are rarely reliant on food or beverages from a single locality and people are mobile as they seek employment, often far from their birthplace, a trend which will undoubtedly become more pronounced after 1992 when the European Economic Community becomes a single, open market. This makes it difficult to relate health directly to the geochemical environment. Simple, unifactorial studies focused on one specific element, disregarding related or interacting elements, are certain to produce inadequate or misleading information. Moreover, much of the research in this field produces only associations, and association is not causation. It is essential that any element or group of elements suspected of being involved in the etiology of a disease must have a demonstrable pathway from the environment to the person (e.g., through the food chain, by inhalation, by direct

TABLE 1
Mortality Data for Major World Regions[1]

Region	Mortality By Cause[a] (Percent of world total)			Years LE
	I & P	Neop	CVS	
Africa	21	5	6.5	49.7
Americas	6.5	17	14.5	66.6
S.E. Asia	40	16	18	57.9
Europe	6	34	32	73.2
E. Mediterranean	10.5	4	4	—
W. Pacific	16	24	25	66.4

[a] I & P = infectious and parasitic diseases, Neop = neoplasms, CVS = cardiovascular diseases, LE = life expectancy.

ingestion of dirt). Additionally, it must be certain that the element is involved in body processes or composition, e.g., as part of some active substance (as cobalt is in vitamin B_{12}) or in some enzyme system (e.g., selenium in glutathione peroxidase) or as a structural component (e.g., fluorine in bone apatite).

There are a number of well-established examples of health being affected by the amounts and properties of elements available from the environment. This section reviews some direct relationships.

A. IODINE

Endemic nonneoplastic goiter has troubled man since time immemorial. It presents as a marked swelling in the neck as the thyroid gland enlarges to compensate for impaired efficiency. The obvious nature of the symptom has led to many local names. Endemic goiter was once so persistent in parts of Britain that it was called Derbyshire Neck, Nithsdale Neck or, in the Forest of Dean, Whin. The earliest known remedy[8] was a concoction of burnt sponge and seaweed, known in the English midlands as the "Coventry Remedy". Besides the visible swelling of the thyroid gland, endemic goiter is often associated with cretinism and deaf-mutism in children.

The link between endemic goiter and environmental iodine was one of the first associations between health and a trace element to be established. By early this century, it was known that iodine is an essential component of the two thyroid hormones, thyroxine and triiodothyronine. Hence, a reduced iodine intake was followed by an impaired ability to synthesize the hormones. Also, we now know that a number of chemical substances impair the utilization of iodine by the thyroid gland, and some occur in vegetables of the cabbage family. This was first recognized by Chesney et al.[9] when they found that rabbits which fed largely on cabbage developed goiter. The active substances have since been identified as goitrin and thyocyanates. Clements[10] has concluded that most people do not consume sufficient quantities of these substances, in their food, to produce goiter.

Generally, the incidence of goiter does not now bear a simple relationship with environmental iodide. Although goiter still occurs, its causation is now complex and any relationship with water, geology, or soil is now obscure. But at the beginning of this century, this was not so in Europe. Murray[11] followed up surveys of goiter in Britain dating to the 1920s and confirmed a high incidence in 12-year-old children in 17 of 39 English counties and 5 out of 10 Welsh counties. Changes in water supply were known to influence goiter. In Derbyshire and Somerset, a change from well water to piped water lowered goiter incidence,[12,13] but increased it in Melton Mowbray and Montgomeryshire.[10] Murray quoted from work done in the U.S. which revealed a simple relationship between goiter rates in World War I recruits and iodide in the water (Table 2).

TABLE 2
Goiter Rates and Water Iodine[10]

I in water (μg/l)	Goiter rate (per 1000 men)
0—0.5	30—15
0.5—2.0	15—3
>3.0	0.1

But in England, a high goiter incidence (48% women) was found where the iodide content of water was appreciable (2.6 to 3.1 μg I per liter). In Scotland, on the other hand, there was a generally lower incidence of thyroid enlargement, compared with England, for waters of similar iodide level.

An association with soil type or parent bedrock has also been observed. In Derbyshire, the condition was associated with limestone terrain and Learmonth[2] reports an association with limestone in the mountainous regions of the Pyrenees and Alps. Fuge and Johnson[14] have published a comprehensive review of the geochemistry of iodine. They noted that the average content in igneous rocks (0.24 mg I per kg) is less than in sedimentary rocks (2.0 mg I per kg) and the concentration in soils is typically 4 to 8 mg I per kg. They suggest that atmospheric inputs to soil are important and quote other authors' work to support a long-recognized association between proximity to the sea and soil iodine content: in France, Germany, and Ireland, soils near the sea contained (mean values) 16.8, 8.5 to 13.3, and 12.7 to 16.1 mg/kg, respectively, whereas mean concentrations inland were 3.2, 1.7, and 3.7 mg I per kg for the respective countries. Låg[15] reported that barley grown in coastal districts of Norway contained (mean) 0.038 mg I per kg dry matter, compared with 0.005 mg I per kg inland.

B. FLUORINE

Interest in fluorine centers primarily on its role in dental health. The relationship between fluoride in drinking water and a reduced incidence of dental decay is probably one of the best established links between geochemistry and disease. The link was first recognized in Colorado by McKay,[16] who observed an unusual mottling of teeth in his patients, yet these teeth seemed resistant to decay. He also discovered the same condition in immigrants coming from volcanic areas of Italy. It emerged that the common factor was a raised fluoride content of drinking water (2 to 13 mg F per l). In Britain, unusual mottling was described[17] for patients in Maldon, Essex, where only 7.9% of the teeth of children in two of the town's schools were carious, compared with 13.1% in all districts examined. Water from the Maldon area contained 4.5 to 5.5 mg F per l, compared with 0.5 mg F per l elsewhere. Similar surveys elsewhere in Britain provided similar results. So strong is the relationship that the addition of fluorides to water supplies (to achieve a concentration of 1 mg F per l) has been undertaken in many countries. But during the period from 1979, especially in western Europe, there has been a marked decline in tooth decay[18] which cannot be attributed to fluoridation of water supplies. Toothpaste containing fluoride is now widely available and generally used, and the importance of oral hygiene combined with regular dental checks is better understood.

Fuge and Johnson[19] have reviewed the occurrence of fluorine in British environments. The average fluoride content of felsic igneous rocks is 735 mg F per kg, but granites in the southwest of England can contain as much as 27,800 mg F per kg and the authors' data indicate that these rocks typically contain 2,600 mg F per kg. Other high fluoride environments include the mineralized carboniferous limestones of the north of England, granites in

Scotland, the Lake District, and Northern Ireland, and Scottish sandstones of permo-Triassic age which contain fluorite cements. But domestic water supplies appeared not to be especially enriched in fluoride. The authors suggested that these areas might be suitable for epidemiological studies.

C. MOLYBDENUM

Dental epidemiology presents some of the most convincing evidence that trace elements can influence health. Adler and Straub[20] compared caries rates in two adjacent Hungarian villages. In one, the incidence in the school population agreed with the rate expected from the fluoride level in the drinking water, but surprisingly low rates were recorded in the other village despite a much lower fluoride content. Since there were no dietary, social, or racial differences between the two study groups, the authors proposed that some other constituent, possibly molybdenum, was responsible. The area around Wells in England (the teart pastures) is arguably the classical site for the study of molybdenosis in sheep and cattle. The area comprises approximately 30,400 ha of flat, low-lying country underlain by molybdenum-rich black shales of the Lower Lias.[21] Anderson[22,23] studied 270 children in the teart district, compared with 163 children from other parts of southern England. Caries rates in the former area were only 69% of the latter (p <0.01). He explained this reduced prevalence by increased dietary intake of molybdenum.

D. LEAD

A report by Barmes and co-workers[24] that environmental lead might be associated with a raised caries incidence stimulated investigations in an old base metal mining area of Britain where soils and vegetables are contaminated by heavy metals.[25] In Devon, no effect on caries prevalence was observed where copper was the major contaminant, but where soils were contaminated mainly by lead, caries was much more prevalent in 12-year-old children. A similar relationship with soil lead was observed in west Wales, but in Shipham, Somerset, cadmium contamination was not associated with any increase in caries prevalence.[26] The general decline in caries incidence in recent years has had the effect of obscuring the environmental-lead/dental health relationship. When Anderson and colleagues[27] revisited the west Wales study area in 1983, 10 years after their earlier surveys, they found that, in their control area, the proportion of children free of caries had risen from 4 to 35%, and in two lead-polluted valleys the caries-free proportions had risen from 8 to 48% and 0 to 40%, respectively. Statistically, there were no longer any differences between the areas.

E. COBALT

In farm animals, deficiencies in cobalt, copper, and selenium are widespread in parts of Europe. But their is little evidence for human health problems related to these elements. The vital importance of cobalt for humans is due to its forming part of the vitamin B_{12} (cyanocobalamin) molecule, which contains one atom of cobalt. Lack of this vitamin leads to pernicious anemia characterized by muscular weakness, easy fatigability, shortness of breath, gastrointestinal disturbances, and, later, disease of the nervous system. But there is no evidence that there is a lack of cobalt in European diets such that pernicious anemia ensues. On the other hand, Scott[28] published some tentative evidence that the prevalence rate varied over Great Britain (Table 3). Scott remarked that there was some association with geological structure, and rates were lowest where sedimentary rocks were found.

There is, however, one curious and unique case study concerning cobalt deficiency. Shuttleworth and colleagues[29] reported the history of a 16-month-old girl born and bred on a hill farm in North Monmouthshire (Wales). The parents complained that whenever she played in the garden she ate soil, and when kept indoors, would eat the soil from flower pots. She was pale, tired, drab-looking, and her hair was matted and very dull. The local

TABLE 3
Pernicious anemia rates in Great Britain[28]

Area	Pernicious anemia rate per 1000
Northern England	1.84
Northwest England	1.69
Northeast England	1.50
Midlands	1.10
Southwest England	1.04
Southeast England	0.85
Wales	1.53
Scotland	1.80

veterinary surgeon had just diagnosed cobalt deficiency in the cows which provided milk for the child. Shuttleworth concluded that cobalt deficiency could be the cause of the child's ill health. For 30 d, the child was given cobaltous chloride (1 mg) daily in a syrup preparation. This treatment was rapidly remedial. The authors commented that low soil cobalt had created a chain of deficiency from herbage to cow to milk to child. Since there had been a delay in weaning, the cobalt deficiency had not been made up by an intake in solid food.

F. SELENIUM

Selenium is an integral part of the enzyme glutathione peroxidase (GSH-Px), which protects cells against oxidative damage by peroxide. Deficiencies in farm animals are prevalent in many areas of many countries and are recognized as "white muscle disease", a degeneration of muscle tissue, especially the heart. Anderson[30] concluded that for 329 farms in Britain, 47% were probably unable to provide grazing livestock with sufficient selenium to maintain blood levels >0.075 µg/l. In Sweden, Carlstrom et al.[31] analyzed 668 blood samples from 136 herds, and 87% of the measured GSH-Px activities were so low as to involve risk of muscle degeneration. In Finland,[32] selenium supplementation markedly reduced the incidence of white muscle disease in cattle.

In Belgium, the selenium content of drinking water has been reported as "low".[33] Barclay and MacPherson[34] analyzed the selenium content of wheat flours used to make bread in Great Britain: inclusion of a higher proportion of home-grown wheats had lowered human dietary intakes markedly and at approximately 43 µg Se per d, they were substantially below recommended daily intakes. The implications of these reports are uncertain.

Very low intakes in China are related to the occurrence of Keshan disease,[35] an endemic cardiomyopathy. But the disease is unknown in Europe. More recently in China, Kaschin-Beck disease has also been linked to low dietary selenium;[36] this is a disease of the bones and resembles rheumatoid arthritis, but it too is not known in Europe.

Diplock[37] has considered whether low intakes of selenium might be associated with an increased incidence of certain forms of cancer, and there was an apparent inverse relationship between blood selenium levels and cancer death rates in the U.S. However, in well-nourished populations, selenium supplements were considered unlikely to be of any benefit against human diseases such as cancer, cardiovascular disease, or cystic fibrosis.

G. ZINC

For zinc, there do not appear to be any simple deficiencies in western Europe which can be linked readily with environmental geochemistry. Moynihan[38] has reviewed trace elements in man: zinc deficiencies have been observed in Africa and Asia, but the only occurrence in children in London was explained by a lack of zinc in the milk formulation used.

H. ALUMINUM

Aluminum is now receiving attention as there is a possibility it might be involved in causing premature senile dementia, Alzheimer's disease. An association with raised soil aluminum was first noticed in Guam. Perl et al.[39] have discussed this report and other associations with environmental aluminum. Since then, it has been reported that the characteristic neurofibrillary tangles which are seen post mortem in patients' brains contain accumulations of aluminum. But whether these accumulations cause the disease or whether the diseased tissue simply concentrates the element is not clear. A symptom of aluminum toxicity is damage to nervous tissue,[40] but there are many dietary sources; tea drinking is likely to be a significant source in Great Britain.[41]

A special category of people at risk from water-borne aluminum are those dependent on renal dialysis units, especially children.[42] Cameron and Ineson[43] have reported marked geographical and seasonal variations in concentrations of aluminum in tap water supply for home-based hemodialysis units within the English Trent Regional Health Authority. High levels were due both to bedrock composition and to the use of coagulents at the water treatment plant.

III. MULTICAUSAL DISEASES

There are important groups of diseases, e.g., cancers, diseases of the central nervous system, and cardiovascular disease, the causes of which are by and large unknown and for which cure and control is uncertain. When the incidence of prevalence of these diseases is mapped, it is commonly observed that there are significant differences from place to place that are not easily explained by genetic traits or social or dietary differences. Environmental influences appear to be operating and a role for environmental geochemistry has been suggested by many authors. But it must be pointed out that most conclusions have been drawn from observed associations and rigorous, reliable epidemiological evidence is rare.

A. CANCERS

Chemotherapy, radiotherapy, and screening programs are reducing the mortality from several cancers. In addition, much progress is being made at a molecular level, through studies of oncogenes, in understanding the causes of cancers. But, at a higher, whole-body level, the causes are still not understood. Ultraviolet radiation, ionizing radiation, many organic substances, and some inorganic substances are known to be carcinogenic. In addition, cancer incidence is geographically very variable. Doll[44] has reported a strong geographical trend in the incidence of gastric cancer across Europe. He noted that people belonging to blood group A have 20% more gastric cancer than people belonging to other groups; but, in general, the variation in cancer experience is seldom determined solely by heredity, and environmental or dietary factors are important.

Over the years, much attention has been paid to the geographical variability of cancers and several authors have speculated that this variability may be influenced by soil or water quality. Legon[45] used the earlier (1936) data of Stocks, which indicated raised mortality ratios for gastric cancer in west and north Wales, and he postulated a link with soil organic content but offered no supporting evidence. Subsequently, Legon[46] extended this hypothesis to the whole of England and Wales. Millar[47] examined data for gastrointestinal cancer in north Montgomeryshire and proposed an association with environmental radioactivity since local black shales were rich in uranium. Again, no direct evidence was adduced to support the hypothesis and his use of cancer statistics was undiscriminating. Tromp and Diehl[48] compared stomach cancer rates in the Netherlands with soil type: sea clay and peats were associated with higher than normal rates, whereas sands and river clays were associated with less than normal rates. But again, although the study related only to men and was age adjusted, the use of the cancer data was less than satisfactory.

A frequently cited study is that of Allen-Price[49] in the Tamar valley of the west of England. Allen-Price remarked that mortality from cancer was unusually low in certain villages and unusually high in others. Within the village of Horrabridge, mortality was linked to the origin of different water supplies: the lowest mortality was associated with reservoir water from Dartmoor, whereas the highest mortality was associated with well or spring water derived from mineralized rock strata. Although this study is again statistically suspect, it stimulated a resurgence of interest in the link between cancer and the environment. Subsequently, a more rigorous study by the Royal College of General Practitioners[50] demonstrated that several Devon practices compared unfavorably with Birmingham practices with respect to malignant neoplasms. In a sequel to this study, Davies[25] reported tht heavy-metal contamination of fields was widespread in the Tamar valley.

A major development in the epidemiology of cancer was the publication by Howe[51] of the National Atlas of Disease Mortality in the U.K. The basic data comprised death certificates as reported to the Registrar General and then published in his Annual Report. The stomach cancer maps (males and females separated) again clearly show an excess of deaths for west and north Wales. A second atlas was published in 1970 and was further interpreted in a paper by Howe.[52] Howe drew attention to the established link of gastric cancer with blood group A and pointed out that this blood group is more common in eastern England where mortality from stomach cancer is lowest. Again, a link with trace elements in the soil, especially heavy metals from mining, was proposed, but no direct evidence was produced to support the hypothesis. More recently, Gardner et al.[53] have published an Atlas of Cancer Mortality for 1968 — 1978 which also identifies parts of north Wales as having unusually high mortalities.

Thus, from the 1930s to the 1970s, published data have consistently shown that certain areas have unusually high rates of deaths from stomach cancer. There seems to be strong evidence for some environmental factor(s) in the etiology of gastric cancer rather than different dietary habits, social class, or lifestyles within individual rural communities. Peeters[54] has written in a recent review (1987): "Soil components definitely act as determinant factors leading to the appearance of certain types of cancer."

In the late 1950s and early 1960s, the British Empire Cancer Campaign supported investigations of cancer in relation to soil factors at the University College of North Wales in Bangor. In the first of two major reports, Stocks and Davies[55] established correlations between garden soil composition and the frequency of stomach cancer in north Wales, Cheshire, and two localities in Devon. Soil organic matter, zinc, and cobalt were related positively with stomach cancer incidence, but not with other intestinal cancer. Chromium was connected with the incidence of both. Vanadium and iron showed inconclusive relations and nickel, titanium, and lead showed no connection. Subsequently, they[56] reported that the average logarithm of the ratio of zinc/copper in garden soils was always higher where a person had just died of stomach cancer after 10 or more years of residence than it was at houses where a person of similar residence had died of a nonmalignant cause. The effect was more pronounced and consistent in soils taken from vegetable gardens and it was not found where the duration of residence was less than 10 years.

It must be accepted that clear conclusions are unlikely to emerge from geochemical investigations. It is not expected that a series of elements responsible for gastric or other cancers will be identified since if trace metals are involved in causing cancer, their role is undoubtedly at the molecular level through some interaction with oncogenes. But the identification of elements or groups of elements as contributary factors would be of considerable importance.

B. MULTIPLE SCLEROSIS

Multiple sclerosis is a relatively uncommon disease which is very difficult to diagnose in its early stages. The numbers of patients in any area are small, which makes it difficult to weight data to produce standardized mortality ratios, and it is usual to quote simple prevalence rates per 100,000. Acheson[57] has suggested that rates in excess of 40/100,000 should be regarded as high and from 0 to 19/100,000 as low. Spillane[58] has noted that rates in excess of 40/100,000 are not known between latitudes 40N and 40S. Rates in western Europe are generally high. There is a general consensus that the disease declines with latitude and is typically one of the cool, temperate climates. Britain is a quite high risk area in general and there are regional differences. Several authors[59-61] have noted that northeast Scotland especially Shetland and the Orkneys, have rates on the order 100 to 150/100,000, compared with the Faroes at approximately 50/100,000.

The disease may be viral in origin and such a virus may lie dormant until triggered by an environmental influence or through a failure in the immunosuppresive system. Mims[62] has argued that there may be no specific virus and that multiple sclerosis results when antibodies to other viruses enter the brain.

A role for lead has been proposed by some authors. Campbell et al.[63] investigated patients who were long-term residents of two villages in Berkshire and Gloucestershire, England. In both cases, the garden soils were contaminated by lead and the lead content of the teeth of multiple sclerosis patients was significantly higher than that of control groups. Warren and colleagues[64-66] have consistently argued a case for lead to be involved in the etiology of multiple sclerosis.

There is some resemblance between multiple sclerosis and swayback in lambs, where the cause can be traced to copper deficiency in the pregnant ewe. Cytochrome oxidase, a copper-containing enzyme, is required for the biosynthesis of the phospholipids that largely constitute the myelin sheaf around nerves. Damage to myelin is common to both swayback and multiple sclerosis. Mills[67] comments that in species such as sheep where neurological sequelae are common in the neonate, such lesions occur subsequent to a decline in nervous tissue cytochrome oxidase activity. In sheep, copper deficiency may be either a simple deficiency or one induced by excess molybdenum in the diet. Raised soil and pasture concentrations of molybdenum are common where black, marine shales form the local soil parent material.[67-70]

Layton and Sutherland[71] have proposed that high risk multiple sclerosis areas are those where molybdenum is retained in the soil against leaching and is more available for plant uptake. Ward et al.[72] reported the elemental concentrations in blood and scalp hair from multiple sclerosis and control subjects from Oxford, England. Their study did not support any association of multiple sclerosis with lead, zinc, or molybdenum. Their data for copper did suggest a possibility that blood-copper concentrations were lower in the multiple sclerosis patients. Also, barium tended to be higher and vanadium lower in multiple sclerosis patients. There is a general acceptance by epidemiologists, based on migration studies, that multiple sclerosis is affected by local factors and further geomedical investigations might be helpful.

C. CARDIOVASCULAR DISEASES

Cardiovascular diseases (coronary heart disease, hypertension and stroke) constitute the major cause of death in adult life in Europe. The data in Table 1 demonstrate the high proportion of these deaths compared with other world regions.

International comparisons suggest that dietary factors, hypertension, cigarette smoking, and alcohol consumption account for much of the variation seen between countries. Nonetheless, one striking factor is the consistent link with drinking water quality, especially hardness, that has been found in many studies. This was first noticed in Japan when Kobayashi[73] found a statistically sound relationship between deaths from cerebral hemorrhage and the sulfate/carbonate ratio in river water, which, in turn, reflected the geochemical nature of the catchment area. In Britain,[74,75] calcium in water was found to correlate inversely with cardiovascular disease, but magnesium did not. Crawford and Crawford[76] analyzed coronary heart tissue from men younger than 40 who had died accidentally and found that those who came from soft-water areas were typified by low-tissue calcium and magnesium: Crawford proposed that hard water exercised some protective effect. Pocock et al.[77] have reported first results of a major British regional heart study. After adjustment for socioeconomic and other factors, cardiovascular mortality in areas with very soft water (calcium carbonate equivalent, 25 mg/l) was estimated to be 10 to 15% higher than that in areas with medium-hard water (170 mg/l).

There is some evidence that hypertension in some areas may be associated with low intakes of calcium.[78] Attention has also been paid to a possible role for magnesium since diseased heart muscle tissue is seen to contain less magnesium than healthy tissue.[79] But it has to be pointed out that hard waters do not necessarily contain raised concentrations of magnesium; this occurs only when the limestones, from which the water is derived, are dolomitized, and most English limestones are not. Another possibility is that soft waters may contain more of a wider variety of trace elements, some of which are injurious.[80]

IV. CONCLUSIONS

It is paradoxical that western Europe is at once the best and the worst region in which to study links between disease and geochemistry. Geology and soils have been extensively mapped in most countries, and much is known about regional geochemical variations. Similarly, health data are relatively accurate and generally available. Public health records exist since earlier last century. This should make the investigative task that much easier when compared with those many countries where even birth and death records are unreliable. But Europeans are mobile. They are rarely reliant on locally produced food; most are urban dwellers who have a nutritious diet. Also, areas which are geochemically anomalous are often thinly populated. Thus, links between environmental geochemistry and health are difficult to substantiate. Nonetheless, the evidence is growing that there may be a link, and exploring it is a challenge to epidemiologists and environmental scientists to work together.

Any studies must go well beyond naive comparisons of geochemical and epidemiological data: dietary or other pathways must be traced and quantified, and a causative role for the element or elements identified in terms of target organs or body processes. Moreover, studies must be predictive. If region A contains soils with too much or too little of element X and this is associated with excess mortality/morbidity ratios of disease Y, then a geochemically similar region, B, should be identified. If the conclusions drawn from region A have any validity, then disease Y should also be unusually prevalent in region B. A negative answer may not invalidate the original conclusions, but attention will then be drawn to the need to identify other factors that are characteristic of region A which are absent in region B, or factors in region B which are obscuring any relationships.

REFERENCES

1. **MARC** (Monitoring and Assessment Research Centre), Prepared for *Environmental Data Report*, United Nations Environmental Programme, Blackwell Scientific, Oxford, 1987, 352.
2. **Learmonth, A.,** *Disease Ecology: An Introduction*, Blackwell Scientific, Oxford, 1988.
3. **Howe, G. M.,** Disease patterns and trace elements, *Spectrum*, 77, 2, 1970.
4. **Mertz, W.,** The essential trace elements, *Science*, 213, 1332, 1981.
5. **Warren, H. V.,** Geology, trace elements and epidemiology, *Geogr. J.*, 130, 525, 1964.
6. **Warren, H. V.,** Some epidemiology, geochemistry and disease relationships, *Ecol. Dis.*, 1, 185, 1982.
7. **Låg, J.,** Some geomedical problems in connection to regional geochemical research, *Norw. Acad. Sci. Lett.*, 329, 10, 1984.

8. **Booth, E.,** The history of the seaweed industry: the iodine industry, *Chem. Ind.,* January 20, 52, 1979.
9. **Chesney, A. M., Clawson, T. A., and Webster, B.,** *Bull. Johns Hopkins Hosp.,* 43, 261, 278, 291, 1982.
10. **Clements, F. W.,** *Br. Med. Bull.,* 16, 133, 1960.
11. **Murray, M. M., Ryle, J. A., Simpson, B. W., and Wilson, D. C.,** Thyroid enlargement and other changes related to the mineral content of drinking water, Medical Research Council Mem. No. 18, Her Majesty's Stationery Office, London, 1984.
12. **Turton, P. H. J.,** Summary of the results of experiments on the prophylaxis and treatment of childhood goitre in Heanor, Derbyshire, *Lancet,* 2, 1170, 1927.
13. **Wilson, D. C.,** Fluorine in the aetiology of endemic goitre. I/F Somerset and goitre, *Lancet,* Feb 15, 211, 1941.
14. **Fuke, R. and Johnson, C. C.,** The geochemistry of iodine — a review, *Environ. Geochem. Health,* 8, 31, 1986.
15. **Låg, J.,** Soil science and geomedicine, *Acta Agric. Scand.,* 22, 150, 1972.
16. **Davies, B. E. and Anderson, R. J.,** The epidemiology of dental caries in relation to environmental trace elements, *Experientia,* 43, 87, 1987.
17. The Incidence of Dental Disease in Children, Medical Research Council, Special Report No. 97, London, 1925.
18. **Diesendorf, M.,** The mystery of declining tooth decay, *Nature,* 322, 125, 1986.
19. **Fuge, R.,** Fluorine in the UK environment, *Environ. Geochemi. Health,* 10, 94, 1988.
20. **Adler, P. and Straus, J.,** Water borne caries protective agent other than fluorine, *Acta Med. Acad. Sci. Hung.,* 4, 221, 1953.
21. **Ferguson, W. S., Lewis, A. H., and Watson, S. J.,** The teart pastures of Somerset. I. The cause and cure of teartness, *J. Agric. Sci.,* 33, 44, 1943.
22. **Anderson, R. J.,** Dental caries prevalence in relation to trace elements, *Br. Dent. J.,* 120, 271, 1966.
23. **Anderson, R. J,** The relationship between dental conditions and the trace element molybdenum, *Caries Res.,* 3, 75, 1969.
24. **Barmes, D. E., Adkins, B. L., and Schamschula, R. G.,** Etiology of caries in Papua-New Guinea: associations in soil, food and water, Bull. *WHO,* 43, 769, 1970.
25. **Davies,, B. E.,** Trace metal content of soils affected by base metal mining in the west of England, *Oikos,* 22, 366, 1971.
26. **Anderson, R. J. and Davies, B. E.,** Dental caries prevalence and trace elements in soil with special reference to lead, *J. Geol. Soc. London,* 137, 547-559, 1980.
27. **Anderson, R. J., Davies, B. E., Healey, J. M., and James, P. M. C.,** Dental caries experience in Ceredigion, Wales, in 1973 and 1983 with special reference to environmental lead, *Commun. Dent. Health,* 3, 193, 1986.
28. **Scott, E.,** Prevalence of pernicious anemia in Great Britain, *J. Coll. Gen. Pract.,* 3, 80, 1960.
29. **Shuttleworth, V. S., Cameron, R. S., Alderman, G., and Davies, H. T.,** A case of cobalt deficiency in a child presenting as 'earth eating', *Practitioner,* 186, 760, 1961.
30. **Anderson, P. H., Berrett, S., and Patterson, D. S. P.,** The biological selection status of livestock in Britain as indicated by sheep erythrocyte glutathione peroxidase activity, *Vet. Rec.,* March 17, 235, 1979.
31. **Carlstrom, G., Jonsson, G., and Pehrson, B. O.,** An evaluation of selenium status of cattle in Sweden by means of glutathion peroxidase, *Swed. J. Agric. Res.,* 9, 43, 1979.
32. **Westermarck, H. W.,** Selenium in long term feeding and the frequency of white muscle disease in Finland during the years 1978-1985, *J. Agric. Sci. Finl.* 59, 47, 1987.
33. **Robberecht, H., van Grieken, R., van Sprundel, M., Vanden Berghe, D., and Deelstfa, H.,** Selenium in environmental and drinking waters of Belgium, *Sci. Total Environ.,* 26, 163, 1983.
34. **Barclay, M. N. I. and MacPherson, A.,** Selenium content of wheat flour used in the UK, *J. Sci. Food Agric.,* 37, 1133, 1986.
35. **Chen, X., Yang, G., Chen, J., Chen, X., Wen, Z., and Ge, K.,** Studies on the relations of selenium and Keshan disease, *Biol. Trace Elements Res.,* 2, 91, 1980.
36. **Xu, G. and Jiang, Y.,** Selenium and the prevalence of Keshan and Kashin-Beck diseases in China, in *Proc. 1st Int. Symp. Geochemistry and Health,* Thornton, I., Ed., Science Reviews, 1986, 192.
37. **Diplock, A. T.,** Biological effects of selenium and relationships with carcinogenesis, *Toxicol. Environ. Chem.,* 8, 305, 1984.
38. **Moynahan, E. J.,** Trace elements in man, *Philos. Trans. R. Soc. London,* B288, 65, 1979.
39. **Perl, D. P., Pendcebury, W., and Munoz-Garcia, D.,** The association of aluminium to Alzheimer's disease and other neurofibrillary tangle-related disorders, in *Proc. 1st Int. Symp. Geochemistry and Health,* Thornton, I., Ed., Science Reviews, 1986, 227.
40. **Lione, A.,** Aluminum toxicology and the aluminum-containing medications, *Pharm. Theory,* 29, 255, 1985.
41. **Coriat, A. M. and Gillard, R. D.,** Beware the cup that cheers, *Nature,* 321, 570, 1986.

42. **Santos, F., Massie, M. D., and Chan, J. C. M.,** Risk factors in aluminum toxicity in the children with chronic renal failure, *Nephron,* 42, 189, 1986.

43. **Cameron, A. P. and Ineson, P. R.,** Hydrogeochemical studies for Al, F and Fi in waters supplying haemodyalysis units in the Trent region, U.K., *Environ. Geochem. Health,* 8, 83, 1968.

44. **Doll, R.,** The geographical distribution of cancer, *Tidskr. Nor. Laegeforen.,* 88, 1160, 1968.

45. **Legon, C. D.,** A note on geographical variations in cancer mortality, with special reference to gastric cancer in Wales, *Br. J. Cancer,* 5, 175, 1951.

46. **Legon, C. D.,** The aetiological significance of geographic variations in cancer mortality, *Br. Med. J.,* Sept. 27, 700, 1952.

47. **Millar, I. B.,** Gastro-intestinal cancer and geochemistry in north Montgameryshire, *Br. J. Cancer,* 15, 176, 1961.

48. **Tromp, S. W. and Diehl, J. C.,** A statistical study of the possible relationship between cancer of the stomach and soil, *Br. J. Cancer,* 9, 349, 1955.

49. **Allen-Price, E. D.,** Uneven distribution of cancer in west Devon, with particular reference to the diverse water supply, *Lancet,* 7136 June, 1235, 1960.

50. **R.C.G.P.,** Some contrasts in morbidity distribution, *J. Coll. Gen. Pract.,* 1, 74, 1966.

51. **Howe, G. M.,** *National Atlas of Disease Mortality in the United Kingdom,* Thames Nelson, London, 1963.

52. **Howe, G. M.,** Mortality from selected malignant neoplasms in the British Isles: the spatial perspective, *Geogr. J.,* 145, 401, 1979.

53. **Gardner, M. J., Winter, P. D., Taylor, C. P., and Acheson, E. D.,** *Atlas of Cancer Mortality,* 1983.

54. **Peeters, E. G.,** The possible influence of the components of the soil and lithosphere on the development and growth of neoplasms, *Experientia,* 43, 74, 1987.

55. **Stocks, P. and Davies, R. I.,** Epidemiological evidence from chemical and spectrographic analyses that soil is concerned in the causation of cancer, *Br. J. Cancer,* 14, 1960.

56. **Stocks, P. and Davies, R. I.,** Zinc and copper content of soils associated with the incidence of cancer of the stomach and other organs, *Br. J. Cancer,* 58, 14, 1964.

57. **Acheson, E. D.,** Epidemiology of multiple sclerosis, *Br. Med. Bull.,* 33, 9, 1977.

58. **Spillane, J. D.,** The geography of neurology, *Br. Med. J.,* 22, 506, 1972.

59. **Allison, R. S.,** Epidemiology of disseminated sclerosis, *Proc. R. Soc. Med.,* 54, 1, 1961.

60. **Shepperd, D. I. and Downie, A. W.,** Prevalence of multiple sclerosis in northeast Scotland, *Br. Med. J.,* 28, 314, 1978.

61. **Sutherland, J. M.,** Observations on the prevalence of multiple sclerosis in northern Scotland, *Brain,* 79, 635, 1956.

62. **Mims, C.,** Multiple sclerosis — the case against viruses, *New Sci.,* 98(1364), 938, 1983.

63. **Campbell, A. M. G., Herdan, G., Tatlow, W. F. T., and Whittle, E. G.,** Lead in relation to disseminated sclerosis, *Brain,* 73, 52, 1950.

64. **Warren, H. V.,** Geology and multiple sclerosis, *Nature* 184, 561, 1969.

65. **Warren, H. V.,** Environmental lead: a survey of its possible physiological significance, *J. Biosoc. Sci.,* 6, 223, 1974.

66. **Warren, H. V.,** Possible correlations between geology and some disease patterns, *Ann. N. Y. Acad. Sci.,* 136, 696, 1967.

67. **Mills, C. F.,** Trace elements in animals, *Philos. Trans. R. Soc. London,* B288, 51, 1979.

68. **Alloway, B. J.,** Copper and molybdenum in swayback pastures, *J. Agric. Sci.,* 80, 521, 1973.

69. **Leech, A. F. and Thornton, I.,** Trace elements in soils and pasture herbage on farms with bovine hypocupraemia, *J. Agric. Sci.,* 108, 591, 1987.

70. **Thomson, I., Thornton, I., and Webb, J. S.,** Molybdenum in black shales and the incidence of bovine hypocuprosis, *J. Sci. Food Agric.,* 23, 879, 1972.

71. **Layton, W.,** Geochemistry and multiple sclerosis — a hypothesis, *Med. J. Austr.,* 1, 73, 1975.

72. **Ward, N. I., Bryce-Smith, D., Minski, M., and Matthews, W. B.,** Multiple sclerosis — multielement survey, *Biol. Trace Elements Res.,* 7, 153, 1985.

73. **Kosayashi, J.,** On geographical relationships between the chemical nature of river water and death rate from apoplexy, *Ber. Ohara Inst. Landwirtsch. Biol. Okayama Univ.,* 9, 12, 1957.

74. **Morris, J. M., Crawford, M. D., and Heady, J. A.,** Hardness of local water supplies and mortality from cardiovascular disease, *Lancet,* April 22, 860, 1961.

75. **Morris, J. M., Crawford, M. D., and Heady, J. A.,** Hardness of local water supplies and mortality from cardiovascular disease, *Lancet,* Sept. 8, 506, 1962.

76. **Crawford, T. and Crawford, M.,** Prevalence and pathological changes of ischaemic heart disease in a hard-water and in a soft-water area, *Lancet,* Feb. 4, 229, 1967.

77. **Pocock, S. J., Shaper, A. G., Cook, D. G., Packham, R. F., Lacey, R. F., Powell, P., and Russell, P. F.,** British regional heart study: geographic variations in cardiovascular mortality and the role of water quality, *Br. Med. J.,* 780, 1243, 1980.

78. **Rolls, B. and Blakeborough, P.,** Calcium and hypertension, *Chem. Ind.,* July 7, 449, 1986.
79. **Shils, M. E.,** Magnesium in health and disease, *Annu. Rev. Nutr.,* 8, 429, 1988.
80. **Masironi, R.,** Trace elements and cardiovascular diseases, *WHO Bull.,* 40, 305, 1969.

INDEX